Smoothing Splines

Methods and Applications

MONOGRAPHS ON STATISTICS AND APPLIED PROBABILITY

General Editors

F. Bunea, V. Isham, N. Keiding, T. Louis, R. L. Smith, and H. Tong

Monographs on Statistics and Applied Probability 121

Smoothing Splines
Methods and Applications

Yuedong Wang

University of California

Santa Barbara, California, USA

CRC Press
Taylor & Francis Group
Boca Raton   London   New York

CRC Press is an imprint of the
Taylor & Francis Group an **informa** business

A CHAPMAN & HALL BOOK

CRC Press
Taylor & Francis Group
6000 Broken Sound Parkway NW, Suite 300
Boca Raton, FL 33487-2742

First issued in paperback 2022

ISBN 13: 978-1-03-247762-6 (pbk)
ISBN 13: 978-1-4200-7755-1 (hbk)

DOI: 10.1201/b10954

This book contains information obtained from authentic and highly regarded sources. Reasonable efforts have been made to publish reliable data and information, but the author and publisher cannot assume responsibility for the validity of all materials or the consequences of their use. The authors and publishers have attempted to trace the copyright holders of all material reproduced in this publication and apologize to copyright holders if permission to publish in this form has not been obtained. If any copyright material has not been acknowledged please write and let us know so we may rectify in any future reprint.

Publisher's Note
The publisher has gone to great lengths to ensure the quality of this reprint but points out that some imperfections in the original copies may be apparent.

Visit the Taylor & Francis Web site at
http://www.taylorandfrancis.com

and the CRC Press Web site at
http://www.crcpress.com

TO

YAN, CATHERINE, AND KEVIN

Contents

List of Tables

List of Figures

Symbol Description

$(x)_+$	$\max\{x, 0\}$
$x \wedge z$	$\min\{x, z\}$
$x \vee z$	$\max\{x, z\}$
$\det^+$	Product of the nonzero eigenvalues
$k_r(x)$	Scaled Bernoulli polynomials
$(\cdot, \cdot)$	Inner product
$\|\cdot\|$:	Norm
$\mathcal{X}$	Domain of a function
$\mathcal{S}$	Unit sphere
$\mathcal{H}$	Function space
$\mathcal{L}$	Linear functional
$\mathcal{N}$	Nonlinear functional
P	Projection
$\mathcal{A}$	Averaging operator
$\mathcal{M}$	Model space
$R(x, z)$	Reproducing kernel
$\mathbb{R}^d$	Euclidean d-space
$NS^{2m}(t_1, \cdots, t_k)$	Natural polynomial spline space
$W_2^m[a, b]$	Sobolev space on $[a, b]$
$W_2^m(per)$	Sobolev space on unit circle
$W_2^m(\mathbb{R}^d)$	Thin-plate spline model space
$W_2^m(\mathcal{S})$	Sobolev space on unit sphere
$\oplus$	Direct sum of function spaces
$\otimes$	Tensor product of function spaces

Preface

Statistical analysis often involves building mathematical models that examine the relationship between dependent and independent variables. This book is about a general class of powerful and flexible modeling techniques, namely, *spline smoothing*.

Research on smoothing spline models has attracted a great deal of attention in recent years, and the methodology has been widely used in many areas. This book provides an introduction to some basic smoothing spline models, including polynomial, periodic, spherical, thin-plate, L-, and partial splines, as well as an overview of more advanced models, including smoothing spline ANOVA, extended and generalized smoothing spline ANOVA, vector spline, nonparametric nonlinear regression, semi-parametric regression, and semiparametric mixed-effects models. Methods for model selection and inference are also presented.

The general forms of nonparametric/semiparametric linear/nonlinear fixed/mixed smoothing spline models in this book provide unified frameworks for estimation, inference, and software implementation. This book draws on the theory of reproducing kernel Hilbert space (RKHS) to present various smoothing spline models in a unified fashion. On the other hand, the subject of smoothing spline in the context of RKHS and regularization is often regarded as technical and difficult. One of my main goals is to make the advanced smoothing spline methodology based on RKHS more accessible to practitioners and students. With this in mind, the book focuses on methodology, computation, implementation, software, and application. It provides a gentle introduction to the RKHS, keeps theory at the minimum level, and provides details on how the RKHS can be used to construct spline models.

User-friendly software is key to the routine use of any statistical method. The `assist` library in R implements methods presented in this book for fitting various nonparametric/semiparametric linear/nonlinear fixed/mixed smoothing spline models. The `assist` library can be obtained at

http://www.r-project.org

Much of the exposition is based on the analysis of real examples. Rather than formal analysis, these examples are intended to illustrate the power and versatility of the spline smoothing methodology. All data analyses are performed in R, and most of them use functions in the `assist` library. Codes for all examples and further developments related to this book will be posted on the web page

http://www.pstat.ucsb.edu/faculty/yuedong/book.html

This book is intended for those wanting to learn about smoothing splines. It can be a reference book for statisticians and scientists who need advanced and flexible modeling techniques. It can also serve as a text for an advanced-level graduate course on the subject. In fact, topics in Chapters 1–4 were covered in a quarter class at the University of California — Santa Barbara, and the University of Science and Technology of China.

I was fortunate indeed to have learned the smoothing spline from Grace Wahba, whose pioneering work has paved the way for much ongoing research and made this book possible. I am grateful to Chunlei Ke, my former student and collaborator, for developing the `assist` package. Special thanks goes to Anna Liu for reading the draft carefully and correcting many mistakes. Several people have helped me over various phases of writing this book: Chong Gu, Wensheng Guo, David Hinkley, Ping Ma, and Wendy Meiring. I must thank my editor, David Grubbes, for his patience and encouragement. Finally, I would like to thank several researchers who kindly shared their data sets for inclusion in this book; they are cited where their data are introduced.

Yuedong Wang
Santa Barbara
December 2010

Chapter 1

Introduction

1.1 Parametric and Nonparametric Regression

Regression analysis builds mathematical models that examine the relationship of a dependent variable to one or more independent variables. These models may be used to predict responses at unobserved and/or future values of the independent variables. In the simple case when both the dependent variable y and the independent variable x are scalar variables, given observations (x_i, y_i) for $i = 1, \ldots, n$, a *regression model* relates dependent and independent variables as follows:

$$y_i = f(x_i) + \epsilon_i, \quad i = 1, \ldots, n, \tag{1.1}$$

where f is the *regression function* and ϵ_i are zero-mean independent random errors with a common variance σ^2. The goal of regression analysis is to construct a model for f and estimate it based on noisy data.

For example, for the Old Faithful geyser in Yellowstone National Park, consider the problem of predicting the waiting time to the next eruption using the length of the previous eruption. Figure 1.1(a) shows the scatter plot of waiting time to the next eruption ($y =$ waiting) against duration of the previous eruption ($x =$ duration) for 272 observations from the Old Faithful geyser. The goal is to build a mathematical model that relates the waiting time to the duration of the previous eruption. A first attempt might be to approximate the regression function f by a straight line

$$f(x) = \beta_0 + \beta_1 x. \tag{1.2}$$

The least squares straight line fit is shown in Figure 1.1(a). There is no apparent sign of lack-of-fit. Furthermore, there is no clear visible trend in the plot of residuals in Figure 1.1(b).

Often f is nonlinear in x. A common approach to dealing with nonlinear relationship is to approximate f by a *polynomial of order m*

$$f(x) = \beta_0 + \beta_1 x + \cdots + \beta_{m-1} x^{m-1}. \tag{1.3}$$

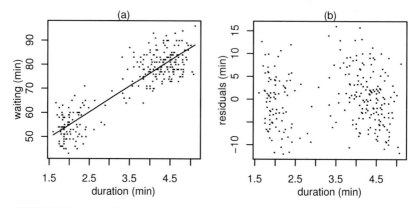

FIGURE 1.1   Geyser data, plots of (a) observations and the least squares straight line fit, and (b) residuals.

Figure 1.2 shows the scatter plot of acceleration ($y = $ `acceleration`) against time after impact ($x = $ `time`) from a simulated motorcycle crash experiment on the efficacy of crash helmets. It is clear that a straight line cannot explain the relationship between acceleration and time. Polynomials with $m = 1, \dots, 20$ are fitted to the data, and Figure 1.2 shows the best fit selected by *Akaike's information criterion* (AIC). There are waves in the fitted curve at both ends of the range. The fit is still not completely satisfactory even when polynomials up to order 20 are considered. Unlike the linear regression model (1.2), except for small m, coefficients in model (1.3) no longer have nice interpretations.

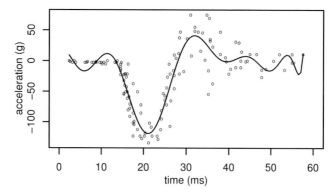

FIGURE 1.2   Motorcycle data, plot of observations, and a polynomial fit.

In general, a *parametric regression* model assumes that the form of f is known except for finitely many unknown parameters. The specific form of f may come from scientific theories and/or approximations to mechanics under some simplified assumptions. The assumptions may be too restrictive and the approximations may be too crude for some applications. An inappropriate model can lead to systematic bias and misleading conclusions. In practice, one should always check the assumed form for the function f.

It is often difficult, if not impossible, to obtain a specific functional form for f. A *nonparametric regression* model does not assume a predetermined form. Instead, it makes assumptions on qualitative properties of f. For example, one may be willing to assume that f is "smooth", which does not reduce to a specific form with finite number of parameters. Rather, it usually leads to some infinite dimensional collections of functions. The basic idea of nonparametric regression is to let the data speak for themselves. That is to let the data decide which function fits the best without imposing any specific form on f. Consequently, nonparametric methods are in general more flexible. They can uncover structure in the data that might otherwise be missed.

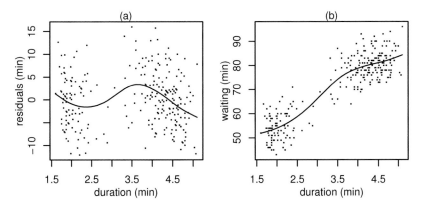

FIGURE 1.3 Geyser data, plots of (a) residuals from the straight line fit and the cubic spline fit to the residuals, and (b) the cubic spline fit to the original data.

For illustration, we fit cubic splines to the geyser data. The cubic spline is a special nonparametric regression model that will be introduced in Section 1.2. A cubic spline fit to residuals from the linear model (1.2) reveals a nonzero trend in Figure 1.3(a). This raises the question of whether a simple linear regression model is appropriate for the geyser data. A cubic spline fit to the original data is shown in Figure 1.3(b). It reveals that there are two clusters in the independent variable, and a different linear model may be required for each cluster. Sections 2.10, 3.8, and 3.9 contain more analysis of the geyser data. A cubic spline fit to the motorcycle data is shown in Figure 1.4. It fits data much better than the polynomial model. Sections 2.10, 3.8, 5.4.1, and 6.4 contain more analysis of the motorcycle data.

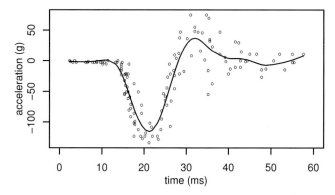

FIGURE 1.4 Motorcycle data, plot of observations, and the cubic spline fit.

The above simple exposition indicates that the nonparametric regression technique can be applied to different steps in regression analysis: data exploration, model building, testing parametric models, and diagnosis. In fact, as illustrated throughout the book, *spline smoothing* is a powerful and versatile tool for building statistical models to exploit structures in data.

1.2 Polynomial Splines

The polynomial (1.3) is a global model which makes it less adaptive to local variations. Individual observations can have undue influence on the fit in remote regions. For example, in the motorcycle data, the behavior of the mean function varies drastically from one region to another.

These local variations led to oscillations at both ends of the range in the polynomial fit. A natural solution to overcome this limitation is to use piecewise polynomials, the basic idea behind *polynomial splines*.

Let $a < t_1 < \cdots < t_k < b$ be fixed points called *knots*. Let $t_0 = a$ and $t_{k+1} = b$. Roughly speaking, polynomial splines are piecewise polynomials joined together smoothly at knots. Formally, a *polynomial spline* of order r is a real-valued function on $[a, b]$, $f(t)$, such that

(i) f is a piecewise polynomial of order r on $[t_i, t_{i+1})$, $i = 0, 1, \ldots, k$;

(ii) f has $r - 2$ continuous derivatives and the $(r - 1)$st derivative is a step function with jumps at knots.

Now consider even orders represented as $r = 2m$. The function f is a *natural polynomial spline* of order $2m$ if, in addition to (i) and (ii), it satisfies the *natural boundary conditions*

(iii) $f^{(j)}(a) = f^{(j)}(b) = 0$, $j = m, \ldots, 2m - 1$.

The natural boundary conditions imply that f is a polynomial of order m on the two outside subintervals $[a, t_1]$ and $[t_k, b]$. Denote the function space of natural polynomial splines of order $2m$ with knots $t_1, \ldots, t_k$ as $NS^{2m}(t_1, \ldots, t_k)$.

One approach, known as *regression spline*, is to approximate f using a polynomial spline or natural polynomial spline. To get a good approximation, one needs to decide the number and locations of knots. This book covers a different approach known as *smoothing spline*. It starts with a well-defined model space for f and introduces a penalty to prevent overfitting. We now describe this approach for polynomial splines.

Consider the regression model (1.1). Suppose f is "smooth". Specifically, assume that $f \in W_2^m[a, b]$ where the *Sobolev space*

$$W_2^m[a, b] = \{ f : f, f', \ldots, f^{(m-1)} \text{ are absolutely continuous,}$$

$$\int_a^b (f^{(m)})^2 dx < \infty \}. \tag{1.4}$$

For any $a \leq x \leq b$, Taylor's theorem states that

$$f(x) = \underbrace{\sum_{\nu=0}^{m-1} \frac{f^{(\nu)}(a)}{\nu!} (x - a)^\nu}_{\text{polynomial of order } m} + \underbrace{\int_a^x \frac{(x - u)^{m-1}}{(m-1)!} f^{(m)}(u) du}_{\text{Rem}(x)}. \tag{1.5}$$

It is clear that the polynomial regression model (1.3) ignores the remainder term $\text{Rem}(x)$ in the hope that it is negligible. It is often difficult to verify this assumption in practice. The idea behind the spline smoothing is to let data decide how large $\text{Rem}(x)$ should be. Since $W_2^m[a,b]$ is an infinite dimensional space, a direct fit to f by minimizing the *least squares* (LS)

$$\frac{1}{n}\sum_{i=1}^{n}(y_i - f(x_i))^2 \tag{1.6}$$

leads to interpolation. Therefore, certain control over $\text{Rem}(x)$ is necessary. One natural approach is to control how far f is allowed to depart from the polynomial model. Under appropriate norms defined later in Sections 2.2 and 2.6, one measure of distance between f and polynomials is $\int_a^b (f^{(m)})^2 dx$. It is then reasonable to estimate f by minimizing the LS (1.6) under the constraint

$$\int_a^b (f^{(m)})^2 dx \le \rho \tag{1.7}$$

for a constant ρ. By introducing a Lagrange multiplier, the constrained minimization problem (1.6) and (1.7) is equivalent to minimizing the *penalized least squares* (PLS):

$$\frac{1}{n}\sum_{i=1}^{n}(y_i - f(x_i))^2 + \lambda \int_a^b (f^{(m)})^2 dx. \tag{1.8}$$

In the remainder of this book, a polynomial spline refers to the solution of the PLS (1.8) in the model space $W_2^m[a,b]$. A *cubic spline* is a special case of the polynomial spline with $m = 2$. Since it measures the roughness of the function f, $\int_a^b (f^{(m)})^2 dx$ is often referred to as a *roughness penalty*. It is obvious that there is no penalty for polynomials of order less than or equal to m. The *smoothing parameter* λ balances the trade-off between goodness-of-fit measured by the LS and roughness of the estimate measured by $\int_a^b (f^{(m)})^2 dx$.

Suppose that $n \ge m$ and $a \le x_1 < x_2 < \cdots < x_n \le b$. Then, for fixed $0 < \lambda < \infty$, (1.8) has a unique minimizer $\hat{f}$ and $\hat{f} \in NS^{2m}(x_1,\ldots,x_n)$ (Eubank 1988). This result indicates that even though we started with the infinite dimensional space $W_2^m[a,b]$ as the model space for f, the solution to the PLS (1.8) belongs to a finite dimensional space. Specifically, the solution is a natural polynomial spline with knots at distinct design points. One approach to computing the polynomial spline estimate is to represent $\hat{f}$ as a linear combination of a basis of $NS^{2m}(x_1,\ldots,x_n)$. Several basis constructions were provided in Section 3.3.3 of Eubank (1988). In particular, the R function `smooth.spline` implements this approach for the cubic spline using the B-spline basis. For example, the cubic spline fit in Figure 1.4 is derived by the following statements:

```
> library(MASS); attach(mcycle)
> smooth.spline(times, accel, all.knots=T)
```

This book presents a different approach. Instead of basis functions, representers of reproducing kernel Hilbert spaces will be used to represent the spline estimate. This approach allows us to deal with many different spline models in a unified fashion. Details of this approach for polynomial splines will be presented in Sections 2.2 and 2.6.

When $\lambda = 0$, there is no penalty, and the natural spline that interpolates observations is the unique minimizer. When $\lambda = \infty$, the unique minimizer is the mth order polynomial. As λ varies from ∞ to 0, we have a family of models ranging from the parametric polynomial model to interpolation. The value of λ decides how far f is allowed to depart from the polynomial model. Thus the choice of λ holds the key to the success of a spline estimate. We discuss how to choose λ based on data in Chapter 3.

1.3   Scope of This Book

Driven by many sophisticated applications and fueled by modern computing power, many flexible nonparametric and semiparametric modeling techniques have been developed to relax parametric assumptions and to exploit possible hidden structure. There are many different nonparametric methods. This book concentrates on one of them, *smoothing spline*. Existing books on this topic include Eubank (1988), Wahba (1990), Green and Silverman (1994), Eubank (1999), Gu (2002), and Ruppert, Wand and Carroll (2003). The goals of this book are to (a) make the advanced smoothing spline methodology based on reproducing kernel Hilbert spaces more accessible to practitioners and students; (b) provide software and examples so that the spline smoothing methods can be routinely used in practice; and (c) provide a comprehensive coverage of recently developed smoothing spline nonparametric/semiparametric linear/nonlinear fixed/mixed models. We concentrate on the methodology, implementation, software, and application. Theoretical results are stated without proofs. All methods will be demonstrated using real data sets and R functions.

The polynomial spline in Section 1.2 concerns the functions defined on the domain $[a, b]$. In many applications, the domain of the regression function is not a continuous interval. Furthermore, the regression function may only be observed indirectly. **Chapter 2** introduces general smoothing spline regression models with reproducing kernel Hilbert spaces on general domains as model spaces. Penalized LS estimation, Kimeldorf–Wahba representer theorem, computation, and the R function `ssr` will be covered. Explicit constructions of model spaces will be discussed in detail for some popular smoothing spline models including polynomial, periodic, thin-plate, spherical, and L-splines.

Chapter 3 introduces methods for selecting the smoothing parameter and making inferences about the regression function. The impact of the smoothing parameter and basic concepts for model selection will be discussed and illustrated using an example. Connections between smoothing spline models and Bayes/mixed-effects models will be established. The unbiased risk, generalized cross-validation, and generalized maximum likelihood methods will be introduced for selecting the smoothing parameter. Bayesian and bootstrap confidence intervals will be introduced for the regression function and its components. The locally most powerful, generalized maximum likelihood and generalized cross-validation tests will also be introduced to test the hypothesis of a parametric model versus a nonparametric alternative.

Analogous to multiple regression, **Chapter 4** constructs models for multivariate regression functions based on smoothing spline analysis of variance (ANOVA) decompositions. The resulting models have hierarchical structures that facilitate model selection and interpretation. Smoothing spline ANOVA decompositions for tensor products of some commonly used smoothing spline models will be illustrated. Penalized LS estimation involving multiple smoothing parameters and componentwise Bayesian confidence intervals will be covered.

Chapter 5 presents spline smoothing methods for heterogeneous and correlated observations. Presence of heterogeneity and correlation may lead to wrong choice of the smoothing parameters and erroneous inference. Penalized weighted LS will be used for estimation. Unbiased risk, generalized cross-validation, and generalized maximum likelihood methods will be extended for selecting the smoothing parameters. Variance and correlation structures will also be discussed.

Analogous to generalized linear models, **Chapter 6** introduces smoothing spline ANOVA models for observations generated from a particular distribution in the exponential family including binomial, Poisson, and gamma distributions. Penalized likelihood will be used for estimation, and methods for selecting the smoothing parameters will be discussed. Nonparametric estimation of variance and spectral density functions will be presented.

Analogous to nonlinear regression, **Chapter 7** introduces spline smoothing methods for nonparametric nonlinear regression models where some unknown functions are observed indirectly through nonlinear functionals. In addition to fitting theoretical and empirical nonlinear nonparametric regression models, methods in this chapter may also be used to deal with constraints on the nonparametric function such as positivity or monotonicity. Several algorithms based on Gauss–Newton, Newton–Raphson, extended Gauss–Newton and Gauss–Seidel methods will be presented for different situations. Computation and the R function `nnr` will be covered.

Chapter 8 introduces semiparametric regression models that involve both parameters and nonparametric functions. The mean function may depend on the parameters and the nonparametric functions linearly or nonlinearly. The semiparametric regression models include many well-known models such as the partial spline, varying coefficients, projection pursuit, single index, multiple index, functional linear, and shape invariant models as special cases. Estimation, inference, computation, and the R function `snr` will also be covered.

Chapter 9 introduces semiparametric linear and nonlinear mixed-effects models. Smoothing spline ANOVA decompositions are extended for the construction of semiparametric mixed-effects models that parallel the classical mixed models. Estimation and inference methods, computation, and the R functions `slm` and `snm` will be covered as well.

1.4   The `assist` Package

The `assist` package was developed for fitting various smoothing spline models covered in this book. It contains five main functions, `ssr`, `nnr`, `snr`, `slm`, and `snm` for fitting various smoothing spline models. The function `ssr` fits smoothing spline regression models in Chapter 2, smoothing spline ANOVA models in Chapter 4, extended smoothing spline ANOVA models with heterogeneous and correlated observations in Chapter 5, generalized smoothing spline ANOVA models in Chapter 6, and semiparametric linear regression models in Chapter 8, Section 8.2. The function `nnr` fits nonparametric nonlinear regression models in Chapter 7. The function `snr` fits semiparametric nonlinear regression models in Chapter 8, Section 8.3. The functions `slm` and `snm` fit semiparametric linear and nonlinear mixed-effects models in Chapter 9. The `assist` package is available at

http://cran.r-project.org

Figure 1.5 shows how the functions in `assist` generalize some of the existing R functions for regression analysis.

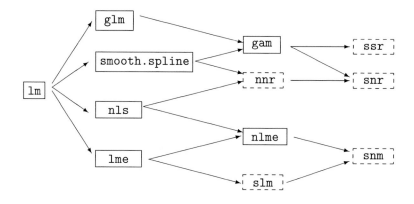

FIGURE 1.5 Functions in assist (dashed boxes) and some existing R functions (solid boxes). An arrow represents an extension to a more general model. lm: linear models. glm: generalized linear models. smooth.spline: cubic spline models. nls: nonlinear regression models. lme: linear mixed-effects models. gam: generalized additive models. nlme: nonlinear mixed-effects models. ssr: smoothing spline regression models. nnr: nonparametric nonlinear regression models. snr: semiparametric nonlinear regression models. slm: semiparametric linear mixed-effects models. snm: semiparametric nonlinear mixed-effects models.

Chapter 2

Smoothing Spline Regression

2.1 Reproducing Kernel Hilbert Space

Polynomial splines concern functions defined on a continuous interval. This is the most common situation in practice. Nevertheless, many applications require modeling functions defined on domains other than a continuous interval. For example, for spatial data with measurements on latitude and longitude, the domain of the function is the Euclidean space $\mathbb{R}^2$. Specific spline models were developed for different applications. It is desirable to develop methodology and software on a general platform such that special cases are dealt with in a unified fashion. *Reproducing Kernel Hilbert Space* (RKHS) provides such a general platform.

This section provides a very brief review of RKHS. Throughout this book, important theoretical results are presented in *italic* without proofs. Details and proofs related to RKHS can be found in Aronszajn (1950), Wahba (1990), Gu (2002), and Berlinet and Thomas-Agnan (2004).

A nonempty set E of elements $f, g, h, \ldots$ forms a *linear space* if there are two operations: (1) *addition*: a mapping $(f, g) \rightarrow f + g$ from $E \times E$ into E; and (2) *multiplication*: a mapping $(\alpha, f) \rightarrow \alpha f$ from $\mathbb{R} \times E$ into E, such that for any $\alpha, \beta \in \mathbb{R}$, the following conditions are satisfied: (a) $f + g = g + f$; (b) $(f + g) + h = f + (g + h)$; (c) for every $f, g \in E$, there exists $h \in E$ such that $f + h = g$; (d) $\alpha(\beta f) = (\alpha\beta)f$; (e) $(\alpha + \beta)f = \alpha f + \beta f$; (f) $\alpha(f + g) = \alpha f + \alpha g$; and (g) $1f = f$. Property (c) implies that there exists a zero element, denoted as 0, such that $f + 0 = f$ for all $f \in E$.

A finite collection of elements $f_1, \ldots, f_k$ in E is called *linearly independent* if the relation $\alpha_1 f_1 + \cdots + \alpha_k f_k = 0$ holds only in the trivial case with $\alpha_1 = \cdots = \alpha_k = 0$. An arbitrary collection of elements A is called linearly independent if every finite subcollection is linearly independent. Let A be a subset of a linear space E. Define

$$\text{span}A \triangleq \{\alpha_1 f_1 + \cdots + \alpha_k f_k : f_1, \ldots, f_k \in A,$$
$$\alpha_1, \ldots, \alpha_k \in \mathbb{R}, \ k = 1, 2, \ldots\}.$$

A set $B \subset E$ is called a *basis* of E if B is linearly independent and span$B = E$.

A nonnegative function $|| \cdot ||$ on a linear space E is called a *norm* if (a) $||f|| = 0$ if and only if $f = 0$; (b) $||\alpha f|| = |\alpha| ||f||$; and (c) $||f + g|| \leq ||f|| + ||g||$. If the function $|| \cdot ||$ satisfies (b) and (c) only, then it is called a *seminorm*. A linear space with a norm is called a *normed linear space*.

Let E be a linear space. A mapping $(\cdot, \cdot) : E \times E \to \mathbb{R}$ is called an *inner product* in E if it satisfies (a) $(f, g) = (g, f)$; (b) $(\alpha f + \beta g, h) = \alpha(f, h) + \beta(g, h)$; and (c) $(f, f) \geq 0$, and $(f, f) = 0$ if and only if $f = 0$. An inner product defines a norm: $||f|| \triangleq \sqrt{(f, f)}$. A linear space with an inner product is called an *inner product space*.

Let E be a normed linear space and f_n be a sequence in E. The sequence f_n is said to *converge* to $f \in E$ if $\lim_{n \to \infty} ||f_n - f|| = 0$, and f is called the *limit point*. The sequence f_n is called a *Cauchy sequence* if $\lim_{l, n \to \infty} ||f_l - f_n|| = 0$. The space E is *complete* if every Cauchy sequence converges to an element in E. A complete inner product space is called a *Hilbert space*.

A *functional* $\mathcal{L}$ on a Hilbert space $\mathcal{H}$ is a mapping from $\mathcal{H}$ to $\mathbb{R}$. $\mathcal{L}$ is a *linear functional* if it satisfies $\mathcal{L}(\alpha f + \beta g) = \alpha \mathcal{L}f + \beta \mathcal{L}g$. $\mathcal{L}$ is said to be *continuous* if $\lim_{n \to \infty} \mathcal{L}f_n = \mathcal{L}f$ when $\lim_{n \to \infty} f_n = f$. $\mathcal{L}$ is said to be *bounded* if there exists a constant M such that $|\mathcal{L}f| \leq M||f||$ for all $f \in \mathcal{H}$. $\mathcal{L}$ is continuous if and only if $\mathcal{L}$ is bounded. For every fixed $h \in \mathcal{H}$, $\mathcal{L}_h f \triangleq (h, f)$ defines a continuous linear functional. Conversely, every continuous linear functional $\mathcal{L}$ can be represented as an inner product with a representer.

Riesz representation theorem
Let $\mathcal{L}$ be a continuous linear functional on a Hilbert space $\mathcal{H}$. There exists a unique $h_{\mathcal{L}}$ such that $\mathcal{L}f = (h_{\mathcal{L}}, f)$ for all $f \in \mathcal{H}$. The element $h_{\mathcal{L}}$ is called the representer of $\mathcal{L}$.

Let $\mathcal{H}$ be a Hilbert space of real-valued functions from $\mathcal{X}$ to $\mathbb{R}$ where $\mathcal{X}$ is an arbitrary set. For a fixed $x \in \mathcal{X}$, the *evaluational functional* $\mathcal{L}_x : \mathcal{H} \to \mathbb{R}$ is defined as

$$\mathcal{L}_x f \triangleq f(x).$$

Note that the evaluational functional $\mathcal{L}_x$ maps a function to a real value while the function f maps a point x to a real value. $\mathcal{L}_x$ applies to all functions in $\mathcal{H}$ with a fixed x. Evaluational functionals are linear since $\mathcal{L}_x(\alpha f + \beta g) = \alpha f(x) + \beta g(x) = \alpha \mathcal{L}_x f + \beta \mathcal{L}_x g$.

Definition A Hilbert space of real-valued functions $\mathcal{H}$ is an RKHS if every evaluational functional is continuous.

Let $\mathcal{H}$ be an RKHS. Then, for each $x \in \mathcal{X}$, the evaluational functional $\mathcal{L}_x f = f(x)$ is continuous. By the Riesz representation theorem, there exists an element R_x in $\mathcal{H}$ such that

$$\mathcal{L}_x f = f(x) = (R_x, f),$$

where the dependence of the representer on x is expressed explicitly as R_x. Consider $R_x(z)$ as a bivariate function of x and z and let $R(x, z) \triangleq R_x(z)$. The bivariate function $R(x, z)$ is called the *reproducing kernel* (RK) of an RKHS $\mathcal{H}$. The term reproducing kernel comes from the fact that $(R_x, R_z) = R(x, z)$. It is easy to check that an RK is *nonnegative definite*. That is, R is symmetric $R(x, z) = R(z, x)$, and for any $\alpha_1, \ldots, \alpha_n \in \mathbb{R}$ and $x_1, \ldots, x_n \in \mathcal{X}$,

$$\sum_{i,j=1}^{n} \alpha_i \alpha_j R(x_i, x_j) \geq 0.$$

Therefore, every RKHS has a unique RK that is nonnegative definite. Conversely, an RKHS can be constructed based on a nonnegative definite function.

Moore–Aronszajn theorem
For every nonnegative definite function R on $\mathcal{X} \times \mathcal{X}$, there exists a unique RKHS on $\mathcal{X}$ with R as its RK.

The above results indicate that there exists an one-to-one correspondence between RKHS's and nonnegative definite functions. For a finite dimensional space $\mathcal{H}$ with an orthonormal basis $\phi_1(x), \ldots, \phi_p(x)$, it is easy to see that

$$R(x, z) \triangleq \sum_{i=1}^{p} \phi_i(x) \phi_i(z)$$

is the RK of $\mathcal{H}$.

The following definitions and results are useful for the construction and decomposition of model spaces. $\mathcal{S}$ is called a *subspace* of a Hilbert space $\mathcal{H}$ if $\mathcal{S} \subset \mathcal{H}$ and $\alpha f + \beta g \in \mathcal{S}$ for every $\alpha, \beta \in \mathbb{R}$ and $f, g \in \mathcal{S}$. A closed subspace $\mathcal{S}$ is a Hilbert space. The *orthogonal complement* of $\mathcal{S}$ is defined as

$$\mathcal{S}^{\perp} \triangleq \{f \in \mathcal{H} : (f, g) = 0 \text{ for all } g \in \mathcal{S}\}.$$

$\mathcal{S}^{\perp}$ is a closed subspace of $\mathcal{H}$. If $\mathcal{S}$ is a closed subspace of a Hilbert space $\mathcal{H}$, then every element $f \in \mathcal{H}$ has a unique decomposition in the form $f = g + h$, where $g \in \mathcal{S}$ and $h \in \mathcal{S}^{\perp}$. Equivalently, $\mathcal{H}$ is decomposed into two subspaces $\mathcal{H} = \mathcal{S} \oplus \mathcal{S}^{\perp}$. This decomposition is called a *tensor sum decomposition*, and elements g and h are called *projections* onto $\mathcal{S}$ and $\mathcal{S}^{\perp}$, respectively. Sometimes the notation $\mathcal{H} \ominus \mathcal{S}$ will be used to denote the subspaces $\mathcal{S}^{\perp}$. Tensor sum decomposition with more than two subspaces can be defined recursively.

All closed subspaces of an RKHS are RKHS's. If $\mathcal{H} = \mathcal{H}_0 \oplus \mathcal{H}_1$ and R, R_0, and R_1 are RKs of $\mathcal{H}$, $\mathcal{H}_0$, and $\mathcal{H}_1$ respectively, then $R = R_0 + R_1$. Suppose $\mathcal{H}$ is a Hilbert space and $\mathcal{H} = \mathcal{H}_0 \oplus \mathcal{H}_1$. If $\mathcal{H}_0$ and $\mathcal{H}_1$ are RKHS's with RKs R_0 and R_1, respectively, then $\mathcal{H}$ is an RKHS with RK $R = R_0 + R_1$.

2.2 Model Space for Polynomial Splines

Before introducing the general smoothing spline models, it is instructive to see how the polynomial splines introduced in Section 1.2 can be derived under the RKHS setup. Again, consider the regression model

$$y_i = f(x_i) + \epsilon_i, \quad i = 1, \dots, n, \tag{2.1}$$

where the domain of the function f is $\mathcal{X} = [a, b]$ and the model space for f is the Sobolev space $W_2^m[a, b]$ defined in (1.4). The smoothing spline estimate $\hat{f}$ is the solution to the PLS (1.8).

Model space construction and decomposition of $W_2^m[a, b]$
The Sobolev space $W_2^m[a, b]$ is an RKHS with the inner product

$$(f, g) = \sum_{\nu=0}^{m-1} f^{(\nu)}(a) g^{(\nu)}(a) + \int_a^b f^{(m)} g^{(m)} dx. \tag{2.2}$$

Furthermore, $W_2^m[a, b] = \mathcal{H}_0 \oplus \mathcal{H}_1$, where

$$\mathcal{H}_0 = span\{1, (x-a), \dots, (x-a)^{m-1}/(m-1)!\},$$
$$\mathcal{H}_1 = \{f: \ f^{(\nu)}(a) = 0, \ \nu = 0, \dots, m-1, \ \int_a^b (f^{(m)})^2 dx < \infty\}, \tag{2.3}$$

are RKHS's with corresponding RKs

$$R_0(x, z) = \sum_{\nu=1}^m \frac{(x-a)^{\nu-1}}{(\nu-1)!} \frac{(z-a)^{\nu-1}}{(\nu-1)!},$$
$$R_1(x, z) = \int_a^b \frac{(x-u)_+^{m-1}}{(m-1)!} \frac{(z-u)_+^{m-1}}{(m-1)!} du. \tag{2.4}$$

The function $(x)_+ = \max\{x, 0\}$.

Details about the foregoing construction can be found in Schumaker (2007). It is clear that $\mathcal{H}_0$ contains the polynomial of order m in the Taylor expansion. Note that the basis listed in (2.3), $\phi_\nu(x) = (x - a)^{\nu-1}/(\nu - 1)!$ for $\nu = 1, \ldots, m$, is an orthonormal basis of $\mathcal{H}_0$. For any $f \in \mathcal{H}_1$, it is easy to check that

$$
\begin{aligned}
f(x) &= \int_a^x f'(u)du = \cdots = \int_a^x dx_1 \int_a^{x_1} dx_2 \cdots \int_a^{x_{m-1}} f^{(m)}(u)du \\
&= \int_a^x dx_1 \int_a^{x_1} dx_2 \cdots \int_a^{x_{m-2}} (x_{m-2} - u)f^{(m)}(u)du = \cdots \\
&= \int_a^x \frac{(x-u)^{m-1}}{(m-1)!} f^{(m)}(u)du.
\end{aligned}
$$

Thus the subspace $\mathcal{H}_1$ contains the remainder term in the Taylor expansion.

Denote P_1 as the orthogonal projection operator onto $\mathcal{H}_1$. From the definition of the inner product, the roughness penalty

$$
\int_a^b (f^{(m)})^2 dx = ||P_1 f||^2. \tag{2.5}
$$

Therefore, $\int_a^b (f^{(m)})^2 dx$ measures the distance between f and the parametric polynomial space $\mathcal{H}_0$. There is no penalty to functions in $\mathcal{H}_0$.

The PLS (1.8) can be rewritten as

$$
\frac{1}{n} \sum_{i=1}^n (y_i - f(x_i))^2 + \lambda ||P_1 f||^2. \tag{2.6}
$$

The solution to (2.6) will be given for the general case in Section 2.4. The above setup for polynomial splines suggests the following ingredients for the construction of a general smoothing spline model:

1. An RKHS $\mathcal{H}$ as the model space for f
2. A decomposition of the model space into two subspaces, $\mathcal{H} = \mathcal{H}_0 \oplus \mathcal{H}_1$, where $\mathcal{H}_0$ consists of functions that are not penalized
3. A penalty $||P_1 f||^2$

Based on prior knowledge and purpose of the study, different choices can be made on the model space, its decomposition, and the penalty. These options make the spline smoothing method flexible and versatile. Choices of these options will be illustrated throughout the book.

2.3   General Smoothing Spline Regression Models

A general *smoothing spline regression* (SSR) model assumes that

$$y_i = f(x_i) + \epsilon_i, \quad i = 1, \ldots, n, \tag{2.7}$$

where y_i are observations of the function f evaluated at design points x_i, and ϵ_i are zero-mean independent random errors with a common variance σ^2. To deal with different situations in a unified fashion, let the domain of the function f be an *arbitrary* set $\mathcal{X}$, and the model space be an RKHS $\mathcal{H}$ on $\mathcal{X}$ with RK $R(x, z)$. The choice of $\mathcal{H}$ depends on several factors including the domain $\mathcal{X}$ and prior knowledge about the function f. Suppose $\mathcal{H}$ can be decomposed into two subspaces,

$$\mathcal{H} = \mathcal{H}_0 \oplus \mathcal{H}_1, \tag{2.8}$$

where $\mathcal{H}_0$ is a finite dimensional space with basis functions $\phi_1(x), \ldots,$ $\phi_p(x)$, and $\mathcal{H}_1$ is an RKHS with RK $R_1(x, z)$. $\mathcal{H}_0$, often referred to as the *null space*, consists of functions that are not penalized. In addition to the construction for polynomial splines in Section 2.2, specific constructions of commonly used model spaces will be discussed in Sections 2.6–2.11. The decomposition (2.8) is equivalent to decomposing the function

$$f = f_0 + f_1, \tag{2.9}$$

where f_0 and f_1 are projections onto $\mathcal{H}_0$ and $\mathcal{H}_1$, respectively. The component f_0 represents a linear regression model in space $\mathcal{H}_0$, and the component f_1 represents systematic variation not explained by f_0. Therefore, the magnitude of f_1 can be used to check or test if the parametric model is appropriate. Projections f_0 and f_1 will be referred to as the "parametric" and "smooth" components, respectively.

Sometimes observations of f are made indirectly through linear functionals. For example, f may be observed in the form $\int_a^b w_i(x)f(x)dx$ where w_i are known functions. Another example is that observations are taken on the derivatives $f'(x_i)$. Therefore, it is useful to consider an even more general SSR model

$$y_i = \mathcal{L}_i f + \epsilon_i, \quad i = 1, \ldots, n, \tag{2.10}$$

where $\mathcal{L}_i$ are bounded linear functionals on $\mathcal{H}$. Model (2.7) is a special case of (2.10) with $\mathcal{L}_i$ being evaluational functionals at design points defined as $\mathcal{L}_i f = f(x_i)$. By the definition of an RKHS, these evaluational functionals are bounded.

2.4 Penalized Least Squares Estimation

The estimation method will be presented for the general model (2.10). The smoothing spline estimate of f, $\hat{f}$, is the minimizer of the PLS

$$\frac{1}{n}\sum_{i=1}^{n}(y_i - \mathcal{L}_i f)^2 + \lambda\|P_1 f\|^2, \qquad (2.11)$$

where λ is a smoothing parameter controlling the balance between the goodness-of-fit measured by the least squares and departure from the null space $\mathcal{H}_0$ measured by $\|P_1 f\|^2$. Functions in $\mathcal{H}_0$ are not penalized since $\|P_1 f\|^2 = 0$ when $f \in \mathcal{H}_0$. Note that $\hat{f}$ depends on λ even though the dependence is not expressed explicitly. Estimation procedures presented in this chapter assume that the λ has been fixed. The impact of the smoothing parameter and methods of selecting it will be discussed in Chapter 3.

Since $\mathcal{L}_i$ are bounded linear functionals, by the Riesz representation theorem, there exists a representer $\eta_i \in \mathcal{H}$ such that $\mathcal{L}_i f = (\eta_i, f)$. For a fixed x, consider $R_x(z) \triangleq R(x, z)$ as a univariate function of z. Then, by properties of the reproducing kernel, we have

$$\eta_i(x) = (\eta_i, R_x) = \mathcal{L}_i R_x = \mathcal{L}_{i(z)} R(x, z), \qquad (2.12)$$

where $\mathcal{L}_{i(z)}$ indicates that $\mathcal{L}_i$ is applied to what follows as a function of z. Equation (2.12) implies that the representer η_i can be obtained by applying the operator to the RK R. Let $\xi_i = P_1 \eta_i$ be the projection of η_i onto $\mathcal{H}_1$. Since $R(x, z) = R_0(x, z) + R_1(x, z)$, where R_0 and R_1 are RKs of $\mathcal{H}_0$ and $\mathcal{H}_1$, respectively, and P_1 is self-adjoint such that $(P_1 g, h) = (g, P_1 h)$ for any $g, h \in \mathcal{H}$, we have

$$\xi_i(x) = (\xi_i, R_x) = (P_1 \eta_i, R_x) = (\eta_i, P_1 R_x) = \mathcal{L}_{i(z)} R_1(x, z). \qquad (2.13)$$

Equation (2.13) implies that the representer ξ_i can be obtained by applying the operator to the RK R_1. Furthermore, $(\xi_i, \xi_j) = \mathcal{L}_{i(x)} \xi_j(x) = \mathcal{L}_{i(x)} \mathcal{L}_{j(z)} R_1(x, z)$. Denote

$$\begin{aligned} T &= \{\mathcal{L}_i \phi_\nu\}_{i=1\ \nu=1}^{n\quad p}, \\ \Sigma &= \{\mathcal{L}_{i(x)} \mathcal{L}_{j(z)} R_1(x, z)\}_{i,j=1}^{n}, \end{aligned} \qquad (2.14)$$

where T is an $n \times p$ matrix, and Σ is an $n \times n$ matrix. For the special case of evaluational functionals $\mathcal{L}_i f = f(x_i)$, we have $\xi_i(x) = R_1(x, x_i)$, $T = \{\phi_\nu(x_i)\}_{i=1\ \nu=1}^{n\quad p}$, and $\Sigma = \{R_1(x_i, x_j)\}_{i,j=1}^{n}$.

Write the estimate $\hat{f}$ as

$$\hat{f}(x) = \sum_{\nu=1}^{p} d_\nu \phi_\nu(x) + \sum_{i=1}^{n} c_i \xi_i(x) + \rho,$$

where $\rho \in \mathcal{H}_1$ and $(\rho, \xi_i) = 0$ for $i = 1, \ldots, n$. Since $\xi_i = P_1 \eta_i$, then η_i can be written as $\eta_i = \zeta_i + \xi_i$, where $\zeta_i \in \mathcal{H}_0$. Therefore,

$$\mathcal{L}_i \rho = (\eta_i, \rho) = (\zeta_i, \rho) + (\xi_i, \rho) = 0. \tag{2.15}$$

Let $\boldsymbol{y} = (y_1, \ldots, y_n)^T$ and $\hat{\boldsymbol{f}} = (\mathcal{L}_1 \hat{f}, \ldots, \mathcal{L}_n \hat{f})^T$ be the vectors of observations and fitted values, respectively. Let $\boldsymbol{d} = (d_1, \ldots, d_p)^T$ and $\boldsymbol{c} = (c_1, \ldots, c_n)^T$. From (2.15), we have

$$\hat{\boldsymbol{f}} = T\boldsymbol{d} + \Sigma\boldsymbol{c}. \tag{2.16}$$

Furthermore, $||P_1 f||^2 = ||\sum_{i=1}^{n} c_i \xi_i + \rho||^2 = \boldsymbol{c}^T \Sigma \boldsymbol{c} + ||\rho||^2$. Then the PLS (2.11) becomes

$$\frac{1}{n}||\boldsymbol{y} - T\boldsymbol{d} - \Sigma\boldsymbol{c}||^2 + \lambda \boldsymbol{c}^T \Sigma \boldsymbol{c} + ||\rho||^2. \tag{2.17}$$

It is obvious that (2.17) is minimized when $\rho = 0$, which leads to the following result in Kimeldorf and Wahba (1971).

Kimeldorf–Wahba representer theorem
Suppose T is of full column rank. Then the PLS (2.11) has a unique minimizer given by

$$\hat{f}(x) = \sum_{\nu=1}^{p} d_\nu \phi_\nu(x) + \sum_{i=1}^{n} c_i \xi_i(x). \tag{2.18}$$

The above theorem indicates that the smoothing spline estimate $\hat{f}$ falls in a finite dimensional space. Equation (2.18) represents the smoothing spline estimate $\hat{f}$ as a linear combination of basis of $\mathcal{H}_0$ and representers in $\mathcal{H}_1$. Coefficients $\boldsymbol{c}$ and $\boldsymbol{d}$ need to be estimated from data. Based on (2.18), the PLS (2.17) reduces to

$$\frac{1}{n}||\boldsymbol{y} - T\boldsymbol{d} - \Sigma\boldsymbol{c}||^2 + \lambda \boldsymbol{c}^T \Sigma \boldsymbol{c}. \tag{2.19}$$

Taking the first derivatives leads to the following equations for $\boldsymbol{c}$ and $\boldsymbol{d}$:

$$\begin{aligned}
(\Sigma + n\lambda I)\Sigma\boldsymbol{c} + \Sigma T\boldsymbol{d} &= \Sigma\boldsymbol{y}, \\
T^T \Sigma \boldsymbol{c} + T^T T \boldsymbol{d} &= T^T \boldsymbol{y},
\end{aligned} \tag{2.20}$$

where I is the identity matrix. Equations in (2.20) are equivalent to

$$\begin{pmatrix} \Sigma + n\lambda I & \Sigma T \\ T^T & T^T T \end{pmatrix} \begin{pmatrix} \Sigma c \\ d \end{pmatrix} = \begin{pmatrix} \Sigma y \\ T^T y \end{pmatrix}.$$

There may be multiple sets of solutions for c when Σ is singular. Nevertheless, all sets of solutions lead to the same estimate of the function $\hat{f}$ (Gu 2002). Therefore, it is only necessary to derive one set of solutions. Consider the following equations

$$(\Sigma + n\lambda I)c + Td = y,$$
$$T^T c = 0. \tag{2.21}$$

It is easy to see that a set of solutions to (2.21) is also a set of solutions to (2.20). The solutions to (2.21) are

$$d = (T^T M^{-1} T)^{-1} T^T M^{-1} y,$$
$$c = M^{-1} \{I - T(T^T M^{-1} T)^{-1} T^T M^{-1}\} y, \tag{2.22}$$

where $M = \Sigma + n\lambda I$.

To compute the coefficients c and d, consider the QR decomposition of T,

$$T = (Q_1 \ Q_2) \begin{pmatrix} R \\ 0 \end{pmatrix},$$

where Q_1, Q_2, and R are $n \times p$, $n \times (n - p)$, and $p \times p$ matrices; $Q = (Q_1 \ Q_2)$ is an orthogonal matrix; and R is upper triangular and invertible. Since $T^T c = R^T Q_1^T c = 0$, we have $Q_1^T c = 0$ and $c = QQ^T c = (Q_1 Q_1^T + Q_2 Q_2^T)c = Q_2 Q_2^T c$. Multiplying the first equation in (2.21) by Q_2^T and using the fact that $Q_2^T T = 0$, we have $Q_2^T M Q_2 Q_2^T c = Q_2^T y$. Therefore,

$$c = Q_2 (Q_2^T M Q_2)^{-1} Q_2^T y. \tag{2.23}$$

Multiplying the first equation in (2.21) by Q_1^T, we have $Rd = Q_1^T(y - Mc)$. Thus,

$$d = R^{-1} Q_1^T (y - Mc). \tag{2.24}$$

Equations (2.23) and (2.24) will be used to compute coefficients c and d.

Based on (2.16), the first equation in (2.21) and equation (2.23), the fitted values

$$\hat{f} = Td + \Sigma c = y - n\lambda c = H(\lambda)y, \tag{2.25}$$

where

$$H(\lambda) \triangleq I - n\lambda Q_2 (Q_2^T M Q_2)^{-1} Q_2^T \tag{2.26}$$

is the so-called *hat* (*influence, smoothing*) matrix. The dependence of the hat matrix on the smoothing parameter λ is expressed explicitly. Note that equation (2.25) provides the fitted values while equation (2.18) can be used to compute estimates at any values of x.

2.5   The `ssr` Function

The R function `ssr` in the `assist` package is designed to fit SSR models. After deciding the model space and the penalty, the estimate $\hat{f}$ is completely decided by y, T, and Σ. Therefore, these terms need to be specified in the `ssr` function. A typical call is

```
ssr(formula, rk)
```

where `formula` and `rk` are required arguments. Together they specify y, T, and Σ. Suppose the vector y and matrices T and Σ have been created in R. Then, `formula` lists y on the left-hand side, and T matrix on the right-hand side of an operator ~. The argument `rk` specifies the matrix Σ.

In the most common situation where $\mathcal{L}_i$ are evaluational functionals, the fitting can be greatly simplified since $\mathcal{L}_i$ are decided by design points x_i. There is no need to compute T and Σ matrices before calling the `ssr` function. Instead, they can be computed internally. Specifically, a direct approach to fit the standard SSR model (2.7) is to list y on the left-hand side and $\phi_1(x), \ldots, \phi_p(x)$ on the right-hand side of an operator ~ in the `formula`, and to specify a function for computing R_1 in the `rk` argument. Functions for computing the RKs of some commonly used RKHS's are available in the `assist` package. Users can easily write their own functions for computing RKs.

There are several optional arguments for the `ssr` function, some of which will be discussed in the following chapters. In particular, methods for selecting the smoothing parameter λ will be discussed in Chapter 3. For simplicity, unless explicitly specified, all examples in this chapter use the default method that selects λ using the generalized cross-validation criterion. Bayesian and bootstrap confidence intervals for fitted functions are constructed based on the methods in Section 3.8.

We now show how to fit polynomial splines to the motorcycle data. Consider the construction of polynomial splines in Section 2.2. For simplicity, we first consider the special cases of polynomial splines with $m = 1$ and $m = 2$, which are called *linear* and *cubic splines*, respectively. Denote $x \wedge z = \min\{x, z\}$ and $x \vee z = \max\{x, z\}$. Based on (2.3) and (2.4), Table 2.1 lists bases for null spaces and RKs of linear and cubic splines for the special domain $\mathcal{X} = [0, b]$.

TABLE 2.1 Bases of null spaces and RKs of linear and cubic splines under the construction in Section 2.2 with $\mathcal{X} = [0, b]$

m	Spline	ϕ_ν	R_0	R_1
1	Linear	1	1	$x \wedge z$
2	Cubic	$1, x$	$1 + xz$	$(x \wedge z)^2 \{3(x \vee z) - x \wedge z\}/6$

Functions `linear2` and `cubic2` in the `assist` package compute evaluations of R_1 in Table 2.1 for linear and cubic splines, respectively. Functions for higher-order polynomial splines are also available. Note that the domain for functions `linear2` and `cubic2` is $\mathcal{X} = [0, b]$ for any fixed $b > 0$. The RK on the general domain $\mathcal{X} = [a, b]$ can be calculated by a translation, for example, `cubic2(x-a)`.

To fit a cubic spline to the motorcycle data, one may create matrices T and Σ first and then call the `ssr` function:

```
> T <- cbind(1, times)
> Sigma <- cubic2(times)
> ssr(accel~T-1, rk=Sigma)
```

The intercept is automatically included in the `formula` statement. Therefore, `T-1` is used to exclude the intercept since it is already included in the T matrix.

Since $\mathcal{L}_i$ are evaluational functionals for the motorcycle example, the `ssr` function can be called directly:

```
> ssr(accel~times, rk=cubic2(times))
```

The inputs for `formula` and `rk` can be modified for fitting polynomial splines of different orders. For example, the following statements fit linear, quintic ($m = 3$), and septic ($m = 4$) splines:

```
> ssr(accel~1, rk=linear2(times))
> ssr(accel~times+I(times^2), rk=quintic2(times))
> ssr(accel~times+I(times^2)+I(times^3), rk=septic2(times))
```

The linear and cubic spline fits are shown in Figure 2.1.

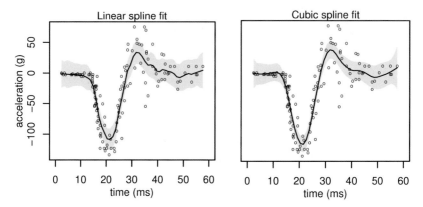

FIGURE 2.1   Motorcycle data, plots of observations (circles), the linear spline fit (left), and the cubic spline fit (right) as solid lines, and 95% Bayesian confidence intervals (shaded regions).

2.6   Another Construction for Polynomial Splines

One construction of RKHS for the polynomial spline was presented in Section 2.2. This section presents an alternative construction for $W_2^m[0, 1]$ on the domain $\mathcal{X} = [0, 1]$. In practice, without loss of generality, a continuous interval $[a, b]$ can always be transformed into $[0, 1]$.

Let $k_r(x) = B_r(x)/r!$ be scaled Bernoulli polynomials where B_r are defined recursively by $B_0(x) = 1$, $B_r'(x) = rB_{r-1}(x)$ and $\int_0^1 B_r(x)dx = 0$ for $r = 1, 2, \ldots$ (Abramowitz and Stegun 1964). The first four scaled Bernoulli polynomials are

$$k_0(x) = 1,$$
$$k_1(x) = x - 0.5,$$
$$k_2(x) = \frac{1}{2}\left\{ k_1^2(x) - \frac{1}{12} \right\}, \qquad (2.27)$$
$$k_4(x) = \frac{1}{24}\left\{ k_1^4(x) - \frac{1}{2}k_1^2(x) + \frac{7}{240} \right\}.$$

Alternative model space construction and decomposition of $W_2^m[0, 1]$

The Sobolev space $W_2^m[0, 1]$ is an RKHS with the inner product

$$(f, g) = \sum_{\nu=0}^{m-1}\left(\int_0^1 f^{(\nu)}dx \right)\left(\int_0^1 g^{(\nu)}dx \right) + \int_0^1 f^{(m)}g^{(m)}dx. \qquad (2.28)$$

Furthermore, $W_2^m[0,1] = \mathcal{H}_0 \oplus \mathcal{H}_1$, where

$$\mathcal{H}_0 = span\{k_0(x), k_1(x), \ldots, k_{m-1}(x)\},$$

$$\mathcal{H}_1 = \{f : \int_0^1 f^{(\nu)} dx = 0, \ \nu = 0, \ldots, m-1, \quad (2.29)$$

$$\int_0^1 (f^{(m)})^2 dx < \infty\},$$

are RKHS's with corresponding RKs

$$R_0(x, z) = \sum_{\nu=0}^{m-1} k_\nu(x) k_\nu(z),$$

$$\quad (2.30)$$

$$R_1(x, z) = k_m(x) k_m(z) + (-1)^{m-1} k_{2m}(|x - z|).$$

The foregoing alternative construction was derived by Craven and Wahba (1979). Note that the inner product (2.28) is different from (2.2). Again, $\mathcal{H}_0$ contains polynomials, and the basis listed in (2.29), $\phi_\nu(x) = k_{\nu-1}(x)$ for $\nu = 1, \ldots, m$, is an orthonormal basis of $\mathcal{H}_0$. Denote P_1 as the orthogonal projection operator onto $\mathcal{H}_1$. From the definition of the inner product, the roughness penalty $\int_0^1 (f^{(m)})^2 dx = ||P_1 f||^2$.

Based on (2.29) and (2.30), Table 2.2 lists bases for the null spaces and RKs of linear and cubic splines under the alternative construction in this section.

TABLE 2.2 Bases of the null spaces and RKs for linear and cubic splines under the construction in Section 2.6 with $\mathcal{X} = [0, 1]$

m	Spline	ϕ_ν	R_0	R_1
1	Linear	1	1	$k_1(x)k_1(z) + k_2(|x - z|)$
2	Cubic	$1, k_1(x)$	$1 + k_1(x)k_1(z)$	$k_2(x)k_2(z) - k_4(|x - z|)$

Functions `linear` and `cubic` in the `assist` package compute evaluations of R_1 in Table 2.2 for linear and cubic splines respectively. Functions for higher-order polynomial splines are also available. Note that the domain under construction in this section is restricted to $[0, 1]$. Thus the `scale` option is needed when the domain is not $[0, 1]$. For example, the following statements fit linear and cubic splines to the motorcycle data:

```
> ssr(accel~1, rk=linear(times), scale=T)
> ssr(accel~times, rk=cubic(times), scale=T)
```

The `scale` option scales the independent variable `times` into the interval $[0, 1]$. It is a good practice to scale a variable first before fitting. For example, the following statements lead to the same cubic spline fit:

```
> x <- (times-min(times))/(max(times)-min(times))
> ssr(accel~x, rk=cubic(x))
```

2.7   Periodic Splines

Many natural phenomena follow a cyclic pattern. For example, many biochemical, physiological, or behavioral processes in living beings follow a daily cycle called circadian rhythm, and many Earth processes follow an annual cycle. In these cases the mean function f is known to be a smooth *periodic* function. Without loss of generality, assume that the domain of the function $\mathcal{X} = [0, 1]$ and f is a periodic function on $[0, 1]$. Since periodic functions can be regarded as functions defined on the unit circle, *periodic splines* are often referred to as *splines on the circle*.

The model space for periodic spline of order m is

$$W_2^m(per) = \left\{ f : \ f^{(j)} \text{ are absolutely continuous, } f^{(j)}(0) = f^{(j)}(1), \right.$$
$$\left. j = 0, \dots, m-1, \ \int_0^1 (f^{(m)})^2 dx < \infty \right\}. \qquad (2.31)$$

Craven and Wahba (1979) derived the following construction.

Model space construction and decomposition of $W_2^m(per)$
The space $W_2^m(per)$ *is an RKHS with inner product*

$$(f, g) = \left(\int_0^1 f dx \right) \left(\int_0^1 g dx \right) + \int_0^1 f^{(m)} g^{(m)} dx.$$

Furthermore, $W_2^m(per) = \mathcal{H}_0 \oplus \mathcal{H}_1$, *where*

$$\mathcal{H}_0 = span\{1\},$$
$$\mathcal{H}_1 = \left\{ f \in W_2^m(per) : \int_0^1 f dx = 0 \right\}, \qquad (2.32)$$

are RKHS's with corresponding RKs

$$R_0(x, z) = 1,$$
$$R_1(x, z) = (-1)^{m-1} k_{2m}(|x - z|). \tag{2.33}$$

Again, the roughness penalty $\int_0^1 (f^{(m)})^2 dx = ||P_1 f||^2$. The function `periodic` in the `assist` library calculates R_1 in (2.33). The order m is specified by the argument `order`. The default is a cubic periodic spline with `order=2`.

We now illustrate how to fit a periodic spline using the Arosa data, which contain monthly mean ozone thickness (Dobson units) in Arosa, Switzerland, from 1926 to 1971. Suppose we want to investigate how ozone thickness changes over months in a year. It is reasonable to assume that the mean ozone thickness is a periodic function of month. Let `thick` be the dependent variable and x be the independent variable `month` scaled into the interval $[0, 1]$. The following statements fit a cubic periodic spline:

```
> data(Arosa); Arosa$x <- (Arosa$month-0.5)/12
> ssr(thick~1, rk=periodic(x), data=Arosa)
```

The fit of the periodic spline is shown in Figure 2.2.

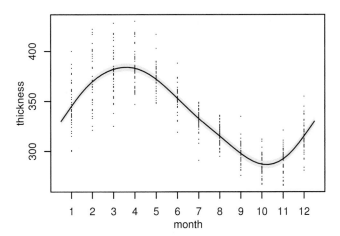

FIGURE 2.2 Arosa data, plot of observations (points), and the periodic spline fits (solid line). The shaded region represents 95% Bayesian confidence intervals.

2.8   Thin-Plate Splines

Suppose f is a function of a multivariate independent variable $\boldsymbol{x} = (x_1, \ldots, x_d) \in \mathbb{R}^d$, where $\mathbb{R}^d$ is the Euclidean d-space. Assume the regression model

$$y_i = f(\boldsymbol{x}_i) + \epsilon_i, \quad i = 1, \ldots, n, \tag{2.34}$$

where $\boldsymbol{x}_i = (x_{i1}, \ldots, x_{id})$ and ϵ_i are zero-mean independent random errors with a common variance σ^2.

Define the model space for a *thin-plate spline* as

$$W_2^m(\mathbb{R}^d) = \{f : \; J_m^d(f) < \infty\}, \tag{2.35}$$

where

$$J_m^d(f) = \sum_{\alpha_1 + \cdots + \alpha_d = m} \frac{m!}{\alpha_1! \ldots \alpha_d!} \int_{-\infty}^{\infty} \cdots \int_{-\infty}^{\infty} \left(\frac{\partial^m f}{\partial x_1^{\alpha_1} \ldots \partial x_d^{\alpha_d}} \right)^2 \prod_{j=1}^{d} dx_j. \tag{2.36}$$

Since $J_m^d(f)$ is invariant under a rotation of the coordinates, the thin-plate spline is especially well suited for spatial data (Wahba 1990, Gu 2002).

Define an inner product as

$$(f, g) = \sum_{\alpha_1 + \cdots + \alpha_d = m} \frac{m!}{\alpha_1! \ldots \alpha_d!} \int_{-\infty}^{\infty} \cdots \int_{-\infty}^{\infty}$$

$$\left(\frac{\partial^m f}{\partial x_1^{\alpha_1} \ldots \partial x_d^{\alpha_d}} \right) \left(\frac{\partial^m g}{\partial x_1^{\alpha_1} \ldots \partial x_d^{\alpha_d}} \right) \prod_{j=1}^{d} dx_j. \tag{2.37}$$

Model space construction of $W_2^m(\mathbb{R}^d)$
With the inner product (2.37), $W_2^m(\mathbb{R}^d)$ is an RKHS if and only if $2m - d > 0$.

Details can be found in Duchon (1977) and Meinguet (1979). A thin-plate spline estimate is the minimizer to the PLS

$$\frac{1}{n} \sum_{i=1}^{n} (y_i - f(\boldsymbol{x}_i))^2 + \lambda J_m^d(f) \tag{2.38}$$

in $W_2^m(\mathbb{R}^d)$. The null space $\mathcal{H}_0$ of the penalty functional $J_m^d(f)$ is the space spanned by polynomials in d variables of total degree up to $m - 1$.

Thus the dimension of the null space $p = \begin{pmatrix} d+m-1 \\ d \end{pmatrix}$. For example, when $d = 2$ and $m = 2$,

$$J_2^2(f) = \int_{-\infty}^{\infty} \int_{-\infty}^{\infty} \left\{ \left(\frac{\partial^2 f}{\partial x_1^2} \right)^2 + 2 \left(\frac{\partial^2 f}{\partial x_1 \partial x_2} \right)^2 + \left(\frac{\partial^2 f}{\partial x_2^2} \right)^2 \right\} dx_1 dx_2,$$

and the null space is spanned by $\phi_1(\boldsymbol{x}) = 1$, $\phi_2(\boldsymbol{x}) = x_1$, and $\phi_3(\boldsymbol{x}) = x_2$.

In general, denote $\phi_1, \dots, \phi_p$ as the p polynomials of total degree up to $m-1$ that span $\mathcal{H}_0$. Denote E_m as the Green function for the m-iterated Laplacian $E_m(\boldsymbol{x}, \boldsymbol{z}) = E(||\boldsymbol{x} - \boldsymbol{z}||)$, where $||\boldsymbol{x} - \boldsymbol{z}||$ is the Euclidean distance and

$$E(u) = \begin{cases} (-1)^{\frac{d}{2}+1+m} |u|^{2m-d} \log |u|, & d \text{ even}, \\ |u|^{2m-d}, & d \text{ odd}. \end{cases}$$

Let $T = \{\phi_\nu(\boldsymbol{x}_i)\}_{i=1}^n{}_{\nu=1}^p$ and $K = \{E_m(\boldsymbol{x}_i, \boldsymbol{x}_j)\}_{i,j=1}^n$. The bivariate function E_m is not the RK of $W_2^m(\mathbb{R}^d)$ since it is not nonnegative definite. Nevertheless, it is *conditionally nonnegative definite* in the sense that $T^T \boldsymbol{c} = \boldsymbol{0}$ implies that $\boldsymbol{c}^T K \boldsymbol{c} \geq 0$. Referred to as a semi-kernel, the function E_m is sufficient for the purpose of estimation. Assume that T is of full column rank. It can be shown that the unique minimizer of the PLS (2.38) is given by (Wahba 1990, Gu 2002)

$$\hat{f}(\boldsymbol{x}) = \sum_{\nu=1}^p d_\nu \phi_\nu(\boldsymbol{x}) + \sum_{i=1}^n c_i \xi_i(\boldsymbol{x}), \tag{2.39}$$

where $\xi_i(\boldsymbol{x}) = E_m(\boldsymbol{x}_i, \boldsymbol{x})$. Therefore, E_m plays the same role as the RK R_1. The coefficients $\boldsymbol{c}$ and $\boldsymbol{d}$ are solutions to

$$\begin{aligned} (K + n\lambda I)\boldsymbol{c} + T\boldsymbol{d} &= \boldsymbol{y}, \\ T^T \boldsymbol{c} &= \boldsymbol{0}. \end{aligned} \tag{2.40}$$

The above equations have the same form as those in (2.21). Therefore, computations in Section 2.4 carry over with Σ being replaced by K. The semi-kernel E_m is calculated by the function `tp.pseudo` in the `assist` package. The order m is specified by the `order` argument with default as `order=2`.

The USA climate data contain average winter temperatures in 1981 from 1214 stations in USA. To investigate how average winter temperature (`temp`) depends on geological locations (`long` and `lat`), we fit a thin-plate spline as follows:

```
> attach(USAtemp)
> ssr(temp~long+lat, rk=tp.pseudo(list(long,lat)))
```

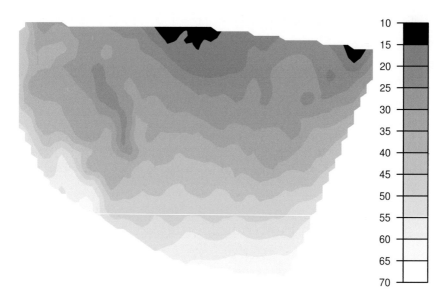

FIGURE 2.3   USA climate data, contour plot of the thin-plate spline fit.

The contour plot of the fit is shown in Figure 2.3.

A genuine RK for $W_2^m(\mathbb{R}^d)$ is needed later in the computation of posterior variances in Chapter 3 and the construction of tensor product splines in Chapter 4. We now discuss briefly how to derive the genuine RK. Define inner product

$$(f, g)_0 = \sum_{j=1}^{J} w_j f(\boldsymbol{u}_j) g(\boldsymbol{u}_j), \qquad (2.41)$$

where $\boldsymbol{u}_j$ are fixed points in $\mathbb{R}^d$, and w_j are fixed positive weights such that $\sum_{j=1}^{J} w_j = 1$. Points $\boldsymbol{u}_j$ and weights w_j are selected in such a way that the matrix $\{(\phi_\nu, \phi_\mu)_0\}_{\nu,\mu=1}^{p}$ is nonsingular. Let $\tilde{\phi}_\nu$, $\nu = 1, \ldots, p$, be an orthonormal basis derived from ϕ_ν with $\tilde{\phi}_1(\boldsymbol{x}) = 1$. Let P_0 be the projection operator onto $\mathcal{H}_0$ defined as $P_0 f = \sum_{\nu=1}^{p}(f, \phi_\nu)_0 \phi_\nu$. Then it can be shown that (Gu 2002)

$$R_0(\boldsymbol{x}, \boldsymbol{z}) = \sum_{\nu=1}^{p} \tilde{\phi}_\nu(\boldsymbol{x}) \tilde{\phi}_\nu(\boldsymbol{z}),$$

$$R_1(\boldsymbol{x}, \boldsymbol{z}) = (I - P_{0(\boldsymbol{x})})(I - P_{0(\boldsymbol{z})}) E(\|\boldsymbol{x} - \boldsymbol{z}\|) \qquad (2.42)$$

are RKs of $\mathcal{H}_0$ and $\mathcal{H}_1 \triangleq W_2^m(\mathbb{R}^d) \ominus \mathcal{H}_0$, where $P_{0(\boldsymbol{x})}$ and $P_{0(\boldsymbol{z})}$ are projections applied to the arguments $\boldsymbol{x}$ and $\boldsymbol{z}$, respectively.

Assume that $T = \{\phi_\nu(\boldsymbol{x}_i)\}_{i=1}^n {}_{\nu=1}^p$ is of full column rank. Let $\Sigma = \{R_1(\boldsymbol{x}_i, \boldsymbol{x}_j)\}_{i,j=1}^n$. One relatively simple approach to compute $\tilde{\phi}_\nu$ and Σ is to let $J = n$, $\boldsymbol{u}_j = \boldsymbol{x}_j$, and $w_j = n^{-1}$. It is easy to see that $\{(\phi_\nu, \phi_\mu)_0\}_{\nu,\mu=1}^p = n^{-1} T^T T$, which is nonsingular. Let

$$T = (Q_1\ Q_2) \begin{pmatrix} R \\ 0 \end{pmatrix}$$

be the QR decomposition of T. Then

$$(\tilde{\phi}_1(\boldsymbol{x}), \ldots, \tilde{\phi}_p(\boldsymbol{x})) = \sqrt{n}(\phi_1(\boldsymbol{x}), \ldots, \phi_p(\boldsymbol{x}))R^{-1}$$

and

$$\Sigma = Q_2 Q_2^T K Q_2 Q_2^T.$$

The function `tp` computes evaluations of R_1 in (2.42) with $J = n$, $\boldsymbol{u}_j = \boldsymbol{x}_j$ and $w_j = n^{-1}$.

2.9 Spherical Splines

Spherical spline, also called *spline on the sphere*, is an extension of both the periodic spline defined on the unit circle and the thin-plate spline defined on $\mathbb{R}^2$. Let the domain be $\mathcal{X} = \mathcal{S}$, where $\mathcal{S}$ is the unit sphere. Any point $\boldsymbol{x}$ on $\mathcal{S}$ can be represented as $\boldsymbol{x} = (\theta, \phi)$, where θ $(0 \leq \theta \leq 2\pi)$ is the longitude and ϕ $(-\pi/2 \leq \phi \leq \pi/2)$ is the latitude. Define

$$J(f) = \begin{cases} \int_0^{2\pi} \int_{-\frac{\pi}{2}}^{\frac{\pi}{2}} (\Delta^{\frac{m}{2}} f)^2 \cos\phi d\phi d\theta, & m \text{ even,} \\ \int_0^{2\pi} \int_{-\frac{\pi}{2}}^{\frac{\pi}{2}} \left\{ \frac{(\Delta^{\frac{m-1}{2}} f)_\theta^2}{\cos^2\phi} + (\Delta^{\frac{m-1}{2}} f)_\phi^2 \right\} \cos\phi d\phi d\theta, & m \text{ odd,} \end{cases}$$

where the notation $(g)_z$ represents the partial derivative of g with respect to z, Δf represents the surface Laplacian on the unit sphere defined as

$$\Delta f = \frac{1}{\cos^2\phi} f_{\theta\theta} + \frac{1}{\cos\phi} (\cos\phi f_\phi)_\phi.$$

Consider the model space

$$W_2^m(\mathcal{S}) = \left\{ f : \left| \int_\mathcal{S} f d\boldsymbol{x} \right| < \infty, \ J(f) < \infty \right\}.$$

Model space construction and decomposition of $W_2^m(\mathcal{S})$
$W_2^m(\mathcal{S})$ *is an RKHS when $m > 1$. Furthermore, $W_2^m(\mathcal{S}) = \mathcal{H}_0 \oplus \mathcal{H}_1$, where*

$$\mathcal{H}_0 = span\{1\},$$

$$\mathcal{H}_1 = \{f \in W_2^m(\mathcal{S}) : \int_{\mathcal{S}} f d\boldsymbol{x} = 0\},$$

are RKHS's with corresponding RKs

$$R_0(\boldsymbol{x}, \boldsymbol{z}) = 1,$$

$$R_1(\boldsymbol{x}, \boldsymbol{z}) = \sum_{i=1}^{\infty} \frac{2i+1}{4\pi} \frac{1}{\{i(i+1)\}^m} G_i(\cos \gamma(\boldsymbol{x}, \boldsymbol{z})),$$

where $\gamma(\boldsymbol{x}, \boldsymbol{z})$ is the angle between $\boldsymbol{x}$ and $\boldsymbol{z}$, and G_i are the Legendre polynomials.

Details of the above construction can be found in Wahba (1981). The penalty $||P_1 f||^2 = J(f)$. The RK R_1 is in the form of an infinite series, which is inconvenient to compute. Closed-form expressions are available only when $m = 2$ and $m = 3$. Wahba (1981) proposed replacing J by a topologically equivalent seminorm Q under which closed-form RKs can be derived. The function `sphere` in the `assist` package calculates R_1 under the seminorm Q for $2 \leq m \leq 6$. The argument `order` specifies m with default as `order=2`.

The world climate data contain average winter temperatures in 1981 from 725 stations around the globe. To investigate how average winter temperature (`temp`) depends on geological locations (`long` and `lat`), we fit a spline on the sphere:

```
> data(climate)
> ssr(temp~1, rk=sphere(cbind(long,lat)), data=climate)
```

The contour plot of the spherical spline fit is shown in Figure 2.4.

2.10   Partial Splines

A *partial spline model* assumes that

$$y_i = \boldsymbol{s}_i^T \boldsymbol{\beta} + \mathcal{L}_i f + \epsilon_i, \quad i = 1, \ldots, n, \tag{2.43}$$

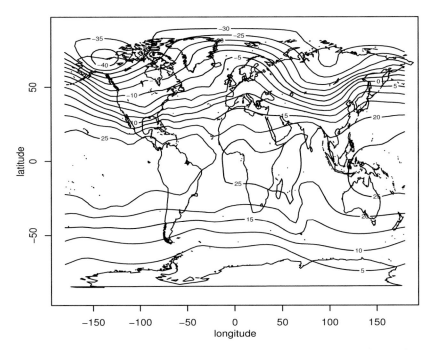

FIGURE 2.4 World climate data, contour plot of the spherical spline fit.

where s is a q-dimensional vector of independent variables, β is a vector of parameters, $\mathcal{L}_i$ are bounded linear functionals, and ϵ_i are zero-mean independent random errors with a common variance σ^2. We assume that $f \in \mathcal{H}$, where $\mathcal{H}$ is an RKHS on an arbitrary domain $\mathcal{X}$. Model (2.43) contains two components: a parametric linear model and a non-parametric function f. The partial spline model is a special case of the *semiparametric linear regression model* discussed in Chapter 8.

Suppose $\mathcal{H} = \mathcal{H}_0 \oplus \mathcal{H}_1$, where $\mathcal{H}_0 = \text{span}\{\phi_1, \ldots, \phi_p\}$ and $\mathcal{H}_1$ is an RKHS with RK R_1. Denote P_1 as the projection onto $\mathcal{H}_1$. The function f and parameters β are estimated as minimizers to the following PLS:

$$\frac{1}{n} \sum_{i=1}^{n} (y_i - s_i^T \beta - \mathcal{L}_i f)^2 + \lambda \|P_1 f\|^2. \tag{2.44}$$

Let $S = (s_1, \ldots, s_n)^T$, $T = \{\mathcal{L}_i \phi_\nu\}_{i=1 \; \nu=1}^{n \quad p}$, $X = (S \; T)$, and $\Sigma = \{\mathcal{L}_{i(x)} \mathcal{L}_{j(z)} R_1(x, z)\}_{i,j=1}^{n}$. Assume that X is of full column rank. Following similar arguments as in Section 2.4, it can be shown that the PLS (2.44) has a unique minimizer, and the solution of f is given in (2.18).

Therefore, the PLS (2.44) reduces to

$$\frac{1}{n}||\boldsymbol{y} - X\boldsymbol{\alpha} - \Sigma\boldsymbol{c}||^2 + \lambda\boldsymbol{c}^T\Sigma\boldsymbol{c},$$

where $\boldsymbol{\alpha} = (\boldsymbol{\beta}^T, \boldsymbol{d}^T)^T$. As in Section 2.4, we can solve $\boldsymbol{\alpha}$ and $\boldsymbol{c}$ from the following equations:

$$
\begin{aligned}
(\Sigma + n\lambda I)\boldsymbol{c} + X\boldsymbol{\alpha} &= \boldsymbol{y}, \\
X^T\boldsymbol{c} &= \boldsymbol{0}.
\end{aligned}
\tag{2.45}
$$

The above equations have the same form as those in (2.21). Thus, computations in Section 2.4 carry over with T and $\boldsymbol{d}$ being replaced by X and $\boldsymbol{\alpha}$, respectively. The `ssr` function can be used to fit partial splines. When $\mathcal{L}_i$ are evaluational functionals, the partial spline model (2.43) can be fitted by adding s variables at the right-hand side of the `formula` argument. When $\mathcal{L}_i$ are not evaluational functionals, matrices X and Σ need to be created and supplied in the `formula` and `rk` arguments.

One interesting application of the partial spline model is to fit a nonparametric regression model with potential change-points. A *change-point* is defined as a discontinuity in the mean function or one of its derivatives. Note that the function $g(x) = (x - t)^k_+$ has a jump in its kth derivative at location t. Therefore, it can be used to model change-points. Specifically, consider the following model

$$y_i = \sum_{j=1}^{J} \beta_j(x_i - t_j)^{k_j}_+ + f(x_i) + \epsilon_i, \quad i = 1, \ldots, n, \tag{2.46}$$

where $x_i \in [a, b]$ are design points, $t_j \in [a, b]$ are change-points, f is a smooth function, and ϵ_i are zero-mean independent random errors with a common variance σ^2. The mean function in model (2.46) has a jump at t_j in its k_jth derivative with magnitude β_j. The choice of model space for f depends on the application. For example, the polynomial or periodic spline space may be used. When t_j and k_j are known, model (2.46) is a special case of partial spline with $q = J$, $s_{ij} = (x_i - t_j)^{k_j}_+$, and $\boldsymbol{s}_i = (s_{i1}, \ldots, s_{iJ})^T$.

We now use the geyser data and motorcycle data to illustrate change-points detection using partial splines. For the geyser data, Figure 1.3(b) indicates that there may be a jump in the mean function between 2.5 and 3.5 minutes. Therefore, we consider the model

$$y_i = \beta(x_i - t)^0_+ + f(x_i) + \epsilon_i, \quad i = 1, \ldots, n, \tag{2.47}$$

where x_i are the duration variable scaled into $[0, 1]$, t is a change-point, and $(x - t)^0_+ = 0$ when $x \leq t$ and 1 otherwise. We assume that $f \in$

$W_2^2[0, 1]$. For a fixed t, say $t = 0.397$, we can fit the partial spline as follows:

```
> attach(faithful)
> x <- (eruptions-min(eruptions))/diff(range(eruptions))
> ssr(waiting~x+(x>.397), rk=cubic(x))
```

The partial spline fit is shown in Figure 2.5(a). No trend is shown in the residual plot in Figure 2.5(b).

The change-point is fixed at $t = 0.397$ in the above fit. Often it is unknown in practice. To search for the location of the change-point t, we compute AIC and GCV (generalized cross-validation) criteria on a grid points between 0.25 and 0.55:

```
> aic <- gcv <- NULL
> for (t in seq(.25,.55,by=.001)) {
  fit <- ssr(waiting~x+(x>t), rk=cubic(x))
  aic <- c(aic, length(x)*log(sum(fit$resi**2))+2*fit$df)
  gcv <- c(gcv, fit$rkpk.obj$score)
  }
```

The vector `fit$resi` contains residuals. The value `fit$df` represents the degrees of freedom $(\mathrm{tr}H(\lambda))$ defined later in Chapter 3, Section 3.2. The GCV criterion is defined in (3.24). Figure 2.5(c) shows the AIC and GCV scores scaled into $[0, 1]$. The scaled scores are identical and reach the minimum in the same region with the middle point at $t = 0.397$.

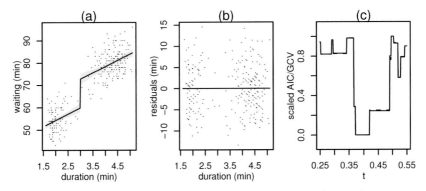

FIGURE 2.5 Geyser data, plots of (a) observations (points), the partial spline fit (solid line) with 95% Bayesian confidence intervals (shaded region), (b) residuals (points) from the partial spline fit and the cubic spline fit (solid line) to the residuals, and (c) the AIC (dashed line) and GCV scores (solid line) scaled into $[0, 1]$.

For the motorcycle data, it is apparent that the mean curve is flat on the left and there is a sharp corner around 15 ms. The linear and cubic splines fit this region with round corners (Figure 2.1). The sharp corner suggests that there may be a change-point in the first derivative of the mean function. Therefore, we consider the model

$$y_i = \beta(x_i - t)_+ + f(x_i) + \epsilon_i, \quad i = 1, \ldots, n, \tag{2.48}$$

where x is the time variable scaled into $[0, 1]$ and t is the change-point in the first derivative. We assume that $f \in W_2^2[0, 1]$. Again, we use the AIC and GCV criteria to search for the location of the change-point t. Figure 2.6(b) shows the scaled AIC and GCV scores. They both reach the minimum at $t = 0.214$. For the fixed $t = 0.214$, the model (2.48) can be fitted as follows:

```
> t <- .214; s <- (x-t)*(x>t)
> ssr(accel~x+s, rk=cubic(x))
```

The partial spline fit is shown in Figure 2.6(a). The sharp corner around 15 ms is preserved. Chapter 3 contains more analysis of potential change-points for the motorcycle data.

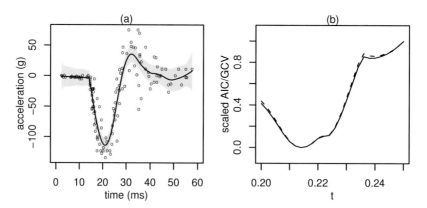

FIGURE 2.6 Motorcycle data, plots of (a) observations (circles), the partial spline fit (solid line) with 95% Bayesian confidence intervals (shaded region), and (b) the AIC (dashed line) and GCV (solid line) scores scaled into $[0, 1]$.

Often, in practice, there is enough knowledge to model some components in the regression function parametrically. For other uncertain components, it may be desirable to leave them unspecified. Combining parametric and nonparametric components, the partial spline (semi-parametric) models are well suited to these situations.

As an illustration, consider the Arosa data. We have investigated how ozone thickness changes over months in a year by fitting a periodic spline (Figure 2.2). Suppose now we want to investigate how ozone thickness changes over time. That is, we need to consider both month (seasonal) effect and year (long-term) effect. Let x_1 and x_2 be the month and year variables scaled into the interval $[0, 1]$. For the purpose of illustration, suppose the seasonal trend can be well approximated by a simple sinusoidal function. The form of long-term trend will be left unspecified. Therefore, we consider the following partial spline model

$$y_i = \beta_1 + \beta_2 \sin(2\pi x_{i1}) + \beta_3 \cos(2\pi x_{i1}) + f(x_{i2}) + \epsilon_i, \qquad (2.49)$$
$$i = 1, \ldots, 518,$$

where y_i represents the average ozone thickness in month x_{i1} of year x_{i2}, and $f \in W_2^2[0, 1] \ominus \{1\}$. Note that the constant functions are removed from the model space for f such that f is identifiable with the constant β_1. The partial spline model (2.49) can be fitted as follows:

```
> x1 <- (Arosa$month-0.5)/12; x2 <- (Arosa$year-1)/45
> ssr(thick~sin(2*pi*x1)+cos(2*pi*x1)+x2, rk=cubic(x2))
```

Estimates of the main effect of month, $\hat{\beta}_2 \sin(2\pi x_1) + \hat{\beta}_3 \cos(2\pi x_1)$, and the main effect of year, $\hat{f}(x_2)$, are shown in Figure 2.7. The simple sinusoidal model for the seasonal trend is too restrictive. More general models for the Arosa data can be found in Chapter 4, Section 4.9.2.

Functional data are observations in a form of functions. The most common forms of functional data are curves defined on a continuous interval and surfaces defined on $\mathbb{R}^2$. A *functional linear model* (FLM) is a linear model that involves functional data as (i) the independent variable, (ii) the dependent variable, or (iii) both independent and dependent variables. Many methods have been developed for fitting FLMs where functional data are curves or surfaces (Ramsay and Silverman 2005). We will use the Canadian weather data to illustrate how to fit FLMs using methods in this book. An FLM corresponding to situation (i) is discussed in this section. FLMs corresponding to situations (ii) and (iii) will be introduced in Chapter 4, Sections 4.9.3, and Chapter 8, Section 8.4.1, respectively. Note that methods illustrated in these sections for functional data apply to general functions defined on arbitrary domains. In particular, they can be used to fit surfaces defined on $\mathbb{R}^2$.

Now consider the Canadian weather data. To investigate the relationship between total annual precipitation and the temperature function,

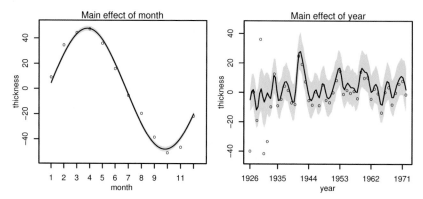

FIGURE 2.7 Arosa data, plots of estimated main effects with 95% Bayesian confidence intervals. A circle in the left panel represents the average thickness for a particular month minus the overall mean. A circle in the right panel represents the average thickness for a particular year minus the overall mean.

consider the following FLM

$$y_i = \beta_1 + \int_0^1 w_i(x)f(x)dx + \epsilon_i, \quad i = 1, \ldots, 35, \qquad (2.50)$$

where y_i is the logarithm of total annual precipitation at station i, β_1 is a constant parameter, x is the variable `month` transformed into $[0, 1]$, $w_i(x)$ is the temperature function at station i, $f(x)$ is an unknown weight function, and ϵ_i are random errors. Model (2.50) is the same as model (15.2) in Ramsay and Silverman (2005). It is an example when the independent variable is a curve. The goal is to estimate the weight function f. It is reasonable to assume that f is a smooth periodic function. Specifically, we model f using cubic periodic spline space $W_2^2(per)$. Write $f(x) = \beta_2 + f_1(x)$ where $f_1 \in W_2^2(per) \ominus \{1\}$. Then model (2.50) can be rewritten as a partial spline model

$$\begin{aligned} y_i &= \beta_1 + \beta_2 \int_0^1 w_i(x)dx + \int_0^1 w_i(x)f_1(x)dx + \epsilon_i \\ &\triangleq \boldsymbol{s}_i^T\boldsymbol{\beta} + \mathcal{L}_i f_1 + \epsilon_i, \end{aligned} \qquad (2.51)$$

where $\boldsymbol{s}_i = (1, \int_0^1 w_i(x)dx)^T$, $\boldsymbol{\beta} = (\beta_1, \beta_2)^T$, and $\mathcal{L}_i f_1 = \int_0^1 w_i(x)f_1(x)dx$. Assume that w_i are square integrable. Then $\mathcal{L}_i$ are bounded linear functionals. Let R_1 be the RK of $W_2^2(per) \ominus \{1\}$. From (2.14), the (i, j)th element of Σ equals

$$
\begin{aligned}
\mathcal{L}_{i(x)}\mathcal{L}_{j(z)}R_1(x,z) &= \int_0^1 \int_0^1 w_i(s)w_j(t)R_1(s,t)dsdt \\
&\approx \frac{1}{144}\sum_{k=1}^{12}\sum_{l=1}^{12} w_i(x_k)w_j(x_l)R_1(x_k,x_l) \\
&= \frac{1}{144}\boldsymbol{w}_i^T R_1(\boldsymbol{x},\boldsymbol{x})\boldsymbol{w}_j,
\end{aligned}
\tag{2.52}
$$

where x_k represents the middle point of month k, $\boldsymbol{x} = (x_1,\ldots,x_{12})^T$, $R_1(\boldsymbol{x},\boldsymbol{x}) = \{R_1(x_k,x_l)\}_{k,l=1}^{12}$, $w_i(x_k)$ is the temperature of month k at station i, and $\boldsymbol{w}_i = (w_i(x_1),\ldots,w_i(x_{12}))^T$. The rectangle rule is used to approximate the integrals. More accurate approximations may be used. Let $W = (\boldsymbol{w}_1,\ldots,\boldsymbol{w}_{35})$. Then $\Sigma \approx W^T R_1(\boldsymbol{x},\boldsymbol{x})W/144$. The following statements fit the partial spline model (2.51):

```
> library(fda); attach(CanadianWeather)
> y <- log(apply(monthlyPrecip,2,sum))
> W <- monthlyTemp
> s <- apply(W,2,mean)
> x <- seq(0.5,11.5,1)/12
> Sigma <- t(W)%*%periodic(x)%*%W/144
> canada.fit1 <- ssr(y~s, rk=Sigma, spar=''m'')
```

where the vector s contains elements $\sum_{j=1}^{12} w_i(x_j)/12$, which are approximations of $\int_0^1 w_i(x)dx$. The generalized maximum likelihood (GML) method in Chapter 3, Section 3.6, is used to select the smoothing parameter since the GCV estimate is too small due to small sample size. This is accomplished by setting the option spar=''m''.

From equation (2.18), the estimate of the weight function is

$$
\hat{f}(x) = d_2 + \sum_{i=1}^{35} c_i \mathcal{L}_{i(z)} R_1(x,z),
$$

where d_2 is an estimate of β_2. To compute $\hat{f}$ at a set of points, say,

$\boldsymbol{x}_0 = (x_{01}, \ldots, x_{0n_0})^T$, we have

$$\hat{f}(\boldsymbol{x}_0) = d_2 \mathbf{1}_{n_0} + \sum_{i=1}^{35} c_i \int_0^1 R_1(\boldsymbol{x}_0, x) w_i(x) dx$$

$$\approx d_2 \mathbf{1}_{n_0} + \sum_{i=1}^{35} c_i \sum_{j=1}^{12} R_1(\boldsymbol{x}_0, x_j) w_i(x_j)/12$$

$$= d_2 \mathbf{1}_{n_0} + \sum_{i=1}^{35} c_i R_1(\boldsymbol{x}_0, \boldsymbol{x}) \boldsymbol{w}_i/12$$

$$= d_2 \mathbf{1}_{n_0} + S\boldsymbol{c},$$

where $\hat{f}(\boldsymbol{x}_0) = (\hat{f}(x_{01}), \ldots, \hat{f}(x_{0n_0}))^T$, $\mathbf{1}_m$ represents a m-vector of all ones, $R_1(\boldsymbol{x}_0, x) = (R_1(x_{01}, x), \ldots, R_1(x_{0n_0}, x))^T$, $R_1(\boldsymbol{x}_0, \boldsymbol{x}) = \{R_1(x_{0i}, x_j)\}_{i=1 \ j=1}^{n_0 \ \ 12}$, $S = R_1(\boldsymbol{x}_0, \boldsymbol{x})W/12$, and $\boldsymbol{c} = (c_1, \ldots, c_n)^T$. We compute $\hat{f}(\boldsymbol{x}_0)$ at 50 equally spaced points in $[0, 1]$ as follows:

```
> S <- periodic(seq(0,1,len=50),x)%*%W/12
> fhat <- canada.fit1$coef$d[2]+S%*%canada.fit1$coef$c
```

Figure 2.8 displays the estimated weight function and 95% bootstrap confidence intervals. The shape of the weight function is similar to that in Figure 15.5 of Ramsay and Silverman (2005). Note that monthly data are used in this book while daily data were used in Ramsay and Silverman (2005).

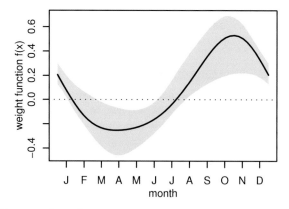

FIGURE 2.8 Canadian weather data, plot of the estimated weight function (solid line), and 95% bootstrap confidence intervals (shaded region).

2.11 *L*-splines

2.11.1 Motivation

In the construction of a smoothing spline model, one needs to decide the penalty functional or, equivalently, the null space $\mathcal{H}_0$ consisting of functions that are not penalized. The squared bias of the spline estimate satisfies the following inequality (Wahba, 1990, p. 59)

$$\frac{1}{n}\sum_{i=1}^{n}\{E(\mathcal{L}_i\hat{f}) - \mathcal{L}_i f\}^2 \leq \lambda ||P_1 f||^2.$$

That is, the squared bias is bounded by the distance of f to the null space $\mathcal{H}_0$. Therefore, selecting a penalty such that $||P_1 f||^2$ is small can lead to low bias in the spline estimate of the function. This is equivalent to selecting a null space $\mathcal{H}_0$ such that it is close to the true function. Ideally, one wants to choose $\mathcal{H}_0$ such that $||P_1 f||^2 = 0$, that is $f \in \mathcal{H}_0$. However, it is usually difficult, if not impossible, to specify such a parametric space in practice. Nevertheless, often there is prior information suggesting that f can be well approximated by a parametric model. That is, f is close to, but not necessarily in, the space $\mathcal{H}_0$. Heckman and Ramsay (2000) called such a space a favored parametric model. *L*-splines allow us to incorporate this kind of indefinite information. *L*-splines can also be used to check or test parametric models (Chapter 3).

Consider functions on the domain $\mathcal{X} = [a,b]$. Let L be the linear differential operator

$$L = D^m + \sum_{j=0}^{m-1} \omega_j(x)D^j, \tag{2.53}$$

where $m \geq 1$, D^j denotes the jth derivative operator, and ω_i are continuous real-valued functions. The minimizer of the following PLS

$$\frac{1}{n}\sum_{i=1}^{n}(y_i - \mathcal{L}_i f)^2 + \lambda \int_a^b (Lf)^2 dx$$

is called an *L-spline*. Note that the penalty is in general different from that of a polynomial or periodic spline. To utilize the general estimation procedure developed in Section 2.4, we need to construct an RKHS such that the penalty $\int_a^b (Lf)^2 dx = ||P_1 f||^2$ with an appropriately defined inner product.

Suppose $f \in W_2^m[a, b]$. Then Lf exists and is square integrable. Let $\mathcal{H}_0 = \{f : Lf = 0\}$ be the kernel of L. Based on results from differential equations (Coddington 1961), there exist real-valued functions $u_1, \ldots, u_m \in W_2^m[a, b]$ such that $u_1, \ldots, u_m$ form a basis of $\mathcal{H}_0$. Furthermore, the Wronskian matrix associated with $u_1, \ldots, u_m$,

$$W(x) = \begin{pmatrix} u_1(x) & u_2(x) & \cdots & u_m(x) \\ u_1'(x) & u_2'(x) & \cdots & u_m'(x) \\ \vdots & \vdots & & \vdots \\ u_1^{(m-1)}(x) & u_2^{(m-1)}(x) & \cdots & u_m^{(m-1)}(x) \end{pmatrix},$$

is invertible for all x. Define inner product

$$(f, g) = \sum_{\nu=0}^{m-1} f^{(\nu)}(a) g^{(\nu)}(a) + \int_a^b (Lf)(Lg)dx. \tag{2.54}$$

Equation (2.54) defines a proper inner product since $f^{(\nu)}(a) = 0$, $\nu = 0, \ldots, m-1$, and $Lf = 0$ leads to $f = 0$. Denote $\boldsymbol{u}(x) = (u_1(x), \ldots, u_m(x))^T$. Let $\boldsymbol{u}^*(x) = (u_1^*(x), \ldots, u_m^*(x))^T$ be the last column of $W^{-1}(x)$ and

$$G(x, s) = \begin{cases} \boldsymbol{u}^T(x)\boldsymbol{u}^*(s), & s \leq x, \\ 0, & s > x, \end{cases}$$

be the Green function associated with L.

Model space construction and decomposition of $W_2^m[a, b]$
The space $W_2^m[a, b]$ is an RKHS with inner product (2.54). Furthermore, $W_2^m[a, b] = \mathcal{H}_0 \oplus \mathcal{H}_1$, where

$$\begin{aligned} \mathcal{H}_0 &= span\{u_1, \ldots, u_m\}, \\ \mathcal{H}_1 &= \{f \in W_2^m[a, b] : f^{(\nu)}(a) = 0, \ \nu = 0, \ldots, m-1\}, \end{aligned} \tag{2.55}$$

are RKHS's with corresponding RKs

$$\begin{aligned} R_0(x, z) &= \boldsymbol{u}^T(x)\{W^T(a)W(a)\}^{-1}\boldsymbol{u}(z), \\ R_1(x, z) &= \int_a^b G(x, s)G(z, s)ds. \end{aligned} \tag{2.56}$$

The above construction of an L-spline is based on a given differential operator L. In practice, rather than a differential operator, prior knowledge may be in a form of basis functions for the null space $\mathcal{H}_0$.

Specifically, suppose prior knowledge suggests that f can be well approximated by a parametric space spanned by a basis $u_1, \ldots, u_m$. Then one can solve the following equations to derive coefficients ω_j in (2.53):

$$(Lu_\nu)(x) = u_\nu^{(m)}(x) + \sum_{j=0}^{m-1} \omega_j(x) u_\nu^{(j)}(x) = 0, \quad \nu = 1, \ldots, m.$$

Let $W(x)$ be the Wronskian matrix, $\boldsymbol{\omega}(x) = (\omega_0(x), \ldots, \omega_{m-1}(x))^T$, and $\boldsymbol{u}^{(m)}(x) = (u_1^{(m)}(x), \ldots, u_m^{(m)}(x))^T$. Then the above equations can be written in a matrix form: $W^T(x)\boldsymbol{\omega}(x) = -\boldsymbol{u}^{(m)}(x)$. Assume that $W(x)$ is invertible. Then $\boldsymbol{\omega}(x) = -W^{-T}(x)\boldsymbol{u}^{(m)}(x)$.

It is clear that the polynomial spline is a special case with $L = D^m$. Some special L-splines are discussed in the following subsections. More details can be found in Schumaker (2007), Wahba (1990), Dalzell and Ramsay (1993), Heckman (1997), Gu (2002), and Ramsay and Silverman (2005).

2.11.2 Exponential Spline

Assume that $f \in W_2^2[a, b]$. Suppose prior knowledge suggests that f can be well approximated by a linear combination of 1 and $\exp(-\gamma x)$ for a fixed $\gamma \neq 0$. Consider the parametric model space

$$\mathcal{H}_0 = \text{span}\{1, \ \exp(-\gamma x)\}.$$

It is easy to see that $\mathcal{H}_0$ is the kernel of the differential operator $L = D^2 + \gamma D$. Nevertheless, we derive this operator following the procedure discussed in Section 2.11.1. Let $u_1(x) = 1$ and $u_2(x) = \exp(-\gamma x)$. The Wronskian matrix

$$W(x) = \begin{pmatrix} 1 & \exp(-\gamma x) \\ 0 & -\gamma \exp(-\gamma x) \end{pmatrix}.$$

Then

$$\begin{pmatrix} \omega_0(x) \\ \omega_1(x) \end{pmatrix} = - \begin{pmatrix} 1 & 0 \\ \frac{1}{\gamma} & -\frac{1}{\gamma}\exp(\gamma x) \end{pmatrix} \begin{pmatrix} 0 \\ \gamma^2 \exp(-\gamma x) \end{pmatrix} = \begin{pmatrix} 0 \\ \gamma \end{pmatrix}.$$

Thus we have

$$L = D^2 + \gamma D.$$

Since

$$\{W(a)^T W(a)\}^{-1} = \begin{pmatrix} 1 + \frac{1}{\gamma^2} & -\frac{1}{\gamma^2}\exp(\gamma a) \\ -\frac{1}{\gamma^2}\exp(\gamma a) & \frac{1}{\gamma^2}\exp(2\gamma a) \end{pmatrix},$$

the RK of $\mathcal{H}_0$

$$R_0(x,z) = 1 + \frac{1}{\gamma^2} - \frac{1}{\gamma^2}\exp(-\gamma x^*) - \frac{1}{\gamma^2}\exp(-\gamma z^*)$$
$$+ \frac{1}{\gamma^2}\exp\{-\gamma(x^* + z^*)\}, \qquad (2.57)$$

where $x^* = x - a$ and $z^* = z - a$.

The Green function

$$G(x,s) = \begin{cases} \frac{1}{\gamma}[1 - \exp\{-\gamma(x-s)\}], & s \le x, \\ 0, & s > x. \end{cases}$$

Thus the RK of $\mathcal{H}_1$

$$R_1(x,z)$$
$$= \int_a^b G(x,s)G(z,s)ds$$
$$= \int_0^{x^* \wedge z^*} \frac{1}{\gamma^2}[1 - \exp\{-\gamma(x^* - s^*)\}][1 - \exp\{-\gamma(z^* - s^*)\}]\,ds^*$$
$$= \frac{1}{\gamma^3}\{\gamma(x^* \wedge z^*) + \exp(-\gamma x^*) + \exp(-\gamma z^*)$$
$$- \exp\{\gamma(x^* \wedge z^* - x^*)\} - \exp\{\gamma(x^* \wedge z^* - z^*)\}$$
$$- \frac{1}{2}\exp\{-\gamma(x^* + z^*)\} + \frac{1}{2}\exp[\gamma\{2(x^* \wedge z^*) - x^* - z^*\}]\}. \qquad (2.58)$$

Evaluations of the RK R_1 for some simple L-splines can be calculated using the `lspline` function in the **assist** package. The argument `type` specifies the type of an L-spline. The option `type=``exp''` computes R_1 in (2.58) for the special case when $a = 0$ and $\gamma = 1$. For general a and $\gamma \ne 0$, the RK R_1 can be computed using a simple transformation $\breve{x} = \gamma(x - a)$.

The weight loss data contain weight measurements of a male obese patient since the start of a weight rehabilitation program. We now illustrate how to fit an exponential spline to the weight loss data. Observations are shown in Figure 2.9(a). Let $y = $ `Weight` and $x = $ `Days`. We first fit a nonlinear regression model as in Venables and Ripley (2002):

$$y_i = \beta_1 + \beta_2 \exp(-\beta_3 x_i) + \epsilon_i, \quad i = 1,\ldots,51. \qquad (2.59)$$

The following statements fit model (2.59):

```
> library(MASS); attach(wtloss)
> y <- Weight; x <- Days
> weight.nls <- nls(y~b1+b2*exp(-b3*x),
                 start=list(b1=81,b2=102,b3=.005))
```

The fit is shown in Figure 2.9(a).

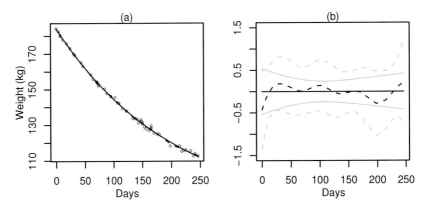

FIGURE 2.9 Weight loss data, plots of (a) observations (circles), the nonlinear regression fit (dotted line), the cubic spline fit (dashed line), and the exponential spline fit (solid line); and (b) the cubic spline fit and 95% Bayesian confidence intervals minus the nonlinear regression fit as dashed lines, and the exponential spline fit and 95% Bayesian confidence intervals minus the nonlinear regression fit as solid lines.

Next we consider the nonparametric regression model (1.1). It is reasonable to assume that the regression function can be well approximated by $\mathcal{H}_0 = \text{span}\{1, \exp(-\gamma x)\}$, where $\gamma = \hat{\beta}_3 = 0.0048$ is the LS estimate of β_3 in the model (2.59). We now fit the exponential spline:

```
> r <- coef(weight.nls)[3]
> ssr(y~exp(-r*x), rk=lspline(r*x,type=''exp''))
```

The exponential spline fit in Figure 2.9(a) is essentially the same as that from nonlinear regression model. The parameter β_3 is fixed as $\hat{\beta}_3$ in the above construction of the exponential spline. One may treat β_3 as a parameter and estimate it using the GCV criterion as in Gu (2002):

```
> gcv.fun <- function(r) ssr(y~exp(-r*x),
    rk=lspline(r*x,type=''exp''))$rkpk.obj$score
> nlm(gcv.fun,.001)$estimate
    0.004884513
```

For comparison, the cubic spline fit is also shown in Figure 2.9(a). To look at the difference between cubic and exponential splines more closely, we plot their fits and 95% Bayesian confidence intervals minus the fit from nonlinear regression in Figure 2.9(b). It is clear that the confidence intervals for the exponential spline are narrower.

Figure 2.9 is essentially the same as Figure 4.3 in Gu (2002). A different approach was used to fit the exponential spline in Gu (2002): it is shown that fitting the exponential spline is equivalent to fitting a cubic spline to the transformed variable $\tilde{x} = 1 - \exp(-\gamma x)$. Thus the following statements lead to the same fit to the exponential spline:

```
> tx <- 1-exp(-r*x)
> ssr(y~tx, rk=cubic2(tx))
```

2.11.3   Logistic Spline

Assume that $f \in W_2^2[a, b]$. Suppose prior knowledge suggests that f can be well approximated by a logistic model $\mathcal{H}_0 = \text{span}\{1/(1 + \delta \exp(-\gamma x))\}$ for some fixed $\delta > 0$ and $\gamma > 0$. It is easy to see that $\mathcal{H}_0$ is the kernel of the differential operator

$$L = D - \frac{\delta\gamma \exp(-\gamma x)}{1 + \delta \exp(-\gamma x)}.$$

The Wronskian is an 1×1 matrix $W(x) = \{1 + \delta \exp(-\gamma x)\}^{-1}$. Since $\{W^T(a)W(a)\}^{-1} = \{1 + \delta \exp(-\gamma a)\}^2$, then the RK of $\mathcal{H}_0$

$$R_0(x, z) = \frac{\{1 + \delta \exp(-\gamma a)\}^2}{\{1 + \delta \exp(-\gamma x)\}\{1 + \delta \exp(-\gamma z)\}}. \tag{2.60}$$

The Green function

$$G(x, s) = \begin{cases} \dfrac{1 + \delta \exp(-\gamma s)}{1 + \delta \exp(-\gamma x)}, & s \le x, \\ 0, & s > x. \end{cases}$$

Thus the RK of $\mathcal{H}_1$

$$\begin{aligned}
R_1(x, z) = & \{1 + \delta \exp(-\gamma x)\}^{-1}\{1 + \delta \exp(-\gamma z)\}^{-1} \\
& \{x \wedge z - a + 2\delta\gamma^{-1}[\exp(-\gamma a) - \exp\{-\gamma(x \wedge z)\}] \\
& + \delta^2(2\gamma)^{-1}[\exp(-2\gamma a) - \exp\{-2\gamma(x \wedge z)\}]\}. \tag{2.61}
\end{aligned}$$

The paramecium caudatum data consist of growth of paramecium caudatum population in the medium of Osterhout. We now illustrate how to fit a logistic spline to the paramecium caudatum data. Observations are shown in Figure 2.10. Let $y = \texttt{density}$ and $x = \texttt{days}$. We first fit the following logistic growth model

$$y_i = \frac{\beta_1}{1 + \beta_2 \exp(-\beta_3 x)} + \epsilon_i, \quad i = 1, \dots, 25, \tag{2.62}$$

using the statements

```
> data(paramecium); attach(paramecium)
> para.nls <- nls(density~b1/(1 + b2*exp(-b3*day)),
               start=list(b1=202,b2=164,b3=0.74))
```

Initial values are the estimates in Neal (2004). The fit is shown in Figure 2.10.

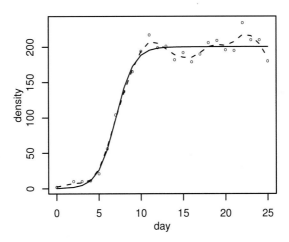

FIGURE 2.10 Paramecium caudatum data, observations (circles), the nonlinear regression fit (dotted line), the cubic spline fit (dashed line), and the logistic spline fit (solid line).

Now we consider the nonparametric regression model (2.7). It is reasonable to assume that the regression function can be well approximated by $\mathcal{H}_0 = \text{span}\{1/(1 + \delta \exp(-\gamma x))\}$, where $\delta = \hat{\beta}_2 = 705.9496$ and $\gamma = \hat{\beta}_3 = 0.9319$ are the LS estimates in model (2.62).

The option type=``logit'' in the lspline function computes R_1 in (2.61) for the special case when $a = 0$, $\delta = 1$, and $\gamma = 1$. It cannot be adapted to compute R_1 for the general situation. We take this opportunity to show how to write a function for computing an RK.

```
> logit.rk <- function(x,a,d,r) {
    tmp1 <- x%o%rep(1,length(x))
    tmp2 <- (tmp1+t(tmp1)-abs(tmp1-t(tmp1)))/2
    tmp3 <- exp(-r*a)-exp(-r*tmp2)
    tmp4 <- exp(-2*r*a)-exp(-2*r*tmp2)
```

```
      tmp5 <- 1/((1+d*exp(-r*x))%o%(1+d*exp(-r*x)))
      (tmp2-a+2*d*tmp3/r+d**2*tmp4/(2*r))*tmp5
    }
> bh <- coef(para.nls)
> ssr(density~I(1/(1+bh[2]*exp(-bh[3]*day)))-1,
      rk=logit.rk(day,0,bh[2],bh[3]),spar=''m'')
```

The function `logit.rk` computes R_1 in (2.61). Since the sample size is small, the GML method is used to select the smoothing parameter. The logistic spline fit is shown in Figure 2.10. The fit is essentially the same as that from the logistic growth model (2.62). For comparison, the cubic spline fit is also shown in Figure 2.10. The logistic spline fit smooths out oscillations after 10 days while the cubic spline fit preserves them.

To include the constant function in $\mathcal{H}_0$, one may consider the operator

$$L = D\left\{D - \frac{\delta\gamma\exp(-\gamma x)}{1 + \delta\exp(-\gamma x)}\right\}.$$

Details of this situation can be found in Gu (2002).

2.11.4   Linear-Periodic Spline

Assume that $f \in W_2^4[a, b]$. Suppose prior knowledge suggests that f can be well approximated by a parametric model

$$\mathcal{H}_0 = \mathrm{span}\{1, \ x, \ \cos x, \ \sin x\}.$$

It is easy to check that $\mathcal{H}_0$ is the kernel of the differential operator

$$L = D^4 + D^2.$$

The Wronskian matrix and its inverse are, respectively,

$$W(x) = \begin{pmatrix} 1 & x & \cos x & \sin x \\ 0 & 1 & -\sin x & \cos x \\ 0 & 0 & -\cos x & -\sin x \\ 0 & 0 & \sin x & -\cos x \end{pmatrix}$$

and

$$W^{-1}(x) = \begin{pmatrix} 1 & -x & 1 & -x \\ 0 & 1 & 0 & 1 \\ 0 & 0 & -\cos x & \sin x \\ 0 & 0 & -\sin x & -\cos x \end{pmatrix}.$$

For simplicity, suppose $a = 0$. Then

$$\{W^T(0)W(0)\}^{-1} = \begin{pmatrix} 2 & 0 & -1 & 0 \\ 0 & 2 & 0 & -1 \\ -1 & 0 & 1 & 0 \\ 0 & -1 & 0 & 1 \end{pmatrix}.$$

Therefore, the RK of $\mathcal{H}_0$

$$R_0(x, z) = 2 - \cos x - \cos z + 2xz - x \sin z - z \sin x$$
$$+ \cos x \cos z + \sin x \sin z. \tag{2.63}$$

The Green function

$$G(x, s) = \begin{cases} x - s - \sin(x - s), & s \leq x, \\ 0, & s > x. \end{cases}$$

Thus the RK of $\mathcal{H}_1$

$$R_1(x, z) = -\frac{1}{6}(x \wedge z)^3 + \frac{1}{2}xz(x \wedge z) - |x - z| - \sin x - \sin z$$
$$+ x \cos z + z \cos x + \frac{1}{2}(x \wedge z) \cos(z - x)$$
$$+ \frac{5}{4}\sin|x - z| - \frac{1}{4}\sin(x + z). \tag{2.64}$$

The option `type=''linSinCos''` in the `lspline` function computes R_1 in (2.64) for the special case when $a = 0$. The translation $x - a$ may be used when $a \neq 0$. When the null space $\mathcal{H}_0 = \text{span}\{1, x, \cos \tau x, \sin \tau x\}$ for a fixed $\tau \neq 0$, the corresponding differential operator is $L = D^4 + \tau^2 D^2$. The linear-periodic spline with a general τ can be fitted using the transformation $\tilde{x} = \tau x$.

The melanoma data contain numbers of melanoma cases per 100,000 in Connecticut from 1936 to 1972. We now illustrate how to fit a linear-periodic spline to the melanoma data. The observations are shown in Figure 2.11. There are two apparent trends: a nearly linear long-term trend over the years, and a cycle of around 10 years corresponding to the sunspot cycle. Let $y = $ `cases` and $x = $ `year`. As in Heckman and Ramsay (2000), we fit a linear-periodic spline with $L = D^4 + \tau^2 D^2$, where $\tau = 0.58$.

```
> library(fda); attach(melanoma)
> x <- year-1936; y <- incidence
> tau <- .58; tx <- tau*x
> ssr(y~tx+cos(tx)+sin(tx),
      rk=lspline(tx,type=''linSinCos''), spar=''m'')
```

Again, since the sample size is small, the GML method is used to select the smoothing parameter. For comparison, the cubic spline fit with smoothing parameter selected by the GML method is also shown in Figure 2.11.

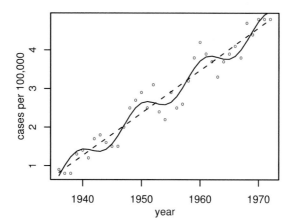

FIGURE 2.11 Melanoma data, observations (circles), the cubic spline fit (dashed line), and the linear-periodic spline fit (solid line).

2.11.5   Trigonometric Spline

Suppose f is a periodic function on $\mathcal{X} = [0, 1]$. Assume that $f \in W_2^m(per)$. Then we have the *Fourier expansion*

$$
\begin{aligned}
f(x) &= a_0 + \sum_{\nu=1}^{\infty} a_\nu \cos 2\pi\nu x + \sum_{\nu=1}^{\infty} b_\nu \sin 2\pi\nu x \\
&= a_0 + \sum_{\nu=1}^{m-1} a_\nu \cos 2\pi\nu x + \sum_{\nu=1}^{m-1} b_\nu \sin 2\pi\nu x + \text{Rem}(x),   \quad (2.65)
\end{aligned}
$$

where the first three elements in (2.65) represent a trigonometric polynomial of degree $m-1$ and $\text{Rem}(x) \triangleq \sum_{\nu=m}^{\infty}(a_\nu \cos 2\pi\nu x + b_\nu \sin 2\pi\nu x)$. The penalty used in Section 2.7 for a periodic spline corresponds to $L = D^m$ with kernel $\mathcal{H}_0 = \text{span}\{1\}$. That is, all nonconstant functions including those in the trigonometric polynomial are penalized. Analogous to the Taylor expansion and polynomial splines, one may want to include lower-degree trigonometric polynomials in the null space. The operator $L = D^m$ does not decompose $W_2^m(per)$ into lower-degree trigonometric polynomials plus the remainder terms.

Now consider the null space

$$
\mathcal{H}_0 = \text{span}\{1, \ \sin 2\pi\nu x, \ \cos 2\pi\nu x, \ \nu = 1, \ldots, m-1\},   \quad (2.66)
$$

which includes trigonometric polynomials with degree up to $m-1$. Assume that $f \in W_2^{m+1}(per)$. It is easy to see that $\mathcal{H}_0$ is the kernel of the differential operator

$$L = D \prod_{\nu=1}^{m-1} \{D^2 + (2\pi\nu)^2\}. \tag{2.67}$$

Model space construction and decomposition of $W_2^{m+1}(per)$
The space $W_2^{m+1}(per)$ is an RKHS with the inner product

$$
\begin{aligned}
(f, g) = & \left(\int_0^1 f dx \right) \left(\int_0^1 g dx \right) \\
& + \sum_{\nu=1}^{m-1} \left(\int_0^1 f \cos 2\pi\nu x dx \right) \left(\int_0^1 g \cos 2\pi\nu x dx \right) \\
& + \sum_{\nu=1}^{m-1} \left(\int_0^1 f \sin 2\pi\nu x dx \right) \left(\int_0^1 g \sin 2\pi\nu x dx \right) \\
& + \int_0^1 (Lf)(Lg) dx,
\end{aligned}
$$

where L is defined in (2.67). Furthermore, $W_2^{m+1}(per) = \mathcal{H}_0 \oplus \mathcal{H}_1$, where $\mathcal{H}_0$ is given in (2.66) and $\mathcal{H}_1 = W_2^{m+1}(per) \ominus \mathcal{H}_0$. $\mathcal{H}_0$ and $\mathcal{H}_1$ are RKHS's with corresponding RKs

$$
R_0(x, z) = 1 + \sum_{\nu=1}^{m-1} \cos\{2\pi(x - z)\},
$$

$$
R_1(x, z) = \sum_{\nu=m}^{\infty} \frac{2}{(2\pi)^{4m+2}} \left\{ \prod_{j=0}^{m-1} (j^2 - \nu^2)^{-2} \right\} \cos 2\pi\nu(x - z). \tag{2.68}
$$

Sometimes f satisfies the constraint $\int_0^1 f dx = 0$. This constraint can be handled easily by removing constant functions in the above construction. Specifically, let

$$\mathcal{H}_0 = \text{span}\{ \sin 2\pi\nu x, \cos 2\pi\nu x, \nu = 1, \ldots, m - 1\}. \tag{2.69}$$

Assume that $f \in W_2^m(per) \ominus \{1\}$. Then $\mathcal{H}_0$ is the kernel of the differential operator

$$L = \prod_{\nu=1}^{m-1} \{D^2 + (2\pi\nu)^2\}. \tag{2.70}$$

Model space construction and decomposition of $W_2^m(per) \ominus \{1\}$
The space $W_2^m(per) \ominus \{1\}$ is an RKHS with the inner product

$$(f, g) = \sum_{\nu=1}^{m-1} \left(\int_0^1 f \cos 2\pi x dx \right) \left(\int_0^1 g \cos 2\pi x dx \right)$$

$$+ \sum_{\nu=1}^{m-1} \left(\int_0^1 f \sin 2\pi x dx \right) \left(\int_0^1 g \sin 2\pi x dx \right)$$

$$+ \int_0^1 (Lf)(Lg) dx,$$

where L is defined in (2.70). Furthermore, $W_2^m(per) = \mathcal{H}_0 \oplus \mathcal{H}_1$, where $\mathcal{H}_0$ is given in (2.69) and $\mathcal{H}_1 = W_2^m(per) \ominus \{1\} \ominus \mathcal{H}_0$. $\mathcal{H}_0$ and $\mathcal{H}_1$ are RKHS's with RKs

$$R_0(x, z) = \sum_{\nu=1}^{m-1} \cos\{2\pi(x - z)\},$$

$$R_1(x, z) = \sum_{\nu=m}^{\infty} \frac{2}{(2\pi)^{4(m-1)}} \left\{ \prod_{j=1}^{m-1} (j^2 - \nu^2)^{-2} \right\} \cos 2\pi\nu(x - z). \tag{2.71}$$

In the `lspline` function, the options `type=``sine1''` and `type=``sine0''` compute RKs R_1 in (2.68) and (2.71), respectively.

We now use the Arosa data to show how to fit a trigonometric spline and illustrate its difference from periodic and partial splines. Suppose we want to investigate how ozone thickness changes over months in a year. We have fitted a cubic periodic spline with $f \in W_2^2(per)$ in Section 2.7. Note that $W_2^2(per) = \mathcal{H}_{10} \oplus \mathcal{H}_{11}$, where $\mathcal{H}_{10} = \{1\}$ and $\mathcal{H}_{11} = W_2^2(per) \ominus \{1\}$. Therefore, we can rewrite the periodic spline model as

$$y_i = f_{10}(x_i) + f_{11}(x_i) + \epsilon_i, \quad i = 1, \dots, n, \tag{2.72}$$

where f_{10} and f_{11} are projections onto $\mathcal{H}_{10}$ and $\mathcal{H}_{11}$, respectively. The estimates of the overall function, $\hat{f}_{10} + \hat{f}_{11}$, and its projections $\hat{f}_{10}$ (parametric) and $\hat{f}_{11}$ (smooth) are shown in the first row of Figure 2.12.

It is apparent that the monthly pattern can be well approximated by a simple sinusoidal function in the model space

$$\mathcal{P} = \text{span}\{1, \ \sin 2\pi x, \ \cos 2\pi x\}. \tag{2.73}$$

Suppose we want to check the departure from the parametric model $\mathcal{P}$. One approach is to add the sine and cosine functions to the null space

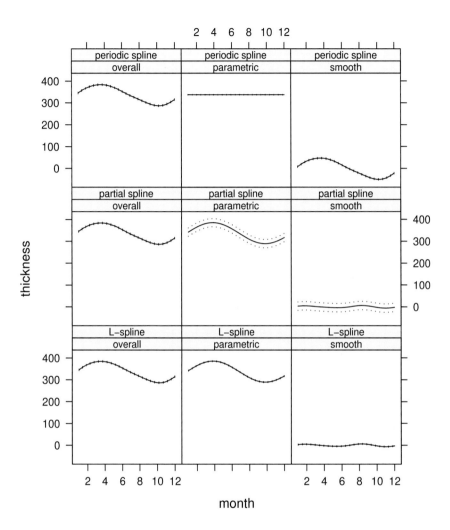

FIGURE 2.12 Arosa data, plots the overall fits (left), the parametric components (middle), and the smooth components (right) of the periodic spline (top), the partial spline (middle), and the L-spline (bottom). Parametric components represent components in the spaces $\mathcal{H}_{10}$, $\mathcal{P}$, and $\mathcal{P}$ for periodic spline, partial spline, and L-spline, respectively. Smooth components represent components in the spaces $\mathcal{H}_{11}$, $\mathcal{H}_{11}$, and $W_2^3(per)\ominus\mathcal{P}$ for periodic spline, partial spline, and L-spline, respectively. Dotted lines are 95% Bayesian confidence intervals.

$\mathcal{H}_{10}$ of the periodic spline. This leads to the following partial spline model

$$y_i = f_{20}(x_i) + f_{21}(x_i) + \epsilon_i, \quad i = 1, \ldots, n, \tag{2.74}$$

where $f_{20} \in \mathcal{P}$ and $f_{21} \in \mathcal{H}_{11}$. The model (2.74) is fitted as follows:

```
> ssr(thick~sin(2*pi*x)+cos(2*pi*x), rk=periodic(x))
```

The estimates of the overall function and its two components $\hat{f}_{20}$ (parametric) and $\hat{f}_{21}$ (smooth) are shown in the second row of Figure 2.12.

Another approach is to fit a trigonometric spline with $m = 2$:

$$y_i = f_{30}(x_i) + f_{31}(x_i) + \epsilon_i, \quad i = 1, \ldots, n, \tag{2.75}$$

where $f_{30} \in \mathcal{P}$ and $f_{31} \in W_2^3(per) \ominus \mathcal{P}$. The model (2.75) is fitted as follows:

```
> ssr(thick~sin(2*pi*x)+cos(2*pi*x),
      rk=lspline(x,type=''sine1''))
```

The estimates of the overall function and its two components $\hat{f}_{30}$ (parametric) and $\hat{f}_{31}$ (smooth) are shown in the third row of Figure 2.12.

All three models have similar overall fits. However, their components are quite different. The smooth component of the periodic spline reveals the departure from a constant function. To check the departure from the simple sinusoidal model space $\mathcal{P}$, we can look at the smooth components from the partial and L-splines. Estimates of the smooth components from the partial and L-splines are similar. However, the confidence intervals based on the L-spline are narrower than those based on the partial spline. This is due to the fact that the two components f_{30} and f_{31} in the L-spline are orthogonal, while the two components f_{20} and f_{21} in the partial spline are not necessarily orthogonal. Therefore, we can expect the inference based on the L-spline to be more efficient in general.

Chapter 3

Smoothing Parameter Selection and Inference

3.1 Impact of the Smoothing Parameter

The penalized least squares (2.11) represents a compromise between the goodness-of-fit and a penalty to the departure from the null space $\mathcal{H}_0$. The balance is controlled by the smoothing parameter λ. As λ varies from 0 to ∞, we have a family of estimates with $\hat{f} \in \mathcal{H}_0$ when $\lambda = \infty$.

To illustrate the impact of the smoothing parameter, consider the Stratford weather data consisting of daily maximum temperatures in Stratford, Texas, during 1990. Observations are shown in Figure 3.1. Consider the regression model (1.1) where $n = 73$ and f represents expected maximum temperature as a function of time in a year. Denote x as the time variable scaled into $[0, 1]$. It is reasonable to assume that f is a smooth periodic function. In particular, we assume that $f \in W_2^2(per)$. For a fixed λ, say 0.001, one can fit the cubic periodic spline as follows:

```
> data(Stratford); attach(Stratford)
> ssr(y~1, rk=periodic(x), limnla=log10(73*.001))
```

where the argument `limnla` specifies a search range for $\log_{10}(n\lambda)$. To see how a spline fit is affected by the choice of λ, periodic spline fits with six different values of λ are shown in Figure 3.1. It is obvious that the fit with $\lambda = \infty$ is a constant, that is, $f_\infty \in \mathcal{H}_0$. The fit with $\lambda = 0$ interpolates data. A larger λ leads to a smoother fit. Both $\lambda = 0.0001$ and $\lambda = 0.00001$ lead to visually reasonable fits.

In practice it is desirable to select the smoothing parameter using an objective method rather than visual inspection. In a sense, a data-driven choice of λ allows data to speak for themselves. Thus, it is not exaggerating to say that the choice of λ is the spirit and soul of nonparametric regression.

We now inspect how λ controls the fit. Again, consider model (1.1) for Stratford weather data. Let us first consider a parametric approach that approximates f using a trigonometric polynomial up to a certain

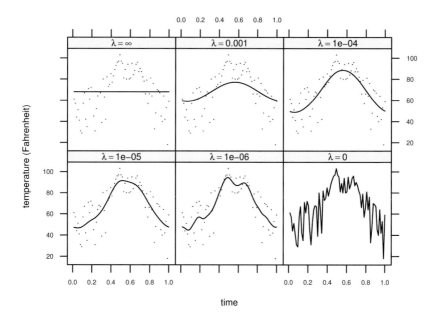

FIGURE 3.1 Stratford weather data, plot of observations, and the periodic spline fits with different smoothing parameters.

degree, say k, where $0 \le k \le K$ and $K = (n-1)/2 = 36$. Denote the corresponding parametric model space for f as

$$\mathcal{M}_k = \text{span}\{1, \sqrt{2}\sin 2\pi\nu x, \sqrt{2}\cos 2\pi\nu x, \ \nu = 1, \ldots, k\}, \qquad (3.1)$$

where $\mathcal{M}_0 = \text{span}\{1\}$. For a fixed k, write the regression model based on $\mathcal{M}_k$ in a matrix form as

$$\boldsymbol{y} = X_k \boldsymbol{\beta}_k + \boldsymbol{\epsilon},$$

where

$$X_k = \begin{pmatrix} 1 & \sqrt{2}\sin 2\pi x_1 & \sqrt{2}\cos 2\pi x_1 & \cdots & \sqrt{2}\sin 2\pi k x_1 & \sqrt{2}\cos 2\pi k x_1 \\ 1 & \sqrt{2}\sin 2\pi x_2 & \sqrt{2}\cos 2\pi x_2 & \cdots & \sqrt{2}\sin 2\pi k x_2 & \sqrt{2}\cos 2\pi k x_2 \\ \vdots & \vdots & \vdots & \cdots & \vdots & \vdots \\ 1 & \sqrt{2}\sin 2\pi x_n & \sqrt{2}\cos 2\pi x_n & \cdots & \sqrt{2}\sin 2\pi k x_n & \sqrt{2}\cos 2\pi k x_n \end{pmatrix}$$

is the design matrix, $x_i = i/n$, $\boldsymbol{\beta}_k = (\beta_1, \ldots, \beta_{2k+1})^T$, and $\boldsymbol{\epsilon} = (\epsilon_1, \ldots, \epsilon_n)^T$. Since design points are equally spaced, we have the following orthogonality relations:

$$\frac{2}{n}\sum_{i=1}^{n}\cos 2\pi\nu x_i \cos 2\pi\mu x_i = \delta_{\nu,\mu}, \quad 1 \le \nu, \ \mu \le K,$$

$$\frac{2}{n}\sum_{i=1}^{n}\sin 2\pi\nu x_i \sin 2\pi\mu x_i = \delta_{\nu,\mu}, \quad 1 \le \nu, \ \mu \le K,$$

$$\frac{2}{n}\sum_{i=1}^{n}\cos 2\pi\nu x_i \sin 2\pi\mu x_i = 0, \quad 1 \le \nu, \ \mu \le K,$$

where $\delta_{\nu,\mu}$ is the Kronecker delta. Therefore, $X_k^T X_k = nI_{2k+1}$, where I_{2k+1} is an identity matrix of size $2k+1$. Note that $X_K/\sqrt{n}$ is an orthogonal matrix. Define the discrete Fourier transformation $\tilde{\boldsymbol{y}} = X_K^T \boldsymbol{y}/n$. Then the LS estimate of $\boldsymbol{\beta}_k$ is $\hat{\boldsymbol{\beta}}_k = (X_k^T X_k)^{-1}X_k^T \boldsymbol{y} = X_k^T \boldsymbol{y}/n = \tilde{\boldsymbol{y}}_k$, where $\tilde{\boldsymbol{y}}_k$ consists of the first $2k+1$ elements of the discrete Fourier transformation $\tilde{\boldsymbol{y}}$. More explicitly,

$$\hat{\beta}_1 = \frac{1}{n}\sum_{i=1}^{n}y_i = \tilde{y}_1,$$

$$\hat{\beta}_{2\nu} = \frac{\sqrt{2}}{n}\sum_{i=1}^{n}y_i \sin 2\pi\nu x_i = \tilde{y}_{2\nu}, \quad 1 \le \nu \le k, \tag{3.2}$$

$$\hat{\beta}_{2\nu+1} = \frac{\sqrt{2}}{n}\sum_{i=1}^{n}y_i \cos 2\pi\nu x_i = \tilde{y}_{2\nu+1}, \quad 1 \le \nu \le k.$$

Now consider modeling f using the cubic periodic spline space $W_2^2(per)$. The exact solution was given in Chapter 2. To simplify the argument, let us consider the following PLS

$$\min_{f \in \mathcal{M}_K}\left\{\frac{1}{n}\sum_{i=1}^{n}(y_i - f(x_i))^2 + \lambda\int_0^1 (f'')^2 dx\right\}, \tag{3.3}$$

where the model space $W_2^2(per)$ is approximated by $\mathcal{M}_K$. The following discussion holds true for the exact solution in $W_2^2(per)$ (Gu 2002). However, the approximation makes the following argument simpler and transparent.

Let

$$\hat{f}(x) = \hat{\alpha}_1 + \sum_{\nu=1}^{K}(\hat{\alpha}_{2\nu}\sqrt{2}\sin 2\pi\nu x + \hat{\alpha}_{2\nu+1}\sqrt{2}\cos 2\pi\nu x)$$

be the solution to (3.3). Then $\hat{\boldsymbol{f}} \triangleq (\hat{f}(x_1), \ldots, \hat{f}(x_n))^T = X_K \hat{\boldsymbol{\alpha}}$, where $\hat{\boldsymbol{\alpha}} = (\hat{\alpha}_1, \ldots, \hat{\alpha}_{2K+1})^T$. The LS

$$\frac{1}{n}||\boldsymbol{y} - \hat{\boldsymbol{f}}||^2 = \frac{1}{n}||\frac{1}{\sqrt{n}}X_K^T(\boldsymbol{y} - \hat{\boldsymbol{f}})||^2$$

$$= ||\frac{1}{n}X_K^T\boldsymbol{y} - \frac{1}{n}X_K^T X_K \hat{\boldsymbol{\alpha}}||^2$$

$$= ||\tilde{\boldsymbol{y}} - \hat{\boldsymbol{\alpha}}||^2.$$

Thus (3.3) reduces to the following ridge regression problem

$$(\hat{\alpha}_1 - \tilde{y}_1)^2 + \sum_{\nu=1}^{K} \left\{ (\hat{\alpha}_{2\nu} - \tilde{y}_{2\nu})^2 + (\hat{\alpha}_{2\nu+1} - \tilde{y}_{2\nu+1})^2 \right\}$$

$$+ \lambda \sum_{\nu=1}^{K} (2\pi\nu)^4 (\hat{\alpha}_{2\nu}^2 + \hat{\alpha}_{2\nu+1}^2). \tag{3.4}$$

The solutions to (3.4) are

$$\hat{\alpha}_1 = \tilde{y}_1,$$

$$\hat{\alpha}_{2\nu} = \frac{\tilde{y}_{2\nu}}{1 + \lambda(2\pi\nu)^4}, \quad \nu = 1, \ldots, K, \tag{3.5}$$

$$\hat{\alpha}_{2\nu+1} = \frac{\tilde{y}_{2\nu+1}}{1 + \lambda(2\pi\nu)^4}, \quad \nu = 1, \ldots, K.$$

Thus the periodic spline is essentially a low-pass filter: components at frequency ν are downweighted by a factor of $1 + \lambda(2\pi\nu)^4$. Figure 3.2 shows how λ controls the nature of the filter: more high frequencies are filtered out as λ increases. Comparing (3.2) and (3.5), it is clear that selecting an order k for the trigonometric polynomial model may be viewed as hard thresholding, and selecting the smoothing parameter λ for the periodic spline may be viewed as soft thresholding.

Now consider the general spline model (2.10). From (2.26), the hat matrix $H(\lambda) = I - n\lambda Q_2 (Q_2^T M Q_2)^{-1} Q_2^T$. Let UEU^T be the eigendecomposition of $Q_2^T \Sigma Q_2$, where $U_{(n-p) \times (n-p)}$ is an orthogonal matrix and $E = \text{diag}(e_1, \ldots, e_{n-p})$. The projection onto the space spanned by T

$$P_T \triangleq T(T^T T)^{-1} T^T = Q_1 R(R^T R)^{-1} R^T Q_1^T = Q_1 Q_1^T.$$

Then

$$H(\lambda) = I - n\lambda Q_2 U(E + n\lambda I)^{-1} U^T Q_2^T \tag{3.6}$$

$$= Q_1 Q_1^T + Q_2 Q_2^T - n\lambda Q_2 U(E + n\lambda I)^{-1} U^T Q_2^T$$

$$= P_T + Q_2 U V U^T Q_2^T, \tag{3.7}$$

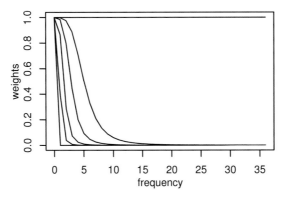

FIGURE 3.2 Weights of the periodic spline filter, $1/(1 + \lambda(2\pi\nu)^4)$, plotted as a function of frequency ν. Six curves from top down correspond to six different λ: 0, 10^{-6}, 10^{-5}, 10^{-4}, 10^{-3}, and ∞.

where $V = \mathrm{diag}(e_1/(e_1 + n\lambda), \ldots, e_{n-p}/(e_{n-p} + n\lambda))$. The hat matrix is divided into two mutually orthogonal matrices: one is the projection onto the space spanned by T, and the other is responsible for shrinking part of the signal that is orthogonal to T. The smoothing parameter shrinks eigenvalues in the form $e_\nu/(e_\nu + n\lambda)$. The choices $\lambda = \infty$ and $\lambda = 0$ lead to the parametric model $\mathcal{H}_0$ and interpolation, respectively.

Equation (3.7) also indicates that the hat matrix $H(\lambda)$ is nonnegative definite. However, unlike the projection matrix for a parametric model, $H(\lambda)$ is usually not idempotent. $H(\lambda)$ has p eigenvalues equal to one and the remaining eigenvalues less than one when $\lambda > 0$.

3.2 Trade-Offs

Before introducing methods for selecting the smoothing parameter, it is helpful to discuss some basic concepts and principles for model selection. In general, model selection boils down to compromises between different aspects of a model. Occam's razor has been the guiding principle for the compromises: the model that *fits observations sufficiently well* in *the least complex way* should be preferred. To be precise on *fits observations sufficiently well*, one needs a quantity that measures how well a model fits the data. One such measure is the LS in (1.6). To be precise on *the least complex way*, one needs a quantity that measures the complexity of a model. For a parametric model, a common measure of model complexity is the number of parameters in the model, often called the *degrees of freedom* (df). For example, the df of model $\mathcal{M}_k$ in (3.1) equals $2k + 1$.

What would be a good measure of model complexity for a nonparametric regression procedure? Consider the general nonparametric regression model (2.10). Let $f_i = \mathcal{L}_i f$, and $\boldsymbol{f} = (f_1, \ldots, f_n)$. Let $\hat{f}$ be an estimate of f based on a modeling procedure $\mathcal{M}$, and $\hat{f}_i = \mathcal{L}_i \hat{f}$. Ye (1998) defined *generalized degrees of freedom* (gdf) of $\mathcal{M}$ as

$$\text{gdf}(\mathcal{M}) \triangleq \sum_{i=1}^{n} \frac{\partial \text{E}_{\mathbf{f}}(\hat{f}_i)}{\partial f_i}. \tag{3.8}$$

The gdf is an extension of the standard degrees of freedom for general modeling procedures. It can be viewed as the sum of the average sensitivities of the fitted values $\hat{f}_i$ to a small change in the response. It is easy to check that (Efron 2004)

$$\text{gdf}(\mathcal{M}) = \frac{1}{\sigma^2} \sum_{i=1}^{n} \text{Cov}(\hat{f}_i, y_i),$$

where $\sum_{i=1}^{n} \text{Cov}(\hat{f}_i, y_i)$ is the so-called *covariance penalty* (Tibshirani and Knight 1999). For spline estimate with a fixed λ, we have $\hat{\boldsymbol{f}} = H(\lambda)\boldsymbol{y}$ based on (2.25). Denote the modeling procedure leading to $\hat{f}$ as $\mathcal{M}_\lambda$ and $H(\lambda) = \{h_{ij}\}_{i,j=1}^{n}$. Then

$$\text{gdf}(\mathcal{M}_\lambda) = \sum_{i=1}^{n} \frac{\partial \text{E}_{\mathbf{f}}(\sum_{j=1}^{n} h_{ij} y_j)}{\partial f_i} = \sum_{i=1}^{n} \frac{\partial \sum_{j=1}^{n} h_{ij} f_j}{\partial f_i} = \text{tr}H(\lambda),$$

where tr represents the trace of a matrix. Even though λ does not have a physical interpretation as k, $\text{tr}H(\lambda)$ is a useful measure of model complexity and will be simply referred to as the *degrees of freedom*.

For Stratford weather data, Figure 3.3(a) depicts how $\text{tr}H(\lambda)$ for the cubic periodic spline depends on the smoothing parameter λ. It is clear that the degrees of freedom decrease as λ increases.

To illustrate the interplay between the LS and model complexity, we fit trigonometric polynomial models from the smallest model with $k = 0$ to the largest model with $k = K$. The square root of residual sum of squares (RSS) are plotted against the degrees of freedom $(2k + 1)$ as circles in Figure 3.3(b). Similarly, we fit the periodic spline with a wide range of values for the smoothing parameter λ. Again, we plot the square root of RSS against the degrees of freedom $(\text{tr}H(\lambda))$ as the solid line in Figure 3.3(b). Obviously, RSS decreases to zero (interpolation) as the degrees of freedom increase to n. The square root of RSS keeps

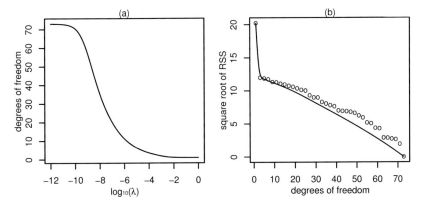

FIGURE 3.3 Stratford data, plots of (a) degrees of freedom of the periodic spline against the smoothing parameter on the logarithm base 10 scale, and (b) square root of RSS from the trigonometric polynomial model (circles) and periodic spline (line) against the degrees of freedom.

declining almost linearly after the initial big drop. It is quite clear that the constant model does not fit data well. However, it is unclear which model *fits observations sufficiently well.*

Figure 3.3(b) shows that the LS and model complexity are two opposite aspects of a model: the approximation error decreases as the model complexity increases. Our goal is to find the "best" model that strikes a balance between these two conflicting aspects. To make the term "best" meaningful, we need a target criterion that quantifies a model's performance. It is clear that the LS cannot be used as the target because it will lead to the most complex model. Even though there is no universally accepted measure, some criteria are widely accepted and used in practice. We now introduce a criterion that is commonly used for regression models.

Consider the *loss* function

$$L(\lambda) = \frac{1}{n} ||\hat{\boldsymbol{f}} - \boldsymbol{f}||^2. \tag{3.9}$$

Define the *risk* function, also called *mean squared error* (MSE), as

$$\mathrm{MSE}(\lambda) \triangleq \mathrm{E}L(\lambda) = \mathrm{E}\left(\frac{1}{n}||\hat{\boldsymbol{f}} - \boldsymbol{f}||^2\right). \tag{3.10}$$

We want the estimate $\hat{f}$ to be as close to the true function f as possible. Obviously, MSE is the expectation of the Euclidean distance between the estimates and the true values. It can be decomposed into two components:

$$\begin{aligned}
\text{MSE}(\lambda) \\
&= \frac{1}{n}\text{E}\|(\text{E}\hat{f} - f) + (\hat{f} - \text{E}\hat{f})\|^2 \\
&= \frac{1}{n}\text{E}\|\text{E}\hat{f} - f\|^2 + \frac{2}{n}\text{E}(\text{E}\hat{f} - f)^T(\hat{f} - \text{E}\hat{f}) + \frac{1}{n}\text{E}\|\hat{f} - \text{E}\hat{f}\|^2 \\
&= \frac{1}{n}\|\text{E}\hat{f} - f\|^2 + \frac{1}{n}\text{E}\|\hat{f} - \text{E}\hat{f}\|^2 \\
&= \frac{1}{n}\|(I - H(\lambda))f\|^2 + \frac{\sigma^2}{n}\text{tr}H^2(\lambda) \\
&\triangleq b^2(\lambda) + v(\lambda),
\end{aligned} \tag{3.11}$$

where b^2 and v represent squared bias and variance, respectively. Note that bias depends on the true function, while the variance does not. Based on notations introduced in Section 3.2, let $\boldsymbol{h} = (h_1, \ldots, h_{n-p})^T \triangleq U^T Q_2^T \boldsymbol{f}$. From (3.6), we have

$$\begin{aligned}
b^2(\lambda) &= \frac{1}{n}\|(I - H(\lambda))\boldsymbol{f}\|^2 \\
&= \frac{1}{n}\boldsymbol{f}^T Q_2 U \text{diag}\left\{\left(\frac{n\lambda}{e_1 + n\lambda}\right)^2, \ldots, \left(\frac{n\lambda}{e_{n-p} + n\lambda}\right)^2\right\} U^T Q_2^T \boldsymbol{f} \\
&= \frac{1}{n}\sum_{\nu=1}^{n-p}\left(\frac{n\lambda h_\nu}{e_\nu + n\lambda}\right)^2.
\end{aligned}$$

From (3.7), we have

$$\begin{aligned}
v(\lambda) &= \frac{\sigma^2}{n}\text{tr}H^2(\lambda) = \frac{\sigma^2}{n}\text{tr}(P_T + Q_2 U V^2 U^T Q_2^T) \\
&= \frac{\sigma^2}{n}\left\{p + \sum_{\nu=1}^{n-p}\left(\frac{e_\nu}{e_\nu + n\lambda}\right)^2\right\}.
\end{aligned}$$

The squared bias measures how well $\hat{f}$ approximates the true function f, and the variance measures how well the function can be estimated. As λ increases from 0 to ∞, $b^2(\lambda)$ increases from 0 to $\sum_{\nu=1}^{n-p} h_\nu^2/n = \|Q_2^T \boldsymbol{f}\|^2/n$, while $v(\lambda)$ decreases from σ^2 to $p\sigma^2/n$. Therefore, the MSE represents a trade-off between bias and variance. Note that $Q_2^T \boldsymbol{f}$ represents the signal that is orthogonal to T.

It is easy to check that $db^2(\lambda)/d\lambda|_{\lambda=0} = 0$ and $dv(\lambda)/d\lambda|_{\lambda=0} < 0$. Thus, $d\text{MSE}(\lambda)/d\lambda|_{\lambda=0} < 0$, and $\text{MSE}(\lambda)$ has at least one minimizer $\lambda^* > 0$. Therefore, when $\text{MSE}(0) \leq \text{MSE}(\infty)$, there exists at least one λ^* such that the corresponding PLS estimate performs better in terms of MSE than the LS estimate in $\mathcal{H}_0$ and the interpolation. When $\mathrm{MSE}(0) > \mathrm{MSE}(\infty)$, considering the MSE as a function of $\delta = 1/\lambda$, we have $db^2(\delta)/d\delta|_{\delta=0} < 0$ and $dv(\delta)/d\delta|_{\delta=0} = 0$. Then, again, there exists at least one δ^* such that the corresponding PLS estimate performs better in terms of MSE than the LS estimate in $\mathcal{H}_0$ and the interpolation.

To calculate MSE, one needs to know the true function f. The following simulation illustrates the bias-variance trade-off. Observations are generated from model (1.1) with $f(x) = \sin(4\pi x^2)$ and $\sigma = 0.5$. The same design points as in the Stratford weather data are used: $x_i = i/n$ for $i = 1, \ldots, n$ and $n = 73$. The true function and one realization of observations are shown in Figure 3.4(a). For a fixed λ, the bias, variance, and MSE can be calculated since the true function is known in the simulation. For the cubic periodic spline, Figure 3.4(b) shows $b^2(\lambda)$, $v(\lambda)$, and $\mathrm{MSE}(\lambda)$ as functions of $\log_{10}(n\lambda)$. Obviously, as λ increases, the squared bias increases and the variance decreases. The MSE represents a compromise between bias and variance.

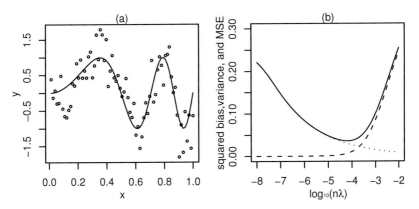

FIGURE 3.4 Plots of (a) true function (line) and observations (circles), and (b) squared bias $b^2(\lambda)$ (dashed), variance $v(\lambda)$ (dotted line), and MSE (solid line) for the cubic periodic spline.

Another closely related target criterion is the average *predictive squared error* (PSE)

$$\mathrm{PSE}(\lambda) = \mathrm{E}\left(\frac{1}{n}||\boldsymbol{y}^+ - \hat{\boldsymbol{f}}||^2\right), \qquad (3.12)$$

where $\boldsymbol{y}^+ = \boldsymbol{f} + \boldsymbol{\epsilon}^+$ are new observations of $\boldsymbol{f}$, $\boldsymbol{\epsilon}^+ = (\epsilon_1^+, \ldots, \epsilon_n^+)^T$ are independent of $\boldsymbol{\epsilon}$, and ϵ_i^+ are independent and identically distributed with mean zero and variance σ^2. PSE measures the performance of a model's prediction for new observations. We have

$$\text{PSE}(\lambda) = \text{E}\left(\frac{1}{n}||(\boldsymbol{y}^+ - \boldsymbol{f}) + (\boldsymbol{f} - \hat{\boldsymbol{f}})||^2\right) = \sigma^2 + \text{MSE}(\lambda).$$

Thus PSE differs from MSE only by a constant σ^2.

Ideally, one would want to select λ that minimizes the MSE (PSE). This is, however, not practical because MSE (PSE) depends on the unknown true function f that one wants to estimate in the first place. Instead, one may estimate MSE (PSE) from the data and then minimize the estimated criterion. We discuss unbiased and cross-validation estimates of PSE (MSE) in Sections 3.3 and 3.4, respectively.

3.3   Unbiased Risk

First consider the case when the error variance σ^2 is known. Since

$$\text{E}\left(\frac{1}{n}||\boldsymbol{y} - \hat{\boldsymbol{f}}||^2\right)$$

$$= \text{E}\left\{\frac{1}{n}||\boldsymbol{y} - \boldsymbol{f}||^2 + \frac{2}{n}(\boldsymbol{y} - \boldsymbol{f})^T(\boldsymbol{f} - \hat{\boldsymbol{f}}) + \frac{1}{n}||\boldsymbol{f} - \hat{\boldsymbol{f}}||^2\right\}$$

$$= \sigma^2 - \frac{2\sigma^2}{n}\text{tr}H(\lambda) + \text{MSE}(\lambda), \tag{3.13}$$

then,

$$\text{UBR}(\lambda) \triangleq \frac{1}{n}||(I - H(\lambda))\boldsymbol{y}||^2 + \frac{2\sigma^2}{n}\text{tr}H(\lambda) \tag{3.14}$$

is an unbiased estimate of PSE(λ). Since PSE differs from MSE only by a constant σ^2, one may expect the minimizer of UBR(λ) to be close to the minimizer of the risk function MSE(λ). In fact, a stronger result holds: under certain regularity conditions, UBR(λ) is a consistent estimate of the *relative loss function* $L(\lambda) + n^{-1}\boldsymbol{\epsilon}^T\boldsymbol{\epsilon}$ (Gu 2002). The function UBR(λ) is referred to as the *unbiased risk* (UBR) criterion, and the minimizer of UBR(λ) is referred to as the UBR estimate of λ. It is obvious that UBR(λ) is an extension of the Mallow's C_p criterion.

The error variance σ^2 is usually unknown in practice. In general, there are two classes of estimators for σ^2: *residual-based* and *difference-based* estimators. The first class of estimators is based on residuals from an estimate of f. For example, analogous to parametric regression, an estimator of σ^2 based on fit $\hat{\boldsymbol{f}} = H(\lambda)\boldsymbol{y}$, is

$$\hat{\sigma}^2 \triangleq \frac{\|(I - H(\lambda))\boldsymbol{y}\|^2}{n - \text{tr}H(\lambda)}. \tag{3.15}$$

The estimator $\hat{\sigma}^2$ is consistent under certain regularity conditions (Gu 2002). However, it depends critically on the smoothing parameter λ. Thus, it cannot be used in the UBR criterion since the purpose of this criterion is to select λ. For choosing the amount of smoothing, it is desirable to have an estimator of σ^2 without needing to fit the function f first.

The difference-based estimators of σ^2 do not require an estimate of the mean function f. The basic idea is to remove the mean function f by taking differences based on some well-chosen subsets of data. Consider the general SSR model (2.10). Let $\mathcal{I}_j = \{i(j,1), \ldots, i(j, K_j)\} \subset \{1, \ldots, n\}$ be a subset of index and $d(j, k)$ be some fixed coefficients such that

$$\sum_{k=1}^{K_j} d^2(j, k) = 1, \quad \sum_{k=1}^{K_j} d(j, k)\mathcal{L}_{i(j,k)} f \approx 0, \quad j = 1, \ldots, J.$$

Since

$$\text{E}\left\{\sum_{k=1}^{K_j} d(j, k)y_{i(j,k)}\right\}^2 \approx \text{E}\left\{\sum_{k=1}^{K_j} d(j, k)\epsilon_{i(j,k)}\right\}^2 = \sigma^2,$$

then

$$\tilde{\sigma}^2 \approx \frac{1}{J} \sum_{j=1}^{J} \left\{\sum_{k=1}^{K_j} d(j, k)y_{i(j,k)}\right\}^2 \tag{3.16}$$

provides an approximately unbiased estimator of σ^2. The estimator $\tilde{\sigma}^2$ is referred to as a differenced-based estimator since $d(j, k)$ are usually chosen to be contrasts such that $\sum_{k=1}^{K_j} d(j, k) = 0$. The specific choices of subsets and coefficients depend on factors including prior knowledge about f and the domain $\mathcal{X}$.

Several methods have been proposed for the common situation when x is a univariate continuous variable, f is a smooth function, and $\mathcal{L}_i$ are evaluational functionals. Suppose design points are ordered such that $x_1 \leq x_2 \leq \cdots \leq x_n$. Since f is smooth, then $f(x_{j+1}) - f(x_j) \approx 0$ when neighboring design points are close to each other. Setting $\mathcal{I}_j = \{j, j+1\}$ and $d(j, 1) = -d(j, 2) = 1/\sqrt{2}$ for $j = 1, \ldots, n-1$, we have the first-order difference-based estimator proposed by Rice (1984):

$$\tilde{\sigma}_{\text{R}}^2 = \frac{1}{2(n-1)} \sum_{i=2}^{n} (y_i - y_{i-1})^2. \tag{3.17}$$

Hall, Kay and Titterington (1990) proposed the mth order difference-based estimator

$$\tilde{\sigma}^2_{\text{HKT}} = \frac{1}{n-m} \sum_{j=1}^{n-m} \left\{ \sum_{k=1}^{m} \delta_k y_{j+k} \right\}^2, \qquad (3.18)$$

where coefficients δ_k satisfy $\sum_{k=1}^{m} \delta_k = 0$, $\sum_{k=1}^{m} \delta_k^2 = 1$, and $\delta_1 \delta_m \neq 0$. Optimal choices of δ_k are studied in Hall et al. (1990). It is easy to see that $\tilde{\sigma}^2_{HKT}$ corresponds to $\mathcal{I}_j = \{j, \ldots, j+m\}$ and $d(j,k) = \delta_k$ for $j = 1, \ldots, n-m$.

Both $\tilde{\sigma}^2_{\text{R}}$ and $\tilde{\sigma}^2_{\text{HKT}}$ require an ordering of design points that could be problematic for multivariate independent variables. Tong and Wang (2005) proposed a different method for a general domain $\mathcal{X}$. Suppose $\mathcal{X}$ is equipped with a norm. Collect squared distances, $d_{ij} = ||x_i - x_j||^2$, for all pairs $\{x_i, x_j\}$, and half squared differences, $s_{ij} = (y_i - y_j)^2/2$, for all pairs $\{y_i, y_j\}$. Then $\text{E}(s_{ij}) = \{f(x_i) - f(x_j)\}^2/2 + \sigma^2$. Suppose $\{f(x_i) - f(x_j)\}^2/2$ can be approximated by βd_{ij} when d_{ij} is small. Then the LS estimate of the intercept in the simple linear model

$$s_{ij} = \alpha + \beta d_{ij} + \epsilon_{ij}, \quad d_{ij} \leq D, \qquad (3.19)$$

provides an estimate of σ^2. Denote such an estimator as $\tilde{\sigma}^2_{\text{TW}}$. Theoretical properties and the choice of bandwidth D were studied in Tong and Wang (2005).

To illustrate the UBR criterion as an estimate of PSE, we generate responses from model (1.1) with $f(x) = \sin(4\pi x^2)$, $\sigma = 0.5$, $x_i = i/n$ for $i = 1, \ldots, n$, and $n = 73$. For the cubic periodic spline, the UBR functions based on 50 replications of simulation data are shown in Figure 3.5 where the true variance is used in (a) and the Rice estimator is used in (b).

3.4   Cross-Validation and Generalized Cross-Validation

Equation (3.13) shows that the RSS underestimates the PSE by the amount of $2\sigma^2 \text{tr} H(\lambda)/n$. The second term in the UBR criterion corrects this bias. The bias in RSS is a consequence of using the same data for model fitting and model evaluation. Ideally, these two tasks should be separated using independent samples. This can be achieved by splitting the whole data into two subsamples: a training (calibration) sample for

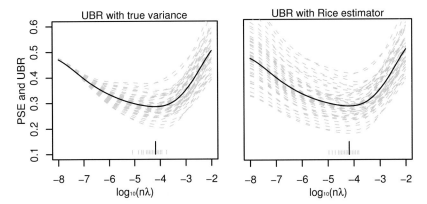

FIGURE 3.5 Plots of the PSE function as solid lines, the UBR functions with true σ^2 as dashed lines (left), and the UBR functions with Rice estimator $\tilde{\sigma}_R^2$. The minimum point of the PSE is marked as long bars at the bottom. The UBR estimates of $\log_{10}(n\lambda)$ are marked as short bars.

model fitting, and a test (validation) sample for model evaluation. This approach, however, is not efficient unless the sample size is large. The idea behind cross-validation is to recycle data by switching the roles of training and test samples. For simplicity, we present leaving-out-one cross-validation only. That is, each time one observation will be left out as the test sample, and the remaining $n-1$ samples will be used as the training sample.

Let $\hat{f}^{[i]}$ be the minimizer of the PLS based on all observations except y_i:

$$\frac{1}{n}\sum_{j\neq i}(y_j - \mathcal{L}_j f)^2 + \lambda||P_1 f||^2. \tag{3.20}$$

The cross-validation estimate of PSE is

$$\text{CV}(\lambda) \triangleq \frac{1}{n}\sum_{i=1}^{n}\left(\mathcal{L}_i \hat{f}^{[i]} - y_i\right)^2. \tag{3.21}$$

$\text{CV}(\lambda)$ is referred to as the *cross-validation* criterion, and the minimizer of $\text{CV}(\lambda)$ is called the cross-validation estimate of the smoothing parameter. Computation of $\hat{f}^{[i]}$ based on (3.21) for each $i = 1, \ldots, n$ would be costly. Fortunately, this is unnecessary due to the following lemma.

Leaving-out-one Lemma
For any fixed i, $\hat{f}^{[i]}$ is the minimizer of

$$\frac{1}{n}\left(\mathcal{L}_i\hat{f}^{[i]} - \mathcal{L}_if\right)^2 + \frac{1}{n}\sum_{j\neq i}(y_j - \mathcal{L}_jf)^2 + \lambda\|P_1f\|^2. \qquad (3.22)$$

[Proof] For any function f, we have

$$\frac{1}{n}\left(\mathcal{L}_i\hat{f}^{[i]} - \mathcal{L}_if\right)^2 + \frac{1}{n}\sum_{j\neq i}(y_j - \mathcal{L}_if)^2 + \lambda\|P_1f\|^2$$

$$\geq \frac{1}{n}\sum_{j\neq i}(y_j - \mathcal{L}_if)^2 + \lambda\|P_1f\|^2$$

$$\geq \frac{1}{n}\sum_{j\neq i}\left(y_j - \mathcal{L}_j\hat{f}^{[i]}\right)^2 + \lambda\|P_1\hat{f}^{[i]}\|^2$$

$$= \frac{1}{n}\left(\mathcal{L}_i\hat{f}^{[i]} - \mathcal{L}_i\hat{f}^{[i]}\right)^2 + \frac{1}{n}\sum_{j\neq i}\left(y_j - \mathcal{L}_j\hat{f}^{[i]}\right)^2 + \lambda\|P_1\hat{f}^{[i]}\|^2,$$

where the second inequality holds since $\hat{f}^{[i]}$ is the minimizer of (3.20).

The above lemma indicates that the solution to the PLS (3.20) without the ith observation, $\hat{f}^{[i]}$, is also the solution to the PLS (2.11) with the ith observation y_i being replaced by the fitted value $\mathcal{L}_i\hat{f}^{[i]}$. Note that the hat matrix $H(\lambda)$ depends on the model space and operators $\mathcal{L}_i$ only. It does not depend on observations of the dependent variable. Therefore, both fits based on (2.11) and (3.22) have the same hat matrix. That is, $\hat{\boldsymbol{f}} = H(\lambda)\boldsymbol{y}$ and $\hat{\boldsymbol{f}}^{[i]} = H(\lambda)\boldsymbol{y}^{[i]}$, where $\hat{\boldsymbol{f}}^{[i]} = (\mathcal{L}_1\hat{f}^{[i]},\ldots,\mathcal{L}_n\hat{f}^{[i]})^T$ and $\boldsymbol{y}^{[i]}$ is the same as $\boldsymbol{y}$ except that the ith element is replaced by $\mathcal{L}_i\hat{f}^{[i]}$. Denote $H(\lambda) = \{h_{ij}\}_{i,j=1}^n$. Then

$$\mathcal{L}_i\hat{f} = \sum_{j=1}^n h_{ij}y_j,$$

$$\mathcal{L}_i\hat{f}^{[i]} = \sum_{j\neq i} h_{ij}y_j + h_{ii}\mathcal{L}_i\hat{f}^{[i]}.$$

Solving for $\mathcal{L}_i\hat{f}^{[i]}$, we have

$$\mathcal{L}_i\hat{f}^{[i]} = \frac{\mathcal{L}_i\hat{f} - h_{ii}y_i}{1 - h_{ii}}.$$

Then

$$\mathcal{L}_i\hat{f}^{[i]} - y_i = \frac{\mathcal{L}_i\hat{f} - h_{ii}y_i}{1 - h_{ii}} - y_i = \frac{\mathcal{L}_i\hat{f} - y_i}{1 - h_{ii}}.$$

Plugging into (3.21), we have

$$\text{CV}(\lambda) = \frac{1}{n} \sum_{i=1}^{n} \frac{(\mathcal{L}_i \hat{f} - y_i)^2}{(1 - h_{ii})^2}. \tag{3.23}$$

Therefore, the cross-validation criterion can be calculated using the fit based on the whole sample and the diagonal elements of the hat matrix.

Replacing h_{ii} by the average of diagonal elements, $\text{tr}H(\lambda)$, we have the *generalized cross-validation* (GCV) criterion

$$\text{GCV}(\lambda) \triangleq \frac{\frac{1}{n} \sum_{i=1}^{n} (\mathcal{L}_i \hat{f} - y_i)^2}{\left\{ \frac{1}{n} \text{tr}(I - H(\lambda)) \right\}^2}. \tag{3.24}$$

The GCV estimate of λ is the minimizer of $\text{GCV}(\lambda)$. Since $\text{tr}H(\lambda)/n$ is usually small in the neighborhood of the optimal λ, we have

$$\text{E}\{\text{GCV}(\lambda)\} \approx \left\{ \frac{1}{n} \|(I - H(\lambda))\boldsymbol{f}\|^2 + \frac{\sigma^2}{n} \text{tr}H^2(\lambda) + \sigma^2 - \frac{2\sigma^2}{n} \text{tr}H(\lambda) \right\}$$
$$\left\{ 1 + \frac{2}{n} \text{tr}H(\lambda) \right\}$$
$$= \text{PSE}(\lambda)\{1 + o(1)\}.$$

The above approximation provides a very crude argument supporting the GCV criterion as a proxy for the PSE. More formally, under certain regularity conditions, $\text{GCV}(\lambda)$ is a consistent estimate of the relative loss function. Furthermore, $\text{GCV}(\lambda)$ is invariant to an orthogonal transformation of $\boldsymbol{y}$. See Wahba (1990) and Gu (2002) for details. One distinctive advantage of the GCV criterion over the UBR criterion is that the former does not require an estimate of σ^2.

To illustrate the $\text{CV}(\lambda)$ and $\text{GCV}(\lambda)$ criteria as estimates of the PSE, we generate responses from model (1.1) with $f(x) = \sin(4\pi x^2)$, $\sigma = 0.5$, $x_i = i/n$ for $i = 1, \ldots, n$, and $n = 73$. For cubic periodic spline, the CV and GCV scores for 50 replications of simulation data are shown in Figure 3.6.

3.5   Bayes and Linear Mixed-Effects Models

Assume a prior for f as

$$F(x) = \sum_{\nu=1}^{p} \zeta_\nu \phi_\nu(x) + \delta^{\frac{1}{2}} U(x), \tag{3.25}$$

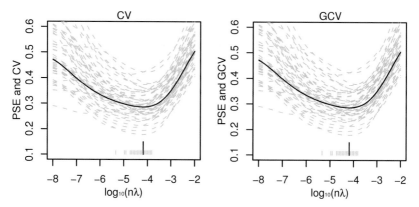

FIGURE 3.6 Plots of the PSE function as solid lines, the CV functions as dashed lines (left), and the GCV functions as dashed lines (right). The minimum point of the PSE is marked as long bars at the bottom. CV and GCV estimates of $\log_{10}(n\lambda)$ are marked as short bars.

where $\zeta_1, \ldots, \zeta_p \overset{iid}{\sim} N(0, \kappa)$, $U(x)$ is a zero-mean Gaussian stochastic process with covariance function $R_1(x, z)$, ζ_ν and $U(x)$ are independent, and κ and δ are positive constants. Note that the bounded linear functionals $\mathcal{L}_i$ are defined for elements in $\mathcal{H}$. Its application to the random process $F(x)$ is yet to be defined. For simplicity, the subscript i in $\mathcal{L}_i$ is ignored in the following definition. Define $\mathcal{L}(\zeta_\nu \phi_\nu) = \zeta_\nu \mathcal{L}\phi_\nu$. The definition of $\mathcal{L}U$ requires the duality between the Hilbert space spanned by a family of random variables and its associated RKHS.

Consider the linear space

$$\mathcal{U} = \left\{ W : W = \sum \alpha_j U(x_j), \ x_j \in \mathcal{X}, \ \alpha_j \in \mathbb{R} \right\}$$

with inner product $(W_1, W_2) = E(W_1 W_2)$. Let $L_2(U)$ be the Hilbert space that is the completion of $\mathcal{U}$. Note that the RK R_1 of $\mathcal{H}_1$ coincides with the covariance function of $U(x)$. Consider a linear map $\Psi : \mathcal{H}_1 \rightarrow L_2(U)$ such that $\Psi\{R_1(x_j, \cdot)\} = U(x_j)$. Since

$$(R_1(x, \cdot), R_1(z, \cdot)) = R_1(x, z) = E\{U(x)U(z)\} = (U(x), U(z)),$$

the map Ψ is inner product preserving. In fact, $\mathcal{H}_1$ is *isometrically isomorphic* to $L_2(U)$. See Parzen (1961) for details. Since $\mathcal{L}$ is a bounded linear functional in $\mathcal{H}_1$, by the Riesz representation theorem, there exists a representer h such that $\mathcal{L}f = (h, f)$. Finally we define

$$\mathcal{L}U \triangleq \Psi h.$$

Note that $\mathcal{L}U$ is a random variable in $L_2(U)$. The application of $\mathcal{L}$ to F is defined as $\mathcal{L}F = \sum_{\nu=1}^{p} \zeta_\nu \mathcal{L}\phi_\nu + \delta^{\frac{1}{2}}\mathcal{L}U$.

When $\mathcal{L}$ is an evaluational functional, say $\mathcal{L}f = f(x_0)$ for a fixed x_0, we have $h(\cdot) = R_1(x_0, \cdot)$. Consequently,

$$\mathcal{L}U = \Psi h = \Psi\{R_1(x_0, \cdot)\} = U(x_0),$$

the evaluation of U at x_0. Therefore, as expected, $\mathcal{L}F = F(x_0)$ when $\mathcal{L}$ is an evaluational functional.

Suppose observations are generated by

$$y_i = \mathcal{L}_i F + \epsilon_i, \quad i = 1, \ldots, n, \tag{3.26}$$

where the prior F is defined in (3.25) and $\epsilon_i \overset{iid}{\sim} N(0, \sigma^2)$. Note that the normality assumption has been made for random errors.

We now compute the posterior mean $E(\mathcal{L}_0 F | \boldsymbol{y})$ for a bounded linear functional $\mathcal{L}_0$ on $\mathcal{H}$. Note that $\mathcal{L}_0$ is arbitrary, which could be quite different from $\mathcal{L}_i$. For example, suppose $f \in W_2^m[a, b]$ and $\mathcal{L}_i$ are evaluational functionals. Setting $\mathcal{L}_0 f = f'(x_0)$ leads to an estimate of f'. Using the correspondence between $\mathcal{H}$ and $L_2(U)$, we have

$$\begin{aligned}
E(\mathcal{L}_i U \mathcal{L}_j U) &= (\mathcal{L}_i U, \mathcal{L}_j U) = (\mathcal{L}_{i(x)} R_1(x, \cdot), \mathcal{L}_{j(z)} R_1(z, \cdot)) \\
&= \mathcal{L}_{i(x)} \mathcal{L}_{j(z)} R_1(x, z), \\
E(\mathcal{L}_0 U \mathcal{L}_i U) &= (\mathcal{L}_0 U, \mathcal{L}_j U) = (\mathcal{L}_{0(x)} R_1(x, \cdot), \mathcal{L}_{j(z)} R_1(z, \cdot)) \\
&= \mathcal{L}_{0(x)} \mathcal{L}_{j(z)} R_1(x, z).
\end{aligned}$$

Let $\boldsymbol{\zeta} = (\zeta_1, \ldots, \zeta_p)^T$, $\boldsymbol{\phi} = (\phi_1, \ldots, \phi_p)^T$ and $\mathcal{L}_0\boldsymbol{\phi} = (\mathcal{L}_0\phi_1, \ldots, \mathcal{L}_0\phi_p)^T$. Then $F(x) = \boldsymbol{\phi}^T(x)\boldsymbol{\zeta} + \delta^{\frac{1}{2}}U(x)$. It is easy to check that

$$\boldsymbol{y} = T\boldsymbol{\zeta} + \delta^{\frac{1}{2}}(\mathcal{L}_1 U, \ldots, \mathcal{L}_n U)^T + \boldsymbol{\epsilon} \sim N(\boldsymbol{0}, \kappa T^T T + \delta \Sigma + \sigma^2 I), \tag{3.27}$$

and

$$\begin{aligned}
\mathcal{L}_0 F &= (\mathcal{L}_0\boldsymbol{\phi})^T \boldsymbol{\zeta} + \delta^{\frac{1}{2}}\mathcal{L}_0 U \\
&\sim N(0, \kappa(\mathcal{L}_0\boldsymbol{\phi})^T \mathcal{L}_0\boldsymbol{\phi} + \delta \mathcal{L}_{0(x)} \mathcal{L}_{0(z)} R_1(x, z)).
\end{aligned} \tag{3.28}$$

Furthermore,

$$E\{(\mathcal{L}_0 F)\boldsymbol{y}\} = \kappa T \mathcal{L}_0\boldsymbol{\phi} + \delta \mathcal{L}_0\boldsymbol{\xi}, \tag{3.29}$$

where

$$\begin{aligned}
\boldsymbol{\xi}(x) &= (\mathcal{L}_{1(z)} R_1(x, z), \ldots, \mathcal{L}_{n(z)} R_1(x, z))^T, \\
\mathcal{L}_0\boldsymbol{\xi} &= (\mathcal{L}_{0(x)} \mathcal{L}_{1(z)} R_1(x, z), \ldots, \mathcal{L}_{0(x)} \mathcal{L}_{n(z)} R_1(x, z))^T.
\end{aligned} \tag{3.30}$$

Let $\lambda = \sigma^2/n\delta$ and $\eta = \kappa/\delta$. Using properties of multivariate normal random variables and equations (3.27), (3.28), and (3.29), we have

$$
\begin{aligned}
&\mathrm{E}(\mathcal{L}_0 F|\boldsymbol{y}) \\
&= (\mathcal{L}_0\boldsymbol{\phi})^T \eta T^T (\eta T T^T + M)^{-1}\boldsymbol{y} + (\mathcal{L}_0\boldsymbol{\xi})^T (\eta T T^T + M)^{-1}\boldsymbol{y}. \tag{3.31}
\end{aligned}
$$

It can be shown (Wahba 1990, Gu 2002) that for any full-column rank matrix T and symmetric and nonsingular matrix M,

$$
\begin{aligned}
\lim_{\eta\to\infty} (\eta T^T T + M)^{-1} &= M^{-1} - M^{-1}T(T^T M^{-1}T)^{-1}T^T M^{-1}, \\
\lim_{\eta\to\infty} \eta T^T (\eta T^T T + M)^{-1} &= T(T^T M^{-1}T)^{-1}T^T M^{-1}. \tag{3.32}
\end{aligned}
$$

Combining results in (3.31), (3.32), and (2.22), we have

$$
\begin{aligned}
\lim_{\kappa\to\infty} \mathrm{E}(\mathcal{L}_0 F|\boldsymbol{y}) &= (\mathcal{L}_0\boldsymbol{\phi})^T T(T^T M^{-1}T)^{-1}T^T M^{-1}\boldsymbol{y} \\
&\quad + (\mathcal{L}_0\boldsymbol{\xi})^T \{M^{-1} - M^{-1}T(T^T M^{-1}T)^{-1}T^T M^{-1}\}\boldsymbol{y} \\
&= (\mathcal{L}_0\boldsymbol{\phi})^T \boldsymbol{d} + (\mathcal{L}_0\boldsymbol{\xi})^T \boldsymbol{c} \\
&= \mathcal{L}_0\hat{f}.
\end{aligned}
$$

The above result indicates that the smoothing spline estimate $\hat{f}$ is a Bayes estimator with a diffuse prior for $\boldsymbol{\zeta}$. From a frequentist perspective, the smoothing spline estimate may be regarded as the *best linear unbiased prediction* (BLUP) estimate of a *linear mixed-effects (LME) model*. We now present three corresponding LME models. The first LME model assumes that

$$
\boldsymbol{y} = T\boldsymbol{\zeta} + \boldsymbol{u} + \boldsymbol{\epsilon}, \tag{3.33}
$$

where $\boldsymbol{\zeta} = (\zeta_1, \ldots, \zeta_p)^T$ are deterministic parameters, $\boldsymbol{u} = (u_1, \ldots, u_n)^T$ are random effects with distribution $\boldsymbol{u} \sim \mathrm{N}(\boldsymbol{0}, \sigma^2\Sigma/n\lambda)$, $\boldsymbol{\epsilon} = (\epsilon_1, \ldots, \epsilon_n)^T$ are random errors with distribution $\boldsymbol{\epsilon} \sim \mathrm{N}(\boldsymbol{0}, \sigma^2 I)$, and $\boldsymbol{u}$ and $\boldsymbol{\epsilon}$ are independent. The second LME model assumes that

$$
\boldsymbol{y} = T\boldsymbol{\zeta} + \Sigma\boldsymbol{u} + \boldsymbol{\epsilon}, \tag{3.34}
$$

where $\boldsymbol{\zeta}$ are deterministic parameters, $\boldsymbol{u}$ are random effects with distribution $\boldsymbol{u} \sim \mathrm{N}(\boldsymbol{0}, \sigma^2\Sigma^+/n\lambda)$, Σ^+ is the Moore–Penrose inverse of Σ, $\boldsymbol{\epsilon}$ are random errors with distribution $\boldsymbol{\epsilon} \sim \mathrm{N}(\boldsymbol{0}, \sigma^2 I)$, and $\boldsymbol{u}$ and $\boldsymbol{\epsilon}$ are independent.

It is inconvenient to use the above two LME models for computation since Σ may be singular. Write $\Sigma = ZZ^T$, where Z is a $n \times m$ matrix with $m = \mathrm{rank}(\Sigma)$. The third LME model assumes that

$$
\boldsymbol{y} = T\boldsymbol{\zeta} + Z\boldsymbol{u} + \boldsymbol{\epsilon}, \tag{3.35}
$$

where $\boldsymbol{\zeta}$ are deterministic parameters, $\boldsymbol{u}$ are random effects with distribution $\boldsymbol{u} \sim \mathrm{N}(\boldsymbol{0}, \sigma^2 I/n\lambda)$, $\boldsymbol{\epsilon}$ are random errors with distribution $\boldsymbol{\epsilon} \sim \mathrm{N}(\boldsymbol{0}, \sigma^2 I)$, and $\boldsymbol{u}$ and $\boldsymbol{\epsilon}$ are independent.

It can be shown that the BLUP estimates for each of the three LME models (3.33), (3.34), and (3.35) are the same as the smoothing spline estimate. See Wang (1998b) and Chapter 9 for more details.

3.6 Generalized Maximum Likelihood

The connection between smoothing spline models and Bayes models can be exploited to develop a likelihood-based estimate for the smoothing parameter. From (3.27), the marginal distribution of $\boldsymbol{y}$ is $\mathrm{N}(\boldsymbol{0}, \delta(\eta T T^T + M))$. Consider the following transformation

$$\begin{pmatrix} \boldsymbol{w}_1 \\ \boldsymbol{w}_2 \end{pmatrix} = \begin{pmatrix} Q_2^T \\ \frac{1}{\sqrt{\eta}} T^T \end{pmatrix} \boldsymbol{y}. \qquad (3.36)$$

It is easy to check that

$$\boldsymbol{w}_1 = Q_2^T \boldsymbol{y} \sim \mathrm{N}(0, \delta Q_2^T M Q_2),$$

$$\mathrm{Cov}(\boldsymbol{w}_1, \boldsymbol{w}_2) = \frac{\delta}{\sqrt{\eta}} Q_2^T (\eta T T^T + M) T \to \boldsymbol{0}, \ \eta \to \infty,$$

$$\mathrm{Var}(\boldsymbol{w}_2) = \frac{\delta}{\eta} T^T (\eta T T^T + M) T \to \delta (T^T T)(T^T T), \ \eta \to \infty.$$

Note that the distribution of $\boldsymbol{w}_2$ is independent of λ. Therefore, we consider the negative marginal log-likelihood of $\boldsymbol{w}_1$

$$l(\lambda, \delta | \boldsymbol{w}_1) = \frac{1}{2} \log |\delta Q_2^T M Q_2| + \frac{1}{2\delta} \boldsymbol{w}_1^T (Q_2^T M Q_2)^{-1} \boldsymbol{w}_1 + C_1, \quad (3.37)$$

where C_1 is a constant. Minimizing $l(\lambda, \delta | \boldsymbol{w}_1)$ with respect to δ, we have

$$\hat{\delta} = \frac{\boldsymbol{w}_1^T (Q_2^T M Q_2)^{-1} \boldsymbol{w}_1}{n - p}. \qquad (3.38)$$

The profile negative log-likelihood

$$\begin{aligned} l_p(\lambda, \hat{\delta} | \boldsymbol{w}_1) &= \frac{1}{2} \log |Q_2^T M Q_2| + \frac{n-p}{2} \log \hat{\delta} + C_2 \\ &= \frac{n-p}{2} \log \frac{\boldsymbol{w}_1^T (Q_2^T M Q_2)^{-1} \boldsymbol{w}_1}{\{\det(Q_2^T M Q_2)^{-1}\}^{\frac{1}{n-p}}} + C_2, \end{aligned} \qquad (3.39)$$

where C_2 is another constant. The foregoing profile negative log-likelihood is equivalent to

$$\mathrm{GML}(\lambda) \triangleq \frac{\boldsymbol{w}_1^T(Q_2^T M Q_2)^{-1}\boldsymbol{w}_1}{\{\det(Q_2^T M Q_2)^{-1}\}^{\frac{1}{n-p}}} = \frac{\boldsymbol{y}^T(I - H(\lambda))\boldsymbol{y}}{[\det^+\{(I - H(\lambda))\}]^{\frac{1}{n-p}}}, \quad (3.40)$$

where the second equality is based on (2.26), and $\det^+$ represents the product of the nonzero eigenvalues. The function $\mathrm{GML}(\lambda)$ is referred to as the *generalized maximum likelihood* (GML) criterion, and the minimizer of $\mathrm{GML}(\lambda)$ is called the GML estimate of the smoothing parameter.

From (3.38), a likelihood-based estimate of σ^2 is

$$\hat{\sigma}^2 \triangleq \frac{n\hat{\lambda}\boldsymbol{w}_1^T(Q_2^T M Q_2)^{-1}\boldsymbol{w}_1}{n - p} = \frac{\boldsymbol{y}^T(I - H(\lambda))\boldsymbol{y}}{n - p}. \quad (3.41)$$

The GML criterion may also be derived from the connection between smoothing spline models and LME models. Consider any one of the three corresponding LME models (3.33), (3.34), and (3.35). The smoothing parameter λ is part of the variance component for the random effects. It is common practice in the mixed-effects literature to estimate the variance components using *restricted likelihood* based on an orthogonal contrast of original observations where the orthogonal contrast is used to eliminate the fixed effects. Note that $\boldsymbol{w}_1$ is one such orthogonal contrast since Q_2 is orthogonal to T. Therefore, $l(\lambda, \delta|\boldsymbol{w}_1)$ in (3.37) is the negative log restricted likelihood, and the GML estimate of the smoothing parameter is the *restricted maximum likelihood* (REML) estimate. Furthermore, the estimate of error variance in (3.41) is the REML estimate of σ^2. The connection between a smoothing spline estimate with GML estimate of the smoothing parameter and a BLUP estimate with REML estimate of the variance component in a corresponding LME model may be utilized to fit a smoothing spline model using software for LME models. This approach will be adopted in Chapters 5, 8, and 9 to fit smoothing spline models for correlated observations.

3.7   Comparison and Implementation

Theoretical properties of the UBR, GCV, and GML criteria can be found in Wahba (1990) and Gu (2002). The UBR criterion requires an estimate of the variance σ^2. No distributional assumptions are required for the UBR and GCV criteria, while the normality assumption is required in the derivation of the GML criterion. Nevertheless, limited simulations suggest that the GML method is quite robust to the departure from the normality assumption.

Theoretical comparisons between UBR, GCV, and GML criteria have been studied using large-sample asymptotics (Wahba 1985, Li 1986, Stein 1990) and finite sample arguments (Efron 2001). Conclusions based on different perspectives do not always agree with each other. In practice, all three criteria usually perform well and lead to similar estimates. Each method has its own strengths and weaknesses. The UBR and GCV criteria occasionally lead to gross undersmooth (interpolation) when sample size is small. Fortunately, this problem diminishes quickly when sample size increases (Wahba and Wang 1995).

The argument `spar` in the `ssr` function specifies which method should be used for selecting the smoothing parameter λ. The options `spar=''v''`, `spar=''m''`, and `spar=''u''` correspond to the GCV, GML, and UBR methods, respectively. The default choice is the GCV method.

We now use the motorcycle data to illustrate how to specify the `spar` option. For simplicity, the variable `times` is first scaled into $[0, 1]$. We first use the Rice method to estimate error variance and use the estimated variance in the UBR criterion:

```
> x <- (times-min(times))/(max(times)-min(times))
> vrice <- mean((diff(accel))**2)/2
> mcycle.ubr.1 <- ssr(accel~x, rk=cubic(x),
                      spar=''u'', varht=vrice)
> summary(mcycle.ubr.1)
Smoothing spline regression fit by UBR method
...


UBR estimate(s) of smoothing parameter(s) : 8.60384e-07
Equivalent Degrees of Freedom (DF):  12.1624
Estimate of sigma:  23.09297
```

The option `varht` specifies the parameter σ^2 required for the UBR method. The `summary` function provides a synopsis including the estimate of the smoothing parameter, the degrees of freedom $\mathrm{tr}H(\lambda)$, and the estimate of standard deviation $\hat{\sigma}$.

Instead of the Rice estimator, we can estimate the error variance using Tong and Wang's estimator $\tilde{\sigma}^2_{\mathrm{TW}}$. Note that there are multiple observations at some time points. We use these replicates to estimate the error variance. That is, we select all pairs with zero distances:

```
> d <- s <- NULL
> for (i in 1:132) {
```

```
   for (j in (i+1):133) {
     d <- c(d, (x[i]-x[j])**2)
     s <- c(s, (accel[i]-accel[j])**2/2)
   }}
> vtw <- coef(lm(s~d,subset=d==0))[1]
> mcycle.ubr.2 <- ssr(accel~x, rk=cubic(x),
                      spar=''u'', varht=vtw)
> summary(mcycle.ubr.2)
Smoothing spline regression fit by UBR method
...


UBR estimate(s) of smoothing parameter(s) : 8.35209e-07
Equivalent Degrees of Freedom (DF):   12.24413
Estimate of sigma:   22.69913
```

Next we use the GCV method to select the smoothing parameter:

```
> mcycle.gcv <- ssr(accel~x, rk=cubic(x), spar=''v'')
> summary(mcycle.gcv)
Smoothing spline regression fit by GCV method
...


GCV estimate(s) of smoothing parameter(s) : 8.325815e-07
Equivalent Degrees of Freedom (DF):   12.25284
Estimate of sigma:   22.65806
```

Finally, we use the GML method to select the smoothing parameter:

```
> mcycle.gml <- ssr(accel~x, rk=cubic(x), spar=''m'')
> summary(mcycle.gml)
Smoothing spline regression fit by GML method
...


GML estimate(s) of smoothing parameter(s) : 4.729876e-07
Equivalent Degrees of Freedom (DF):   13.92711
Estimate of sigma:   22.57701
```

For the motorcycle data, all three methods lead to similar estimates of the smoothing parameter and the function f.

3.8 Confidence Intervals

3.8.1 Bayesian Confidence Intervals

Consider the Bayes model (3.25) and (3.26). The computation in Section 3.5 can be carried out one step further to derive posterior distributions. In the following arguments, as in Section 3.5, a diffuse prior is assumed for ζ with $\kappa \to \infty$. For simplicity of notation, the limit is not expressed explicitly.

Let $F_{0\nu} = \zeta_\nu \phi_\nu$ for $\nu = 1, \ldots, p$, and $F_1 = \delta^{\frac{1}{2}} U$. Let $\mathcal{L}_0$, $\mathcal{L}_{01}$, and $\mathcal{L}_{02}$ be bounded linear functionals. Since $F_{0\nu}$, F_1, and ϵ_i are all normal random variables, then the posterior distributions of $\mathcal{L}_0 F_{0\nu}$ and $\mathcal{L}_0 F_1$ are normal with the following mean and covariances.

Posterior means and covariances
For ν, $\mu = 1, \ldots, p$, the posterior means are

$$E(\mathcal{L}_0 F_{0\nu}|\boldsymbol{y}) = (\mathcal{L}_0 \phi_\nu) \boldsymbol{e}_\nu^T \boldsymbol{d},$$
$$E(\mathcal{L}_0 F_1|\boldsymbol{y}) = (\mathcal{L}_0 \boldsymbol{\xi})^T \boldsymbol{c}, \tag{3.42}$$

and the posterior covariances are

$$\delta^{-1} Cov(\mathcal{L}_{01} F_{0\nu}, \mathcal{L}_{02} F_{0\mu}|\boldsymbol{y}) = (\mathcal{L}_{01}\phi_\nu)(\mathcal{L}_{02}\phi_\mu)\boldsymbol{e}_\nu^T A \boldsymbol{e}_\mu,$$
$$\delta^{-1} Cov(\mathcal{L}_{01} F_{0\nu}, \mathcal{L}_{02} F_1|\boldsymbol{y}) = -(\mathcal{L}_{01}\phi_\nu)\boldsymbol{e}_\nu^T B(\mathcal{L}_{02}\boldsymbol{\xi}), \tag{3.43}$$
$$\delta^{-1} Cov(\mathcal{L}_{01} F_1, \mathcal{L}_{02} F_1|\boldsymbol{y}) = \mathcal{L}_{01}\mathcal{L}_{02} R_1 - (\mathcal{L}_{01}\boldsymbol{\xi})^T C(\mathcal{L}_{02}\boldsymbol{\xi}),$$

where $\boldsymbol{e}_\nu$ is a vector of dimension p with the νth element being one and all other elements being zero, the vectors $\boldsymbol{c}$ and $\boldsymbol{d}$ are given in (2.22), and the matrices $A = (T^T M^{-1} T)^{-1}$, $B = A T^T M^{-1}$, and $C = M^{-1}(I - B)$.

The vectors $\boldsymbol{\xi}$ and $\mathcal{L}\boldsymbol{\xi}$ are defined in (3.30). Proofs can be found in Wahba (1990) and Gu (2002). Note that $\mathcal{H} = \text{span}\{\phi_1, \ldots, \phi_p\} \oplus \mathcal{H}_1$. Then any $f \in \mathcal{H}$ can be represented as

$$f = f_{01} + \cdots + f_{0p} + f_1, \tag{3.44}$$

where $f_{0\nu} \in \text{span}\{\phi_\nu\}$ for $\nu = 1, \ldots, p$, and $f_1 \in \mathcal{H}_1$. The estimate $\hat{f}$ can also be decomposed similarly

$$\hat{f} = \hat{f}_{01} + \cdots + \hat{f}_{0p} + \hat{f}_1, \tag{3.45}$$

where $\hat{f}_{0\nu} = \phi_\nu d_\nu$ for $\nu = 1, \ldots, p$, and $\hat{f}_1 = \boldsymbol{\xi}^T \boldsymbol{c}$.

The functionals $\mathcal{L}_0$, $\mathcal{L}_{01}$, and $\mathcal{L}_{02}$ are arbitrary as long as they are well defined. Equations in (3.42) indicate that the posterior means of components in F equal their corresponding components in the spline estimate $\hat{f}$. Equations (3.42) and (3.43) can be used to compute posterior means and variances for any combinations of components of F. Specifically, consider the linear combination

$$F_{\boldsymbol{\gamma}}(x) = \sum_{\nu=1}^{p} \gamma_\nu F_{0\nu}(x) + \gamma_{p+1} F_1(x), \tag{3.46}$$

where γ_ν equals 1 when the corresponding component in F is to be included and 0 otherwise, and $\boldsymbol{\gamma} = (\gamma_1, \ldots, \gamma_{p+1})^T$. Then, for any linear functional $\mathcal{L}_0$,

$$\mathrm{E}(\mathcal{L}_0 F_{\boldsymbol{\gamma}}|\boldsymbol{y}) = \sum_{\nu=1}^{p} \gamma_\nu (\mathcal{L}_0 \phi_\nu) d_\nu + \gamma_{p+1} (\mathcal{L}_0 \boldsymbol{\xi})^T \boldsymbol{c},$$

$$\mathrm{Var}(\mathcal{L}_0 F_{\boldsymbol{\gamma}}|\boldsymbol{y}) = \sum_{\nu=1}^{p} \sum_{\mu=1}^{p} \gamma_\nu \gamma_\mu \mathrm{Cov}(\mathcal{L}_0 F_{0\nu}, \mathcal{L}_0 F_{0\mu}|\boldsymbol{y})$$

$$+ \sum_{\nu=1}^{p} \gamma_\nu \gamma_{p+1} \mathrm{Cov}(\mathcal{L}_0 F_{0\nu}, \mathcal{L}_0 F_1|\boldsymbol{y}) \tag{3.47}$$

$$+ \gamma_{p+1} \mathrm{Cov}(\mathcal{L}_0 F_1, \mathcal{L}_0 F_1|\boldsymbol{y}).$$

For various reasons it is often desirable to have interpretable confidence intervals for the function f and its components. For example, one may want to decide whether a nonparametric model is more suitable than a particular parametric model. A parametric regression model may be considered not suitable if a larger portion of its estimate is outside the confidence intervals of a smoothing spline estimate.

Consider a collection of points $x_{0j} \in \mathcal{X}$, $j = 1, \ldots, J$. For each j, posterior mean $\mathrm{E}\{F_{\boldsymbol{\gamma}}(x_{0j})|\boldsymbol{y}\}$ and variance $\mathrm{Var}\{F_{\boldsymbol{\gamma}}(x_{0j})|\boldsymbol{y}\}$ can be calculated using equations in (3.47) by setting $\mathcal{L}_0 F = F(x_{0j})$. Then, $100(1 - \alpha)\%$ Bayesian confidence intervals for

$$f_{\boldsymbol{\gamma}}(x_{0j}) = \sum_{\nu=1}^{p} \gamma_\nu f_{0\nu}(x_{0j}) + \gamma_{p+1} f_1(x_{0j}), \quad j = 1, \ldots, J \tag{3.48}$$

are

$$\mathrm{E}\{F_{\boldsymbol{\gamma}}(x_{0j})|\boldsymbol{y}\} \pm z_{\frac{\alpha}{2}} \sqrt{\mathrm{Var}\{F_{\boldsymbol{\gamma}}(x_{0j})|\boldsymbol{y}\}}, \quad j = 1, \ldots, J, \tag{3.49}$$

where $z_{\frac{\alpha}{2}}$ is the $1 - \alpha/2$ percentile of a standard normal distribution.

In particular, let $\boldsymbol{F} = (\mathcal{L}_1 F, \ldots, \mathcal{L}_n F)^T$. Applying (3.42) and (3.43), we have

$$\begin{aligned}
\mathrm{E}(\boldsymbol{F}|\boldsymbol{y}) &= H(\lambda)\boldsymbol{y}, \\
\mathrm{Cov}(\boldsymbol{F}|\boldsymbol{y}) &= \sigma^2 H(\lambda).
\end{aligned} \tag{3.50}$$

Therefore, the posterior variances of the fitted values $\mathrm{Var}(\mathcal{L}_i F|\boldsymbol{y}) = \sigma^2 h_{ii}$, where h_{ii} are diagonal elements of the matrix $H(\lambda)$. When $\mathcal{L}_i$ are evaluational functionals $\mathcal{L}_i f = f(x_i)$, Wahba (1983) proposed the following $100(1 - \alpha)\%$ confidence intervals

$$\hat{f}(x_i) \pm z_{\frac{\alpha}{2}} \hat{\sigma} \sqrt{h_{ii}}, \tag{3.51}$$

where $\hat{\sigma}$ is an estimates of σ. Note that confidence intervals for a linear combination of components of f can be constructed similarly.

Though based on the Bayesian argument, it has been found that the Bayesian confidence intervals have good frequentist properties provided that the smoothing parameter has been estimated properly. They must be interpreted as "across-the-function" rather than pointwise. More precisely, define the *average coverage probability* (ACP) as

$$\mathrm{ACP} = \frac{1}{n} \sum_{i=1}^{n} P\{f(x_i) \in C(\alpha, x_i)\}$$

for some $(1 - \alpha)100\%$ confidence intervals $\{C(\alpha, x_i), \ i = 1, \ldots, n\}$. Rather than considering a confidence interval for $f(\tau)$, where $f(\cdot)$ is the realization of a stochastic process and τ is fixed, one may consider confidence intervals for $f(\tau_n)$, where f is now a fixed function and τ_n is a point randomly selected from $\{x_i, \ i = 1, \ldots, n\}$. Then $\mathrm{ACP} = P\{f(\tau_n) \in C(\alpha, \tau_n)\}$. Note that the ACP coverage property is weaker than the pointwise coverage property. For polynomial splines and $C(\alpha, x_i)$ being the Bayesian confidence intervals defined in (3.51), under certain regularity conditions, Nychka (1988) showed that $\mathrm{ACP} \approx 1 - \alpha$.

The `predict` function in the `assist` package computes the posterior mean and standard deviation of $F_{\gamma}(x)$ in (3.46). The option `terms` specifies the coefficients γ, and the option `newdata` specifies a data frame consisting of the values at which predictions are required. We now use the geyser, motorcycle, and Arosa data to illustrate how to use the `predict` function to compute the posterior means and standard deviations.

For the geyser data, we have fitted a cubic spline in Chapter 1, Section 1.1 and a partial spline in Chapter 2, Section 2.10. In the following we fit a cubic spline using the `ssr` function, compute posterior means and standard deviations for the estimate of the smooth component $P_1 f$ using the `predict` function, and plot the estimate of the smooth components and 95% Bayesian confidence intervals:

```
> geyser.cub.fit <- ssr(waiting~x, rk=cubic(x))
> grid <- seq(0,1,len=200)
> geyser.cub.pred <- predict(geyser.cub.fit, pstd=T,
    terms=c(0,0,1), newdata=data.frame(x=grid))
> grid1 <- grid*diff(range(eruptions))+min(eruptions)
> plot(eruptions, waiting, xlab=''duration (mins)'',
    ylim=c(-6,6), ylab=''smooth component (mins)'',
    type=''n'')
> polygon(c(grid1,rev(grid1)),
    c(geyser.cub.pred$fit-1.96*geyser.cub.pred$pstd,
      rev(geyser.cub.pred$fit+1.96*geyser.cub.pred$pstd)),
    col=gray(0:8/8)[8], border=NA)
> lines(grid1, geyser.cub.pred$fit)
> abline(0,0,lty=2)
```

where the option `pstd` specifies whether the posterior standard deviations should be calculated. Note that the option `pstd=T` can be dropped in the above statement since it is the default. There are in total three components in the cubic spline fit: two basis functions ϕ_1 (constant) and ϕ_2 (linear) for the null space, and the smooth component in the space $\mathcal{H}_1$. In the order in which they appear in the `ssr` function, these three components correspond to the intercept (`~1`, which is automatically included), the linear basis specified by `~x`, and the smooth component specified by `rk=cubic(x)`. Therefore, the option `terms=c(0,0,1)` was used to compute the posterior means and standard deviations for the smooth component f_1 in the space $\mathcal{H}_1$. The estimate of the smooth components and 95% Bayesian confidence intervals are shown in Figure 3.7(a). A large portion of the zero constant line is outside the confidence intervals, indicating the lack-of-fit of a linear model (the null space of the cubic spline).

For the partial spline model (2.43), consider the following Bayes model

$$y_i = s_i^T \beta + \mathcal{L}_i F + \epsilon_i, \quad i = 1, \ldots, n, \qquad (3.52)$$

where the prior for β is assumed to be $N(\mathbf{0}, \kappa I_q)$, the prior F is defined in (3.25), and $\epsilon_i \overset{iid}{\sim} N(0, \sigma^2)$. Again, it can be shown that the PLS estimates of components in β and f based on (2.44) equal the posterior means of their corresponding components in the Bayes model as $\kappa \to \infty$. Posterior covariances and Bayesian confidence intervals for β and f_γ can be calculated similarly.

We now refit the partial spline model (2.47) and compute posterior means and standard deviations for the estimate of the smooth component $P_1 f$:

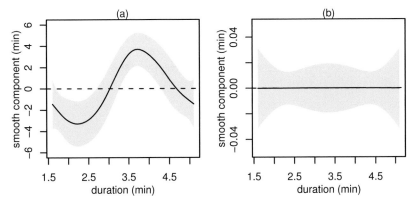

FIGURE 3.7 Geyser data, plots of estimates of the smooth components, and 95% Bayesian confidence intervals for (a) the cubic spline and (b) the partial spline models. The constant zero is marked as the dotted line in each plot.

```
> t <- .397; s <- 1*(x>t)
> geyser.ps.fit <- ssr(waiting~x+s, rk=cubic(x))
> geyser.ps.pred <- predict(geyser.ps.fit,
    terms=c(0,0,0,1),
    newdata=data.frame(x=grid,s=1*(grid>t)))
```

Since there are in total four components in the partial spline fit, the option `terms=c(0,0,0,1)` was used to compute the posterior means and standard deviations for the smooth component f_1.

Figure 3.7(b) shows the estimate of the smooth components and 95% Bayesian confidence intervals for the partial spline. The zero constant line is well inside the confidence intervals, indicating that this smooth component may be dropped in the partial spline model. That is, a simple linear change-point model may be appropriate for this data.

For the motorcycle data, based on visual inspection, we have searched a potential change-point t to the first derivative in the interval $[0.2, 0.25]$ for the variable x in Section 2.10. To search for all possible change-points to the first derivative, we fit the partial spline model (2.48) repeatedly with t taking values on a grid point in the interval $[0.1, 0.9]$. We then calculate the posterior mean and standard deviation for β. Define a t-statistic at point t as $E(\beta|\boldsymbol{y})/\sqrt{\text{Var}(\beta|\boldsymbol{y})}$. The t-statistics were calculated as follows:

```
> tgrid <- seq(0.05,.95,len=200); tstat <- NULL
> for (t in tgrid) {
    s <- (x-t)*(x>t)
```

```
tmp <- ssr(accel~x+s, rk=cubic(x))
tmppred <- predict(tmp, terms=c(0,0,1,0),
  newdata=data.frame(x=.5,s=1))
tstat <- c(tstat, tmppred$fit/tmppred$pstd)
}
```

Note that `term=c(0,0,1,0)` requests the posterior mean and standard deviation for the component $\beta \times s$, where $s = (x - t)_+$, and s is set to one in the `newdata` argument. The t-statistics are shown in the bottom panel of Figure 3.8.

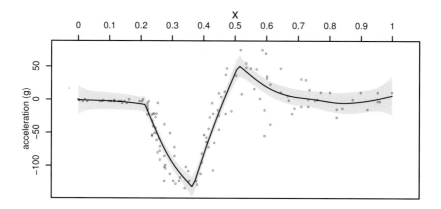

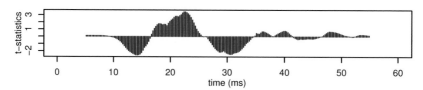

FIGURE 3.8 Motorcycle data, plots of t-statistics (bottom panel), and the partial spline fit with 95% Bayesian confidence intervals (top panel). The horizontal axis at the top shows the transformed scale.

There are three clusters of large t-statistics that suggests three potential change points in the first derivative at $t_1 = 0.2128$, $t_2 = 0.3666$, and $t_3 = 0.5113$, respectively (see also Speckman (1995)). So we fit the following partial spline model

$$y_i = \sum_{j=1}^{3} \beta_j (x_i - t_j)_+ + f(x_i) + \epsilon_i, \quad i = 1, \dots, n, \tag{3.53}$$

where x is the variable `times` scaled into $[0, 1]$, t_j are the change-points in the first derivative, and $f \in W_2^2[0, 1]$. We fit model (3.53) and compute posterior means and standard deviations as follows:

```
> t1 <- .2128; t2 <- .3666; t3 <- .5113
> s1 <- (x-t1)*(x>t1); s2 <- (x-t2)*(x>t2)
> s3 <- (x-t3)*(x>t3)
> mcycle.ps.fit2 <- ssr(accel~x+s1+s2+s3, rk=cubic(x))
> grid <- seq(0,1,len=100)
> mcycle.ps.pred2 <- predict(mcycle.ps.fit2,
    newdata=data.frame(x=grid, s1=(grid-t1)*(grid>t1),
    s2=(grid-t2)*(grid>t2), s3=(grid-t3)*(grid>t3)))
```

The fit and Bayesian confidence intervals are shown in Figure 3.8.

For the Arosa data, Figure 2.12 in Chapter 2 shows the estimates and 95% confidence intervals for the overall function and its decomposition. In particular, the posterior means and standard deviations for the periodic spline were calculated as follows:

```
> arosa.per.fit <- ssr(thick~1, rk=periodic(x))
> grid <- data.frame(x=seq(.5/12,11.5/12,length=50))
> arosa.per.pred <- predict(arosa.per.fit,grid,
    terms=matrix(c(1,0,0,1,1,1),nrow=3,byrow=T))
```

where the input for the `term` argument is a 3×2 matrix with the first row (`1,0`) specifying the parametric component, the second row (`0,1`) specifying the smooth component, and the third row (`1,1`) specifying the overall function.

3.8.2 Bootstrap Confidence Intervals

Consider the general SSR model (2.10). Let $\hat{f}$ and $\hat{\sigma}^2$ be the estimates of f and σ^2, respectively. Let

$$y_{i,b}^* = \mathcal{L}_i \hat{f} + \epsilon_{i,b}^*, \quad i = 1, \dots, n; \ b = 1, \dots, B \tag{3.54}$$

be B bootstrap samples where $\epsilon_{i,b}^* \overset{iid}{\sim} N(0, \hat{\sigma}^2)$. The random errors $\epsilon_{i,b}^*$ may also be drawn from residuals with replacement when the normality assumption is undesirable. Let $\hat{f}_{\gamma,b}^*$ be the estimate of f_γ in (3.48) based on the bth bootstrap sample $\{y_{i,b}^*,\ i = 1, \ldots, n\}$. For any well-defined functional $\mathcal{L}_0$, there are B bootstrap estimates $\mathcal{L}_0 \hat{f}_{\gamma,b}^*$ for $b = 1, \ldots, B$. Then the $100(1 - \alpha)\%$ *percentile bootstrap confidence interval* of $\mathcal{L}_0 f_\gamma$ is

$$(\mathcal{L}_0 \hat{f}_{\gamma,L}, \mathcal{L}_0 \hat{f}_{\gamma,U}), \tag{3.55}$$

where $\mathcal{L}_0 \hat{f}_{\gamma,L}$ and $\mathcal{L}_0 \hat{f}_{\gamma,U}$ are the lower and upper $\alpha/2$ quantiles of $\{\mathcal{L}_0 \hat{f}_{\gamma,b}^*,\ b = 1, \ldots, B\}$.

The percentile-t bootstrap confidence intervals can be constructed as follows. Let $\tau = (\mathcal{L}_0 \hat{f}_\gamma - \mathcal{L}_0 f_\gamma)/\hat{\sigma}$, where division by $\hat{\sigma}$ is introduced to reduce the dependence on σ. Let $\tau_b^* = (\mathcal{L}_0 \hat{f}_{\gamma,b}^* - \mathcal{L}_0 \hat{f}_\gamma)/\hat{\sigma}_b^*$ be the bootstrap estimates of τ for $b = 1, \ldots, B$, where $\hat{\sigma}_b^*$ is the estimate of σ based on the bth bootstrap sample. Then the $100(1 - \alpha)\%$ *percentile-t bootstrap confidence interval* of $\mathcal{L}_0 f_\gamma$ is

$$(\mathcal{L}_0 \hat{f}_\gamma - q_{1-\frac{\alpha}{2}} \hat{\sigma}, \mathcal{L}_0 \hat{f}_\gamma - q_{\frac{\alpha}{2}} \hat{\sigma}), \tag{3.56}$$

where $q_{\frac{\alpha}{2}}$ and $q_{1-\frac{\alpha}{2}}$ are the lower and upper $\alpha/2$ quantiles of $\{\tau_b^*,\ b = 1, \ldots, B\}$. Note that the bounded linear condition is not required for $\mathcal{L}_0$ in the above construction of bootstrap confidence intervals. Other forms of bootstrap confidence intervals and comparison between Bayesian and bootstrap approaches can be found in Wang and Wahba (1995).

For example, the 95% percentile bootstrap confidence intervals in Figure 2.8 in Chapter 2 were computed as follows:

```
> nboot <- 9999
> fb <- NULL
> for (i in 1:nboot) {
    yb <- canada.fit1$fit +
      sample(canada.fit1$resi, 35, replace=T)
    bfit <- ssr(yb~s, rk=Sigma, spar=''m'')
    fb <- cbind(fb, bfit$coef$d[2]+S%*%bfit$coef$c)
  }
> lb <- apply(fb, 1, quantile, prob=.025)
> ub <- apply(fb, 1, quantile, prob=.975)
```

where random errors for bootstrap samples were drawn from residuals with replacement, and `lb` and `ub` represent lower and upper bounds.

We use the following simulation to show the performance of Bayesian and bootstrap confidence intervals. Observations are generated from model (1.1) with $f(x) = \exp\{-64(x - 0.5)^2\}$, $\sigma = 0.1$, $x_i = i/n$ for $i = 1, \ldots, n$, and $n = 100$. We fit a cubic spline and construct 95% Bayesian, percentile bootstrap (denoted as Per), and percentile-t bootstrap (denoted as T) confidence intervals. We set $B = 1000$ and repeat the simulation 100 times. Figures 3.9(a)(b)(c) show average pointwise coverages for the Bayesian, percentile bootstrap, and percentile-t bootstrap confidence intervals, respectively. The absolute value of f'' is also plotted to show the curvature of the function. The pointwise coverage is usually smaller than the nominal value at high curvature points. Boxplots of the across-the-function coverages for these three methods are shown in Figure 3.9(d). The average and median ACPs are close to the nominal value.

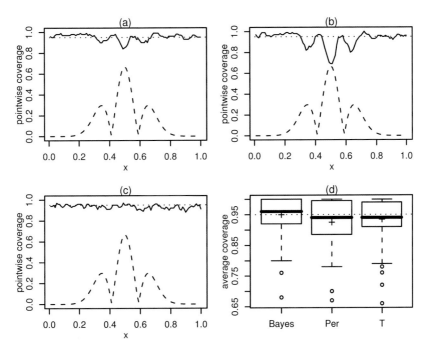

FIGURE 3.9 Plots of pointwise coverages (solid line) and nominal value (dotted line) for (a) the Bayesian confidence intervals, (b) the percentile bootstrap confidence intervals, and (c) the percentile-t bootstrap confidence intervals. Dashed lines in (a), (b), and (c) represent a scaled version of $|f''|$. Boxplots of ACPs are shown in (d) with mean coverages marked as pluses and nominal value plotted as a dotted line.

3.9   Hypothesis Tests

3.9.1   The Hypothesis

One of the most useful applications of the nonparametric regression models is to check or suggest a parametric model. When appropriate, parametric models, especially linear models, are preferred in practice because of their simplicity and interpretability. One important step in building a parametric model is to investigate potential departure from a specified model. Tests with specific alternatives are often performed in practice. This kind of tests would not perform well for different forms of departure from the parametric model, especially those orthogonal to the specific alternative. For example, to detect departure from a straight line model, one may consider a quadratic polynomial as the alternative. Then departure in the form of higher-order polynomials may be missed. It is desirable to have tools that can detect general departures from a specific parametric model.

One approach to check a parametric model is to construct confidence intervals for f or its smooth component in an SSR model using methods in Section 3.8. The parametric model may be deemed unsuitable if a larger portion of its estimate is outside the confidence intervals of f. When the null space $\mathcal{H}_0$ corresponds to the parametric model under consideration, one may check the magnitude of the estimate of the smooth component $P_1 f$ since it represents the remaining systematic variation not explained by the parametric model under consideration. If a larger portion of the confidence intervals for $P_1 f$ does not contain zero, the parametric model may be deemed unsuitable. This approach was illustrated using the geyser data in Section 3.8.1.

Often the confidence intervals are all one needs in practice. Nevertheless, sometimes it may be desirable to conduct a formal test on the departure from a parametric model. In this section we consider the following hypothesis

$$\mathrm{H}_0: \ f \in \mathcal{H}_0, \qquad \mathrm{H}_1: \ f \in \mathcal{H} \text{ and } f \notin \mathcal{H}_0, \qquad (3.57)$$

where $\mathcal{H}_0$ corresponds to the parametric model. Note that when the parametric model satisfies $Lf = 0$ with a differential operator L given in (2.53), the L-spline may be used to check or test the parametric model.

The alternative is equivalent to $||P_1 f|| > 0$. Note that $\lambda = \infty$ in (2.11), or equivalently, $\delta = 0$ in the corresponding Bayes model (3.25) leads to $f \in \mathcal{H}_0$. Thus the hypothesis (3.57) can be reexpressed as

$$\mathrm{H}_0: \ \lambda = \infty, \qquad \mathrm{H}_1: \ \lambda < \infty, \qquad (3.58)$$

or

$$H_0 : \delta = 0, \qquad H_1 : \delta > 0. \tag{3.59}$$

3.9.2 Locally Most Powerful Test

Consider the hypothesis (3.59). As in Section 3.8, we use the marginal log-likelihood of $\boldsymbol{w}_1$, where $\boldsymbol{w}_1 = Q_2^T \boldsymbol{y} \sim N(0, \delta Q_2^T M Q_2)$. Since Q_2 is orthogonal to T, it is clear that the transformation $Q_2^T \boldsymbol{y}$ eliminates contribution from the model under the null hypothesis. Thus $\boldsymbol{w}_1$ reflects signals, if any, from $\mathcal{H}_1$. Note that $M = \Sigma + n\lambda I$ and $Q_2^T \Sigma Q_2 = U E U^T$, where notations Q_2, U, and E were defined in Section 3.2. Let $\boldsymbol{z} \triangleq U^T \boldsymbol{w}_1$ and denote $\boldsymbol{z} = (z_1, \dots, z_{n-p})^T$. Then $\boldsymbol{z} \sim N(0, \delta E + \sigma^2 I)$. First assume that σ^2 is known. Note that $\lambda = \sigma^2 / n\delta$. The negative log-likelihood of $\boldsymbol{z}$

$$l(\delta | \boldsymbol{z}) = \frac{1}{2} \sum_{\nu=1}^{n-p} \log(\delta e_\nu + \sigma^2) + \frac{1}{2} \sum_{\nu=1}^{n-p} \frac{z_\nu^2}{\delta e_\nu + \sigma^2} + C_1, \tag{3.60}$$

where C_1 is a constant. Let $U_\delta(\delta)$ and $I_{\delta\delta}(\delta)$ be the score and Fisher information of δ. It is not difficult to check that the score test statistic (Cox and Hinkley 1974)

$$t_{\text{score}} \triangleq \frac{U_\delta(0)}{\sqrt{I_{\delta\delta}(0)}} = C_2 \left(\sum_{\nu=1}^{n-p} e_\nu z_\nu^2 + C_3 \right),$$

where C_2 and C_3 are constants. Therefore, the score test is equivalent to the following test statistic

$$t_{\text{LMP}} = \sum_{\nu=1}^{n-p} e_\nu z_\nu^2. \tag{3.61}$$

For polynomial splines, Cox, Koh, Wahba and Yandell (1988) showed that there is no uniformly most powerful test and that t_{LMP} is the *locally most powerful test*.

The variance is usually unknown in practice. Replacing σ^2 by its MLE under the null hypothesis (3.59), $\hat{\sigma}_0^2 = \sum_{\nu=1}^{n-p} z_\nu^2 / (n-p)$, leads to the approximate LMP test statistic

$$t_{\text{appLMP}} = \frac{\sum_{\nu=1}^{n-p} e_\nu z_\nu^2}{\sum_{\nu=1}^{n-p} z_\nu^2}. \tag{3.62}$$

The null hypothesis is rejected for large values of t_{appLMP}. The test statistic t_{appLMP} does not follow a simple distribution under H_0. Nevertheless, it is straightforward to simulate the null distribution. Under

H_0, $z_\nu \overset{iid}{\sim} N(0, \sigma^2)$. Without loss of generality, σ^2 can be set to one in the simulation for the null distribution since both the numerator and the denominator depend on z_ν^2. Specifically, samples $z_{\nu,j} \overset{iid}{\sim} N(0, 1)$ for $\nu = 1, \ldots, n - p$ and $j = 1, \ldots, N$ are generated, and the statistics $t_{\text{appLMP},j} = \sum_{\nu=1}^{n-p} e_\nu z_{\nu,j}^2 / \sum_{\nu=1}^{n-p} z_{\nu,j}^2$ are computed. Note that $t_{\text{appLMP},j}$ are N realizations of the statistic under H_0. Then the proportion that $t_{\text{appLMP},j}$ is greater than t_{appLMP} provides an estimate of the p-value. This approach usually requires a very large N. The p-value can also be calculated numerically using the algorithm in Davies (1980). The approximation method is very fast and agrees with the results from the Monte Carlo method (Liu and Wang 2004).

3.9.3　Generalized Maximum Likelihood Test

Consider the hypothesis (3.58). Since $z \sim N(0, \delta(E + n\lambda I))$, the MLE of δ is

$$\hat{\delta} = \frac{1}{n - p} \sum_{\nu=1}^{n-p} \frac{z_\nu^2}{e_\nu + n\lambda}.$$

From (3.39), the profile likelihood of λ is

$$L_p(\lambda, \hat{\delta}|z) = C \left\{ \frac{\sum_{\nu=1}^{n-p} z_\nu^2/(e_\nu + n\lambda)}{\prod_{\nu=1}^{n-p}(e_\nu + n\lambda)^{-\frac{1}{n-p}}} \right\}^{-\frac{n-p}{2}}, \tag{3.63}$$

where C is a constant. The GML estimate of λ, $\hat{\lambda}_{\text{GML}}$, is the maximizer of (3.63). The GML test statistic for the hypothesis (3.58) is

$$
\begin{aligned}
t_{\text{GML}} &\triangleq \left\{ \frac{L_p(\hat{\lambda}_{\text{GML}}|z)}{L_p(\infty|z)} \right\}^{-\frac{2}{n-p}} \\
&= \frac{\sum_{\nu=1}^{n-p} z_\nu^2/(e_\nu + n\hat{\lambda}_{\text{GML}})}{\prod_{\nu=1}^{n-p}(e_\nu + n\hat{\lambda}_{\text{GML}})^{-\frac{1}{n-p}}} \frac{1}{\sum_{\nu=1}^{n-p} z_\nu^2}.
\end{aligned}
\tag{3.64}
$$

It is clear that the GML test is equivalent to the ratio of restricted likelihoods. The null hypothesis is rejected when t_{GML} is too small. The standard theory for likelihood ratio tests does not apply because the parameter λ locates on the boundary of the parameter space under the null hypothesis. Thus, it is difficult to derive the null distribution for t_{GML}. The Monte Carlo method described for the LMP test can be adapted to compute an estimate of the p-value. Note that t_{GML} involves the GML estimate of the smoothing parameter. Therefore, $\hat{\lambda}_{\text{GML}}$ needs to be estimated for each simulation sample, which makes this approach computationally intensive.

The null distribution of $-(n-p)\log t_{\text{GML}}$ can be well approximated by a mixture of χ_1^2 and χ_0^2, denoted by $r\chi_1^2 + (1-r)\chi_0^2$. However, the ratio r is not fixed. It is difficult to derive a formula for r since it depends on many factors. One can approximate the ratio r first and then calculate the p-value based on the mixture of χ_1^2 and χ_0^2 with the approximated r. A relatively small sample size is required to approximate r. See Liu and Wang (2004) for details.

3.9.4 Generalized Cross-Validation Test

Consider the hypothesis (3.58). Let $\hat{\lambda}_{\text{GCV}}$ be the GCV estimate of λ. Similar to the GML test statistic, the GCV test statistic is defined as the ratio between GCV scores

$$
\begin{aligned}
t_{\text{GCV}} &\triangleq \frac{\text{GCV}(\hat{\lambda}_{\text{GCV}})}{\text{GCV}(\infty)} \\
&= (n-p)^2 \frac{\sum_{\nu=1}^{n-p} z_\nu^2/(1+e_\nu/n\hat{\lambda}_{\text{GCV}})^2}{\{\sum_{\nu=1}^{n-p} 1/(1+e_\nu/n\hat{\lambda}_{\text{GCV}})\}^2} \frac{1}{\sum_{\nu=1}^{n-p} z_\nu^2}.
\end{aligned}
\tag{3.65}
$$

H_0 is rejected when t_{GCV} is too small. Again, similar to the GML test, the Monte Carlo method can be used to compute an estimate of the p-value.

3.9.5 Comparison and Implementation

The LMP and GML tests derived based on Bayesian arguments perform well under deterministic models. In terms of eigenvectors of $Q_2^T \Sigma Q_2$, the LMP test is more powerful in detecting departure in the direction of the first eigenvector, the GML test is more powerful in detecting departure in low-frequencies, and the GCV test is more powerful in detecting departure in high frequencies. See Liu and Wang (2004) for more details.

These tests can be carried out using the `anova` function in the `assist` package. We now use the geyser and Arosa data to illustrate how to use this function.

For the geyser data, first consider the following hypotheses

$$
H_0 : f \in \text{span}\{1, x\}, \qquad H_1 : f \in W_2^2[0,1] \text{ and } f \notin \text{span}\{1, x\}.
$$

We have fitted cubic spline model where the null space corresponds to the model under H_0. Therefore, we can test the hypothesis as follows:

```
> anova(geyser.cub.fit, simu.size=500)
```

```
Testing H_0: f in the NULL space

      test.value simu.size simu.p-value
LMP   0.02288057         500              0
GCV   0.00335470         500              0
```

where the option `simu.size` specifies the Monte Carlo sample size N. The simple linear model is rejected. Next we consider the hypothesis

$$H_0 : f \in \text{span}\{1, x, s\}, \qquad H_1 : f \in W_2^2[0, 1] \text{ and } f \notin \text{span}\{1, x, s\},$$

where $s = (x - 0.397)_+^0$. We have fitted a partial spline model where the null space corresponds to the model under H_0. Therefore, we can test the hypothesis as follows:

```
> anova(geyser.ps.fit, simu.size=500)

Testing H_0: f in the NULL space

      test.value simu.size simu.p-value
LMP   0.000351964        500         0.602
GCV   0.003717477        500         0.602
```

The null hypothesis is not rejected. The conclusions are the same as those based on Bayesian confidence intervals. To apply the GML test, we need to first fit using the GML method to select the smoothing parameter:

```
> geyser.ps.fit.m <- ssr(waiting~x+s, rk=cubic(x),
    spar=''m'')
> anova(geyser.ps.fit.m, simu.size=500)

Testing H_0: f in the NULL space

      test.value simu.size simu.p-value approximate.p-value
LMP   0.000352          500         0.634
GML   1.000001          500         0.634                  0.5
```

where the `approximate.`p`-value` was computed using the mixture of two Chi-square distributions.

For the Arosa data, consider the hypothesis

$$H_0 : f \in \mathcal{P}, \qquad H_1 : f \in W_2^2(per) \text{ and } f \notin \mathcal{P},$$

where $\mathcal{P} = \text{span}\{1, \sin 2\pi x, \cos 2\pi x\}$ is the model space for the sinusoidal model. Two approaches can be used to test the above hypothesis: fit a partial spline and fit an L-spline:

```
> arosa.ps.fit <- ssr(thick~sin(2*pi*x)+cos(2*pi*x),
                       rk=periodic(x), data=Arosa)
> anova(arosa.ps.fit,simu.size=500)

Testing H_0: f in the NULL space

     test.value simu.size simu.p-value
LMP 0.001262064       500            0
GCV 0.001832394       500            0

> arosa.ls.fit <- ssr(thick~sin(2*pi*x)+cos(2*pi*x),
                       rk=lspline(x,type=''sine1''))
> anova(arosa.ls.fit,simu.size=500)

Testing H_0: f in the NULL space

     test.value simu.size simu.p-value
LMP 2.539163e-06       500            0
GCV  0.001828071       500            0
```

The test based on the *L*-spline is usually more powerful since the parametric and the smooth components are orthogonal.

Chapter 4

Smoothing Spline ANOVA

4.1   Multiple Regression

Consider the problem of building regression models that examine the relationship between a dependent variable y and *multiple* independent variables $x_1, \ldots, x_d$. For generality, let the domain of each x_k be an arbitrary set $\mathcal{X}_k$. Denote $\boldsymbol{x} = (x_1, \ldots, x_d)$. Given observations $(\boldsymbol{x}_i, y_i)$ for $i = 1, \ldots, n$, where $\boldsymbol{x}_i = (x_{i1}, \ldots, x_{id})$, a *multiple regression model* relates the dependent variable and independent variables as follows:

$$y_i = f(\boldsymbol{x}_i) + \epsilon_i, \quad i = 1, \ldots, n, \tag{4.1}$$

where f is a *multivariate regression function*, and ϵ_i are zero-mean independent random errors with a common variance σ^2. The goal is to construct a model for f and estimate it based on noisy data.

There exist many different methods to construct a model space for f, parametrically, semiparametrically, or nonparametrically. For example, a thin-plate spline model may be used when all x_k are univariate continuous variables, and partial spline models may be used if a linear parametric model can be assumed for all but one variable. This chapter introduces a nonparametric approach called *smoothing spline analysis of variance (smoothing spline ANOVA or SS ANOVA) decomposition* for constructing model spaces for the multivariate function f.

The multivariate function f is defined on the product domain $\mathcal{X} = \mathcal{X}_1 \times \mathcal{X}_2 \times \cdots \times \mathcal{X}_d$. Note that each $\mathcal{X}_k$ is arbitrary: it may be a continuous interval, a discrete set, a unit circle, a unit sphere, or $\mathbb{R}^d$. Construction of model spaces for a single variable was introduced in Chapter 2. Let $\mathcal{H}^{(k)}$ be an RKHS on $\mathcal{X}_k$. The choice of the marginal space $\mathcal{H}^{(k)}$ depends on the domain $\mathcal{X}_k$ and prior knowledge about f as a function of x_k. To model the joint function f, we start with the tensor product of these marginal spaces defined in the following section.

4.2   Tensor Product Reproducing Kernel Hilbert Spaces

First consider the simple case when $d = 2$. Denote RKs for $\mathcal{H}^{(1)}$ and $\mathcal{H}^{(2)}$ as $R^{(1)}$ and $R^{(2)}$, respectively. It is known that the product of non-negative definite functions is nonnegative definite (Gu 2002). As RKs, both $R^{(1)}$ and $R^{(2)}$ are nonnegative definite. Therefore, the bivariate function on $\mathcal{X} = \mathcal{X}_1 \times \mathcal{X}_2$

$$R((x_1, x_2), (z_1, z_2)) \triangleq R^{(1)}(x_1, z_1) R^{(2)}(x_2, z_2)$$

is nonnegative definite. By Moore–Aronszajn theorem, there exists a unique RKHS $\mathcal{H}$ on $\mathcal{X} = \mathcal{X}_1 \times \mathcal{X}_2$ such that R is its RK. The resulting RKHS $\mathcal{H}$ is called the *tensor product RKHS* and is denoted as $\mathcal{H}^{(1)} \otimes \mathcal{H}^{(2)}$. For $d > 2$, the tensor product RKHS of $\mathcal{H}^{(1)}, \ldots, \mathcal{H}^{(d)}$ on the product domain $\mathcal{X} = \mathcal{X}_1 \times \mathcal{X}_2 \times \cdots \times \mathcal{X}_d$, $\mathcal{H}^{(1)} \otimes \mathcal{H}^{(2)} \otimes \cdots \otimes \mathcal{H}^{(d)}$, is defined recursively. Note that the RK for a tensor product space equals the product of RKs of marginal spaces. That is, the RK of $\mathcal{H}^{(1)} \otimes \mathcal{H}^{(2)} \otimes \cdots \otimes \mathcal{H}^{(d)}$ equals

$$R(\boldsymbol{x}, \boldsymbol{z}) = R^{(1)}(x_1, z_1) R^{(2)}(x_2, z_2) \cdots R^{(d)}(x_d, z_d),$$

where $\boldsymbol{x} \in \mathcal{X}$, $\boldsymbol{z} = (z_1, \ldots, z_d) \in \mathcal{X}$, $\mathcal{X} = \mathcal{X}_1 \times \mathcal{X}_2 \times \cdots \times \mathcal{X}_d$, and $R^{(k)}$ is the RK of $\mathcal{H}^{(k)}$ for $k = 1, \ldots, d$.

For illustration, consider the ultrasound data consisting of tongue shape measurements over time from ultrasound imaging. The data set contains observations on the response variable `height` (y) and three independent variables: `environment` (x_1), `length` (x_2), and `time` (x_3). The variable x_1 is a factor with three levels: $x_1 = 1, 2, 3$ corresponding to *2words*, *cluster*, and *schwa*, respectively. Both continuous variables x_2 and x_3 are scaled into $[0, 1]$. Interpolations of the raw data are shown in Figure 4.1.

In linguistic studies, researchers want to determine (1) how tongue shapes for an articulation differ under different environments, (2) how the tongue shape changes as a function of time, and (3) how changes over time differ under different environments. To address the first question at a fixed time point, we need to model a bivariate regression function $f(x_1, x_2)$. Assume marginal spaces $\mathbb{R}^3$ and $W_2^m[0, 1]$ for variables x_1 and x_2, respectively. Then we may consider the tensor product space $\mathbb{R}^3 \otimes W_2^m[0, 1]$ for the bivariate function f. To address the second question, for a fixed environment, we need to model a bivariate regression function $f(x_2, x_3)$. Assume marginal spaces $W_2^{m_1}[0, 1]$ and $W_2^{m_2}[0, 1]$ for variables

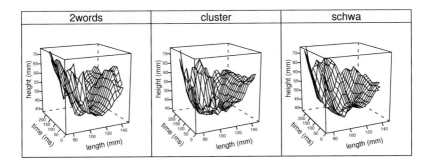

FIGURE 4.1 Ultrasound data, 3-d plots of observations.

x_2 and x_3, respectively. Then we may consider the tensor product space $W_2^{m_1}[0,1] \otimes W_2^{m_2}[0,1]$ for the bivariate function. To address the third question, we need to model a trivariate regression function $f(x_1, x_2, x_3)$. We may consider the tensor product space $\mathbb{R}^3 \otimes W_2^{m_1}[0,1] \otimes W_2^{m_2}[0,1]$ for the trivariate function. Analysis of the ultrasound data are given in Section 4.9.1.

The SS ANOVA decomposition decomposes a tensor product space into subspaces with a hierarchical structure similar to the main effects and interactions in the classical ANOVA. The resulting hierarchical structure facilitates model selection and interpretation. Sections 4.3, 4.4, and 4.5 present SS ANOVA decompositions for a single space, tensor product of two spaces, and tensor product of d spaces. More SS ANOVA decompositions can be found in Sections 4.9, 5.4.4, 6.3, and 9.2.4.

4.3 One-Way SS ANOVA Decomposition

SS ANOVA decompositions of tensor product RKHS's are based on decompositions of marginal spaces for each independent variable. Therefore, decompositions for a single space are introduced first in this section. Spline models for a single independent variable have been introduced in Chapter 2. Denote the independent variable as x and the regression function as f. It was assumed that f belongs to an RKHS $\mathcal{H}$. The function f was decomposed into a parametric and a smooth component, $f = f_0 + f_1$, or in terms of model space, $\mathcal{H} = \mathcal{H}_0 \oplus \mathcal{H}_1$. This can be regarded as one form of the SS ANOVA decomposition. We now introduce general decompositions based on averaging operators. Consider a function space $\mathcal{H}$ on the domain $\mathcal{X}$. An operator $\mathcal{A}$ is called an *averaging operator* if $\mathcal{A} = \mathcal{A}^2$. Instead of the more appropriate term as an *idempotent operator*, the term averaging operator is used since it is motivated by averaging in the classical ANOVA decomposition. Note that an averaging operator does not necessarily involve averaging. As we will see in the following subsections, the commonly used averaging operators are *projection operators*. Thus they are idempotent.

Suppose model space $\mathcal{H} = \mathcal{H}_0 \oplus \mathcal{H}_1$, where $\mathcal{H}_0$ is a finite dimensional space with orthogonal basis $\phi_1(x), \ldots, \phi_p(x)$. Let $\mathcal{A}_\nu$ be the projection operator onto the subspace $\{\phi_\nu(x)\}$ for $\nu = 1, \ldots, p$, and $\mathcal{A}_{p+1}$ be the projection operator onto $\mathcal{H}_1$. Then the function can be decomposed as

$$f = (\mathcal{A}_1 + \cdots + \mathcal{A}_p + \mathcal{A}_{p+1})f \triangleq f_{01} + \cdots + f_{0p} + f_1. \qquad (4.2)$$

Correspondingly, the model space is decomposed into

$$\mathcal{H} = \{\phi_1(x)\} \oplus \cdots \oplus \{\phi_p(x)\} \oplus \mathcal{H}_1.$$

For simplicity, $\{\cdot\}$ represents the space spanned by the basis functions inside the bracket. Some averaging operators $\mathcal{A}_\nu$ (subspaces) can be combined. For example, combining $\mathcal{A}_1, \ldots, \mathcal{A}_p$ leads to the decomposition $f = f_0 + f_1$, a parametric component plus a smooth component. When $\phi_1(x) = 1$, combining $\mathcal{A}_2, \ldots, \mathcal{A}_{p+1}$ leads to the decomposition $f = f_{01} + \tilde{f}_1$, where f_{01} is a constant independent of x, and $\tilde{f}_1 = f_{02} + \cdots + f_{0p} + f_1$ collects all components that depend on x. Therefore, $f = f_{01} + \tilde{f}_1$ decomposes the function into a constant plus a nonconstant function.

For the same model space, different SS ANOVA decompositions may be constructed for different purposes. In general, we denote the one-way SS ANOVA decomposition as

$$f = \mathcal{A}_1 f + \cdots + \mathcal{A}_r f, \qquad (4.3)$$

where $\mathcal{A}_1 + \cdots + \mathcal{A}_r = I$, and I is the identity operator. The above equality always holds since $f = If$. Equivalently, in terms of the model space, the one-way SS ANOVA decomposition is denoted as

$$\mathcal{H} = \mathcal{H}_{(1)} \oplus \cdots \oplus \mathcal{H}_{(r)}.$$

The following subsections provide one-way SS ANOVA decompositions for some special model spaces.

4.3.1 Decomposition of $\mathbb{R}^a$: One-Way ANOVA

Suppose x is a discrete variable with a levels. The classical one-way *mean model* assumes that

$$y_{ik} = \mu_i + \epsilon_{ik}, \quad i = 1, \dots, a; \ k = 1, \dots, n_i, \tag{4.4}$$

where y_{ik} represents the observation of the kth replication at level i of x, μ_i represents the mean at level i, and ϵ_{ik} represent random errors.

Regarding μ_i as a function of i and expressing explicitly as $f(i) \triangleq \mu_i$, then f is a function defined on the discrete domain $\mathcal{X} = \{1, \dots, a\}$. It is easy to see that the model space for f is the Euclidean a-space $\mathbb{R}^a$.

Model space construction and decomposition of $\mathbb{R}^a$
The space $\mathbb{R}^a$ is an RKHS with the inner product $(f, g) = f^T g$. Furthermore, $\mathbb{R}^a = \mathcal{H}_0 \oplus \mathcal{H}_1$, where

$$\mathcal{H}_0 = \{f : \ f(1) = \cdots = f(a)\},$$
$$\mathcal{H}_1 = \{f : \ \sum_{i=1}^{a} f(i) = 0\}, \tag{4.5}$$

are RKHS's with corresponding RKs

$$R_0(i, j) = \frac{1}{a},$$
$$R_1(i, j) = \delta_{i,j} - \frac{1}{a}, \tag{4.6}$$

and $\delta_{i,j}$ is the Kronecker delta.

Details about the above construction can be found in Gu (2002). Define an averaging operator $\mathcal{A}_1 : \ \mathbb{R}^a \to \mathbb{R}^a$ such that

$$\mathcal{A}_1 f = \frac{1}{a} \sum_{i=1}^{a} f(i).$$

The operator $\mathcal{A}_1$ maps f to the constant function that equals the average over all indices. Let $\mathcal{A}_2 = I - \mathcal{A}_1$. The one-way ANOVA *effect model* is based on the following decomposition of the function f:

$$f = \mathcal{A}_1 f + \mathcal{A}_2 f \triangleq \mu + \alpha_i, \tag{4.7}$$

where μ is the overall mean, and α_i is the effect at level i. From the definition of the averaging operator, α_i satisfy the sum-to-zero side condition $\sum_{i=1}^{a} \alpha_i = 0$. It is clear that $\mathcal{A}_1$ and $\mathcal{A}_2$ are the projection operators from $\mathbb{R}^a$ onto $\mathcal{H}_0$ and $\mathcal{H}_1$ defined in (4.5). Thus, they divide $\mathbb{R}^a$ into $\mathcal{H}_0$ and $\mathcal{H}_1$.

Under the foregoing construction, $||P_1 f||^2 = \sum_{i=1}^{a}\{f(i) - \bar{f}\}^2$, where $\bar{f} = \sum_{i=1}^{a} f(i)/a$. For balanced designs with $n_i = n$, the solution to the PLS (2.11) is

$$\hat{f}(i) = \bar{y}_{i\cdot} - \frac{a\lambda}{1 + a\lambda}(\bar{y}_{i\cdot} - \bar{y}_{\cdot\cdot}), \qquad (4.8)$$

where $\bar{y}_{i\cdot} = \sum_{k=1}^{n} y_{ik}/n$, and $\bar{y}_{\cdot\cdot} = \sum_{i=1}^{a} \bar{y}_{i\cdot}/a$. It is easy to check that when $n = 1$, $\sigma^2 = 1$ and $\lambda = (a - 3)/a(\sum_{i=1}^{a} \bar{y}_{i\cdot}^2 - a + 3)$, the spline estimate $\hat{f}$ is the *James–Stein estimator* (shrink toward mean). Therefore, in a sense, the Stein's shrinkage estimator can be regarded as spline estimate on a discrete domain.

There exist other ways to decompose $\mathbb{R}^a$. For example, the averaging operator $\mathcal{A}_1 f = f(1)$ leads to the same decomposition as (4.7) with the set-to-zero side condition: $\alpha_1 = 0$ (Gu 2002).

4.3.2   Decomposition of $W_2^m[a, b]$

Under the construction in Section 2.2, let

$$\mathcal{A}_\nu f(x) = f^{(\nu-1)}(a)\frac{(x - a)^{\nu-1}}{(\nu - 1)!}, \qquad \nu = 1, \ldots, m. \qquad (4.9)$$

It is easy to see that $\mathcal{A}_v^2 = \mathcal{A}_v$. Thus, they are averaging operators. In fact, $\mathcal{A}_\nu$ is the projection operator onto $\{(x - a)^{\nu-1}/(\nu - 1)!\}$. Let $\mathcal{A}_{m+1} = I - \mathcal{A}_1 - \cdots - \mathcal{A}_m$. The decomposition

$$f = \mathcal{A}_1 f + \cdots + \mathcal{A}_m f + \mathcal{A}_{m+1} f$$

corresponds to the Taylor expansion (1.5). It decomposes the model space $W_2^m[a, b]$ into

$$W_2^m[a, b] = \{1\} \oplus \{x - a\} \oplus \cdots \oplus \{(x - a)^{m-1}/(m - 1)!\} \oplus \mathcal{H}_1,$$

where $\mathcal{H}_1$ is given in (2.3). For $f \in \mathcal{H}_1$, conditions $f^{(\nu)}(a) = 0$ for $\nu = 0, \ldots, m-1$ are analogous to the set-to-zero condition in the classical one-way ANOVA model.

Under the construction in Section 2.6 for $W_2^m[0, 1]$, let

$$\mathcal{A}_\nu f(x) = \left\{\int_0^1 f^{(\nu-1)}(u)du\right\} k_\nu(x), \qquad \nu = 1, \ldots, m.$$

Again, $\mathcal{A}_\nu$ is an averaging (projection) operator extracting the polynomial of order ν. In particular, $\mathcal{A}_1 f = \int_0^1 f(u)du$ is a natural extension of the averaging in the discrete domain. Let $\mathcal{A}_{m+1} = I - \mathcal{A}_1 - \cdots - \mathcal{A}_m$. The decomposition

$$f = \mathcal{A}_1 f + \cdots + \mathcal{A}_m f + \mathcal{A}_{m+1} f$$

decomposes the model space $W_2^m[0,1]$ into

$$W_2^m[0,1] = \{1\} \oplus \{k_1(x)\} \oplus \cdots \oplus \{k_{m-1}(x)\} \oplus \mathcal{H}_1,$$

where $\mathcal{H}_1$ is given in (2.29). For $f \in \mathcal{H}_1$, conditions $\int_0^1 f^{(\nu)} dx = 0$ for $\nu = 0, \ldots, m-1$ are analogous to the sum-to-zero condition in the classical one-way ANOVA model.

4.3.3 Decomposition of $W_2^m(per)$

Under the construction in Section 2.7, let

$$\mathcal{A}_1 f = \int_0^1 f \, du$$

be an averaging (projection) operator. Let $\mathcal{A}_2 = I - \mathcal{A}_1$. The decomposition

$$f = \mathcal{A}_1 f + \mathcal{A}_2 f$$

decomposes the model space

$$W_2^m(per) = \{1\} \oplus \{f \in W_2^m(per) : \int_0^1 f \, du = 0\}.$$

4.3.4 Decomposition of $W_2^m(\mathbb{R}^d)$

For simplicity, consider the special case with $d = 2$ and $m = 2$. Decompositions for general d and m can be derived similarly. Let $\phi_1(\boldsymbol{x}) = 1$, $\phi_2(\boldsymbol{x}) = x_1$, and $\phi_3(\boldsymbol{x}) = x_2$ be polynomials of total degree less than $m = 2$. Let $\tilde{\phi}_1 = 1$, and $\tilde{\phi}_2$ and $\tilde{\phi}_3$ be an orthonormal basis such that $(\tilde{\phi}_\nu, \tilde{\phi}_\mu)_0 = \delta_{\nu,\mu}$ based on the norm (2.41). Define two averaging operators

$$\mathcal{A}_1 f(\boldsymbol{x}) = \sum_{j=1}^J w_j f(\boldsymbol{u}_j),$$

$$\mathcal{A}_2 f(\boldsymbol{x}) = \sum_{j=1}^J w_j f(\boldsymbol{u}_j) \{ \tilde{\phi}_2(\boldsymbol{u}_j) \tilde{\phi}_2(\boldsymbol{x}) + \tilde{\phi}_3(\boldsymbol{u}_j) \tilde{\phi}_3(\boldsymbol{x}) \},$$

(4.10)

where $\boldsymbol{u}_j$ are fixed points in $\mathbb{R}^2$, and w_j are fixed positive weights such that $\sum_{j=1}^{J} w_j = 1$. It is clear that $\mathcal{A}_1$ and $\mathcal{A}_2$ are projection operators onto spaces $\{\tilde{\phi}_1\}$ and $\{\tilde{\phi}_2, \tilde{\phi}_3\}$, respectively. To see how they generalize averaging operators, define a probability measure μ on $\mathcal{X} = \mathbb{R}^2$ by assigning probability w_j to the point $\boldsymbol{u}_j$, $j = 1, \ldots, J$. Then $\mathcal{A}_1 f(\boldsymbol{x}) = \int_{\mathbb{R}^2} f \tilde{\phi}_1 d\mu$ and $\mathcal{A}_2 f(\boldsymbol{x}) = (\int_{\mathbb{R}^2} f \tilde{\phi}_2 d\mu) \tilde{\phi}_2(\boldsymbol{x}) + (\int_{\mathbb{R}^2} f \tilde{\phi}_3 d\mu) \tilde{\phi}_3(\boldsymbol{x})$. Therefore, $\mathcal{A}_1$ and $\mathcal{A}_2$ take averages with respect to the discrete probability measure μ. In particular, μ puts mass $1/n$ on design points $\boldsymbol{x}_j$ when $J = n$, $w_j = 1/n$ and $\boldsymbol{u}_j = \boldsymbol{x}_j$. A continuous density on $\mathbb{R}^2$ may be used. However, the resulting integrals usually do not have closed forms, and approximations such as a quadrature formula would have to be used. It is then essentially equivalent to using an approximate discrete probability measure.

Let $\mathcal{A}_3 = I - \mathcal{A}_1 - \mathcal{A}_2$. The decomposition

$$f = \mathcal{A}_1 f + \mathcal{A}_2 f + \mathcal{A}_3 f$$

divides the model space

$$W_2^m(\mathbb{R}^2) = \{1\} \oplus \{\tilde{\phi}_2, \tilde{\phi}_3\} \oplus \{f \in W_2^m(\mathbb{R}^2) : \ J_2^2(f) = 0\},$$

where $J_m^d(f)$ is defined in (2.36).

4.4   Two-Way SS ANOVA Decomposition

Suppose there are two independent variables, $x_1 \in \mathcal{X}_1$ and $x_2 \in \mathcal{X}_2$. Consider the tensor product space $\mathcal{H}^{(1)} \otimes \mathcal{H}^{(2)}$ on the product domain $\mathcal{X}_1 \times \mathcal{X}_2$. For f as a marginal function of x_k, assume the following one-way decomposition based on Section 4.3,

$$f = \mathcal{A}_1^{(k)} f + \cdots + \mathcal{A}_{r_k}^{(k)} f, \quad k = 1, 2, \tag{4.11}$$

where $\mathcal{A}_j^{(k)}$ are averaging operators on $\mathcal{H}^{(k)}$ and $\sum_{j=1}^{r_k} \mathcal{A}_j^{(k)} = I$. Then for the joint function, we have

$$f = \{\mathcal{A}_1^{(1)} + \cdots + \mathcal{A}_{r_1}^{(1)}\}\{\mathcal{A}_1^{(2)} + \cdots + \mathcal{A}_{r_2}^{(2)}\} f = \sum_{j_1=1}^{r_1} \sum_{j_2=1}^{r_2} \mathcal{A}_{j_1}^{(1)} \mathcal{A}_{j_2}^{(2)} f. \tag{4.12}$$

The above decomposition of the bivariate function f is referred to as the *two-way SS ANOVA decomposition*.

Denote

$$\mathcal{H}^{(k)} = \mathcal{H}_{(1)}^{(k)} \oplus \cdots \oplus \mathcal{H}_{(r_k)}^{(k)}, \quad k = 1, 2,$$

as the one-way decomposition to $\mathcal{H}^{(k)}$ associated with (4.11). Then, (4.12) decomposes the tensor product space

$$\mathcal{H}^{(1)} \otimes \mathcal{H}^{(2)} = \left\{ \mathcal{H}_{(1)}^{(1)} \oplus \cdots \oplus \mathcal{H}_{(r_1)}^{(1)} \right\} \otimes \left\{ \mathcal{H}_{(1)}^{(2)} \oplus \cdots \oplus \mathcal{H}_{(r_2)}^{(2)} \right\}$$
$$= \sum_{j_1=1}^{r_1} \sum_{j_2=1}^{r_2} \mathcal{H}_{(j_1)}^{(1)} \otimes \mathcal{H}_{(j_2)}^{(2)}.$$

Consider the special case when $r_k = 2$ for $k = 1, 2$. Assume that $\mathcal{A}_1^{(k)} f$ is independent of x_k, or equivalently, $\mathcal{H}_0^{(k)} = \{1\}$. Then the decomposition (4.12) can be written as

$$f = \mathcal{A}_1^{(1)} \mathcal{A}_1^{(2)} f + \mathcal{A}_2^{(1)} \mathcal{A}_1^{(2)} f + \mathcal{A}_1^{(1)} \mathcal{A}_2^{(2)} f + \mathcal{A}_2^{(1)} \mathcal{A}_2^{(2)} f$$
$$\triangleq \mu + f_1(x_1) + f_2(x_2) + f_{12}(x_1, x_2), \tag{4.13}$$

where μ represents the grand mean, $f_1(x_1)$ and $f_2(x_2)$ represent the main effects of x_1 and x_2, respectively, and $f_{12}(x_1, x_2)$ represents the interaction between x_1 and x_2.

For general r_k, assuming $\mathcal{A}_1^{(k)} f$ is independent of x_k, the decomposition (4.13) can be derived by combining operators $\mathcal{A}_2^{(k)}, \ldots, \mathcal{A}_{r_k}^{(k)}$ into one averaging operator $\tilde{\mathcal{A}}_2^{(k)} = \mathcal{A}_2^{(k)} + \cdots + \mathcal{A}_{r_k}^{(k)}$. Therefore, decomposition (4.13) combines components in (4.12) and reorganizes them into the overall main effects and interactions.

The following subsections provide two-way SS ANOVA decompositions for combinations of some special model spaces.

4.4.1 Decomposition of $\mathbb{R}^a \otimes \mathbb{R}^b$: Two-Way ANOVA

Suppose both x_1 and x_2 are discrete variables with a and b levels, respectively. The classical two-way *mean model* assumes that

$$y_{ijk} = \mu_{ij} + \epsilon_{ijk}, \quad i = 1, \ldots, a; \ j = 1, \ldots, b; \ k = 1, \ldots, n_{ij},$$

where y_{ijk} represents the observation of the kth replication at level i of x_1, and level j of x_2, μ_{ij} represents the mean at level i of x_1 and level j of x_2, and ϵ_{ijk} represent random errors.

Regarding μ_{ij} as a bivariate function of (i, j) and letting $f(i, j) \triangleq \mu_{ij}$, then f is a bivariate function defined on the product domain $\mathcal{X} = \{1, \ldots, a\} \times \{1, \ldots, b\}$. The model space for f is the tensor product space

$\mathbb{R}^a \otimes \mathbb{R}^b$. Define two averaging operators $\mathcal{A}_1^{(1)}$ and $\mathcal{A}_2^{(2)}$ such that

$$\mathcal{A}_1^{(1)} f = \frac{1}{a} \sum_{i=1}^{a} f(i, j),$$

$$\mathcal{A}_1^{(2)} f = \frac{1}{b} \sum_{j=1}^{b} f(i, j).$$

$\mathcal{A}_1^{(1)}$ and $\mathcal{A}_1^{(2)}$ map f to univariate functions by averaging over all levels of x_1 and x_2, respectively. Let $\mathcal{A}_2^{(1)} = I - \mathcal{A}_1^{(1)}$ and $\mathcal{A}_2^{(2)} = I - \mathcal{A}_1^{(2)}$. Then the classical two-way ANOVA *effect model* is based on the following SS ANOVA decomposition of the function f:

$$
\begin{aligned}
f &= (\mathcal{A}_1^{(1)} + \mathcal{A}_2^{(1)})(\mathcal{A}_1^{(2)} + \mathcal{A}_2^{(2)}) f \\
&= \mathcal{A}_1^{(1)} \mathcal{A}_1^{(2)} f + \mathcal{A}_2^{(1)} \mathcal{A}_1^{(2)} f + \mathcal{A}_1^{(1)} \mathcal{A}_2^{(2)} f + \mathcal{A}_2^{(1)} \mathcal{A}_2^{(2)} f \\
&\triangleq \mu + \alpha_i + \beta_j + (\alpha\beta)_{ij},
\end{aligned}
\tag{4.14}
$$

where μ represents the overall mean, α_i represents the main effect of x_1, β_j represents the main effect of x_2, and $(\alpha\beta)_{ij}$ represents the interaction between x_1 and x_2. The sum-to-zero side conditions are satisfied from the definition of the averaging operators.

Based on one-way ANOVA decomposition in Section 4.3.1, we have $\mathbb{R}^a = \mathcal{H}_0^{(1)} \oplus \mathcal{H}_1^{(1)}$ and $\mathbb{R}^b = \mathcal{H}_0^{(2)} \oplus \mathcal{H}_1^{(2)}$, where $\mathcal{H}_0^{(1)}$ and $\mathcal{H}_0^{(2)}$ are subspaces containing constant functions, and $\mathcal{H}_1^{(1)}$ and $\mathcal{H}_1^{(2)}$ are orthogonal complements of $\mathcal{H}_0^{(1)}$ and $\mathcal{H}_0^{(2)}$, respectively. The classical two-way ANOVA model decomposes the tensor product space

$$
\begin{aligned}
\mathbb{R}^a \otimes \mathbb{R}^b &= \left\{ \mathcal{H}_0^{(1)} \oplus \mathcal{H}_1^{(1)} \right\} \otimes \left\{ \mathcal{H}_0^{(2)} \oplus \mathcal{H}_1^{(2)} \right\} \\
&= \left\{ \mathcal{H}_0^{(1)} \otimes \mathcal{H}_0^{(2)} \right\} \oplus \left\{ \mathcal{H}_1^{(1)} \otimes \mathcal{H}_0^{(2)} \right\} \\
&\quad \oplus \left\{ \mathcal{H}_0^{(1)} \otimes \mathcal{H}_1^{(2)} \right\} \oplus \left\{ \mathcal{H}_1^{(1)} \otimes \mathcal{H}_1^{(2)} \right\}.
\end{aligned}
\tag{4.15}
$$

The four subspaces in (4.15) contain components μ, α_i, β_j, and $(\alpha\beta)_{ij}$ in (4.14), respectively.

4.4.2 Decomposition of $\mathbb{R}^a \otimes W_2^m[0, 1]$

Suppose x_1 is a discrete variable with a levels, and x_2 is a continuous variable in $[0, 1]$. A natural model space for x_1 is $\mathbb{R}^a$ and a natural model space for x_2 is $W_2^m[0, 1]$. Therefore, we consider the tensor product space $\mathbb{R}^a \otimes W_2^m[0, 1]$ for the bivariate regression function $f(x_1, x_2)$. For simplicity, we derive SS ANOVA decompositions for $m = 1$ and $m = 2$ only. SS ANOVA decompositions for higher-order m can be derived similarly. In this and the remaining sections, the construction in Section 2.6 for the marginal space $W_2^m[0,1]$ will be used. Similar SS ANOVA decompositions can be derived under the construction in Section 2.2.

Consider the tensor product space $\mathbb{R}^a \otimes W_2^1[0,1]$ first. Define two averaging operators $\mathcal{A}_1^{(1)}$ and $\mathcal{A}_1^{(2)}$ as

$$\mathcal{A}_1^{(1)} f = \frac{1}{a} \sum_{x_1=1}^{a} f,$$

$$\mathcal{A}_1^{(2)} f = \int_0^1 f \, dx_2,$$

where $\mathcal{A}_1^{(1)}$ and $\mathcal{A}_1^{(2)}$ extract the constant term out of all possible functions for each variable. Let $\mathcal{A}_2^{(1)} = I - \mathcal{A}_1^{(1)}$ and $\mathcal{A}_2^{(2)} = I - \mathcal{A}_1^{(2)}$. Then

$$\begin{aligned}
f &= \{\mathcal{A}_1^{(1)} + \mathcal{A}_2^{(1)}\}\{\mathcal{A}_1^{(2)} + \mathcal{A}_2^{(2)}\} f \\
&= \mathcal{A}_1^{(1)} \mathcal{A}_1^{(2)} f + \mathcal{A}_2^{(1)} \mathcal{A}_1^{(2)} f + \mathcal{A}_1^{(1)} \mathcal{A}_2^{(2)} f + \mathcal{A}_2^{(1)} \mathcal{A}_2^{(2)} f \\
&\triangleq \mu + f_1(x_1) + f_2(x_2) + f_{12}(x_1, x_2).
\end{aligned} \qquad (4.16)$$

Obviously, (4.16) is a natural extension of the classical two-way ANOVA decomposition (4.14) from the product of two discrete domains to the product of one discrete and one continuous domain. As in the classical ANOVA model, components in (4.16) have nice interpretations: μ represents the overall mean, $f_1(x_1)$ represents the main effect of x_1, $f_2(x_2)$ represents the main effect of x_2, and $f_{12}(x_1, x_2)$ represents the interaction. Components also have nice interpretation collectively: $\mu + f_2(x_2)$ represents the mean curve among all levels of x_1, and $f_1(x_1) + f_{12}(x_1, x_2)$ represents the departure from the mean curve at level x_1. Write $\mathbb{R}^a = \mathcal{H}_0^{(1)} \oplus \mathcal{H}_1^{(1)}$ and $W_2^1[0,1] = \mathcal{H}_0^{(2)} \oplus \mathcal{H}_1^{(2)}$, where $\mathcal{H}_0^{(1)}$ and $\mathcal{H}_1^{(1)}$ are given in (4.5), $\mathcal{H}_0^{(2)} = \{1\}$ and $\mathcal{H}_1^{(2)} = \{f \in W_2^1[0,1] : \int_0^1 f \, du = 0\}$. Then, in terms of the model space, (4.16) decomposes

$$\begin{aligned}
&\mathbb{R}^a \otimes W_2^1[0,1] \\
&= \{\mathcal{H}_0^{(1)} \oplus \mathcal{H}_1^{(1)}\} \otimes \{\mathcal{H}_0^{(2)} \oplus \mathcal{H}_1^{(2)}\} \\
&= \{\mathcal{H}_0^{(1)} \otimes \mathcal{H}_0^{(2)}\} \oplus \{\mathcal{H}_1^{(1)} \otimes \mathcal{H}_0^{(2)}\} \oplus \{\mathcal{H}_0^{(1)} \otimes \mathcal{H}_1^{(2)}\} \oplus \{\mathcal{H}_1^{(1)} \otimes \mathcal{H}_1^{(2)}\} \\
&\triangleq \mathcal{H}^0 \oplus \mathcal{H}^1 \oplus \mathcal{H}^2 \oplus \mathcal{H}^3.
\end{aligned} \qquad (4.17)$$

To fit model (4.16), we need to find basis for $\mathcal{H}^0$ and RKs for $\mathcal{H}^1$, $\mathcal{H}^2$, and $\mathcal{H}^3$. It is clear that $\mathcal{H}^0$ contains all constant functions. Thus, $\mathcal{H}^0$

is an one-dimensional space with the basis $\phi(\boldsymbol{x}) = 1$. The RKs of $\mathcal{H}_0^{(1)}$ and $\mathcal{H}_1^{(1)}$ are given in (4.6), and the RKs of $\mathcal{H}_0^{(2)}$ and $\mathcal{H}_1^{(2)}$ are given in Table 2.2. The RKs of $\mathcal{H}^1$, $\mathcal{H}^2$, and $\mathcal{H}^3$ can be calculated using the fact that the RK of a tensor product space equals the product of RKs of the involved marginal spaces. For example, the RK of $\mathcal{H}^3 = \mathcal{H}_1^{(1)} \otimes \mathcal{H}_1^{(2)}$ equals $(\delta_{x_1,z_1} - 1/a)\{k_1(x_2)k_1(z_2) + k_2(|x_2 - z_2|)\}$.

Now suppose we want to model the effect of x_2 using the cubic spline space $W_2^2[0, 1]$. Consider the tensor product space $\mathbb{R}^a \otimes W_2^2[0, 1]$. Define three averaging operators $\mathcal{A}_1^{(1)}$, $\mathcal{A}_1^{(2)}$, and $\mathcal{A}_2^{(2)}$ as

$$\mathcal{A}_1^{(1)} f = \frac{1}{a} \sum_{x_1=1}^{a} f,$$

$$\mathcal{A}_1^{(2)} f = \int_0^1 f dx_2,$$

$$\mathcal{A}_2^{(2)} f = \left(\int_0^1 f' dx_2 \right) (x_2 - 0.5),$$

where $\mathcal{A}_1^{(1)}$ and $\mathcal{A}_1^{(2)}$ extract the constant function out of all possible functions for each variable, and $\mathcal{A}_2^{(2)}$ extracts the linear function for x_2. Let $\mathcal{A}_2^{(1)} = I - \mathcal{A}_1^{(1)}$ and $\mathcal{A}_3^{(2)} = I - \mathcal{A}_1^{(2)} - \mathcal{A}_2^{(2)}$. Then

$$
\begin{aligned}
f &= \{\mathcal{A}_1^{(1)} + \mathcal{A}_2^{(1)}\}\{\mathcal{A}_1^{(2)} + \mathcal{A}_2^{(2)} + \mathcal{A}_3^{(2)}\} f \\
&= \mathcal{A}_1^{(1)} \mathcal{A}_1^{(2)} f + \mathcal{A}_1^{(1)} \mathcal{A}_2^{(2)} f + \mathcal{A}_1^{(1)} \mathcal{A}_3^{(2)} f \\
&\quad + \mathcal{A}_2^{(1)} \mathcal{A}_1^{(2)} f + \mathcal{A}_2^{(1)} \mathcal{A}_2^{(2)} f + \mathcal{A}_2^{(1)} \mathcal{A}_3^{(2)} f \\
&\triangleq \mu + \beta \times (x_2 - 0.5) + f_2^s(x_2) \\
&\quad + f_1(x_1) + \gamma_{x_1} \times (x_2 - 0.5) + f_{12}^{ss}(x_1, x_2),
\end{aligned}
\tag{4.18}
$$

where μ represents the overall mean, $f_1(x_1)$ represents the main effect of x_1, $\beta \times (x_2 - 0.5)$ represents the linear main effect of x_2, $f_2^s(x_2)$ represents the smooth main effect of x_2, $\gamma_{x_1} \times (x_2 - 0.5)$ represents the smooth–linear interaction, and $f_{12}^{ss}(x_1, x_2)$ represents the smooth–smooth interaction. The overall main effect of x_2

$$f_2(x_2) = \beta \times (x_2 - 0.5) + f_2^s(x_2),$$

and the overall interaction between x_1 and x_2

$$f_{12}(x_1, x_2) = \gamma_{x_1} \times (x_2 - 0.5) + f_{12}^{ss}(x_1, x_2).$$

It is obvious that f_2 and f_{12} are the results of combining averaging operators $\mathcal{A}_2^{(2)}$ and $\mathcal{A}_2^{(3)}$. One may look at the components in the overall main effects and interactions to decide whether to include them in the model. The first three terms in (4.18) represent the mean curve among all levels of x_1, and the last three terms represent the departure from the mean curve. The simple ANCOVA (analysis of covariance) model with x_2 being modeled by a straight line is a special case of (4.18) with $f_2^s = f_{12}^{ss} = 0$. Thus, checking whether f_2^s and f_{12}^{ss} are negligible provides a diagnostic tool for the ANCOVA model. Write $\mathbb{R}^a = \mathcal{H}_0^{(1)} \oplus \mathcal{H}_1^{(1)}$ and $W_2^2[0,1] = \mathcal{H}_0^{(2)} \oplus \mathcal{H}_1^{(2)} \oplus \mathcal{H}_2^{(2)}$, where $\mathcal{H}_0^{(1)}$ and $\mathcal{H}_1^{(1)}$ are given in (4.5), $\mathcal{H}_0^{(2)} = \{1\}$, $\mathcal{H}_1^{(2)} = \{x_2 - 0.5\}$, and $\mathcal{H}_2^{(2)} = \{f \in W_2^2[0,1], \int_0^1 f du = \int_0^1 f' du = 0\}$. Then, in terms of the model space, (4.18) decomposes

$$\mathbb{R}^a \otimes W_2^2[0,1]$$
$$= \left\{\mathcal{H}_0^{(1)} \oplus \mathcal{H}_1^{(1)}\right\} \otimes \left\{\mathcal{H}_0^{(2)} \oplus \mathcal{H}_1^{(2)} \oplus \mathcal{H}_2^{(2)}\right\}$$
$$= \left\{\mathcal{H}_0^{(1)} \otimes \mathcal{H}_0^{(2)}\right\} \oplus \left\{\mathcal{H}_0^{(1)} \otimes \mathcal{H}_1^{(2)}\right\} \oplus \left\{\mathcal{H}_0^{(1)} \otimes \mathcal{H}_2^{(2)}\right\}$$
$$\oplus \left\{\mathcal{H}_1^{(1)} \otimes \mathcal{H}_0^{(2)}\right\} \oplus \left\{\mathcal{H}_1^{(1)} \otimes \mathcal{H}_1^{(2)}\right\} \oplus \left\{\mathcal{H}_1^{(1)} \otimes \mathcal{H}_2^{(2)}\right\}$$
$$\triangleq \mathcal{H}^0 \oplus \mathcal{H}^1 \oplus \mathcal{H}^2 \oplus \mathcal{H}^3 \oplus \mathcal{H}^4, \tag{4.19}$$

where $\mathcal{H}^0 = \{\mathcal{H}_0^{(1)} \otimes \mathcal{H}_0^{(2)}\} \oplus \{\mathcal{H}_0^{(1)} \otimes \mathcal{H}_1^{(2)}\}$, $\mathcal{H}^1 = \mathcal{H}_0^{(1)} \otimes \mathcal{H}_2^{(2)}$, $\mathcal{H}^2 = \mathcal{H}_1^{(1)} \otimes \mathcal{H}_0^{(2)}$, $\mathcal{H}^3 = \mathcal{H}_1^{(1)} \otimes \mathcal{H}_1^{(2)}$, and $\mathcal{H}^4 = \mathcal{H}_1^{(1)} \otimes \mathcal{H}_2^{(2)}$. It is easy to see that $\mathcal{H}^0$ is a two-dimensional space with basis functions $\phi_1(\boldsymbol{x}) = 1$, and $\phi_2(\boldsymbol{x}) = x_2 - 0.5$. RKs of $\mathcal{H}^1$, $\mathcal{H}^2$, $\mathcal{H}^3$, and $\mathcal{H}^4$ can be calculated from RKs of $\mathcal{H}_0^{(1)}$ and $\mathcal{H}_1^{(1)}$ given in (4.6) and the RKs of $\mathcal{H}_0^{(2)}$, $\mathcal{H}_1^{(2)}$, and $\mathcal{H}_2^{(2)}$ given in Table 2.2.

4.4.3 Decomposition of $W_2^{m_1}[0,1] \otimes W_2^{m_2}[0,1]$

Suppose both x_1 and x_2 are continuous variables in $[0,1]$. $W_2^m[0,1]$ is a natural model space for both effects of x_1 and x_2. Therefore, we consider the tensor product space $W_2^{m_1}[0,1] \otimes W_2^{m_2}[0,1]$. For simplicity, we will derive SS ANOVA decompositions for combinations $m_1 = m_2 = 1$ and $m_1 = m_2 = 2$ only. SS ANOVA decompositions for other combinations of m_1 and m_2 can be derived similarly.

Consider the tensor product space $W_2^1[0,1] \otimes W_2^1[0,1]$ first. Define two averaging operators as

$$\mathcal{A}_1^{(k)} f = \int_0^1 f dx_k, \quad k = 1, 2,$$

where $\mathcal{A}_1^{(k)}$ extracts the constant term out of all possible functions of

x_k. Let $\mathcal{A}_2^{(k)} = I - \mathcal{A}_1^{(k)}$ for $k = 1, 2$. Then

$$
\begin{aligned}
f &= \{\mathcal{A}_1^{(1)} + \mathcal{A}_2^{(1)}\}\{\mathcal{A}_1^{(2)} + \mathcal{A}_2^{(2)}\}f \\
&= \mathcal{A}_1^{(1)}\mathcal{A}_1^{(2)}f + \mathcal{A}_2^{(1)}\mathcal{A}_1^{(2)}f + \mathcal{A}_1^{(1)}\mathcal{A}_2^{(2)}f + \mathcal{A}_2^{(1)}\mathcal{A}_2^{(2)}f \\
&\triangleq \mu + f_1(x_1) + f_2(x_2) + f_{12}(x_1, x_2).
\end{aligned}
\tag{4.20}
$$

Obviously, (4.20) is a natural extension of the classical two-way ANOVA decomposition (4.14) from the product of two discrete domains to the product of two continuous domains. Components μ, $f_1(x_1)$, $f_2(x_2)$, and $f_{12}(x_1, x_2)$ represent the overall mean, the main effect of x_1, the main effect of x_2, and the interaction between x_1 and x_2, respectively. In terms of the model space, (4.20) decomposes

$$
\begin{aligned}
&W_2^1[0, 1] \otimes W_2^1[0, 1] \\
&= \left\{\mathcal{H}_0^{(1)} \oplus \mathcal{H}_1^{(1)}\right\} \otimes \left\{\mathcal{H}_0^{(2)} \oplus \mathcal{H}_1^{(2)}\right\} \\
&= \left\{\mathcal{H}_0^{(1)} \otimes \mathcal{H}_0^{(2)}\right\} \oplus \left\{\mathcal{H}_1^{(1)} \otimes \mathcal{H}_0^{(2)}\right\} \oplus \left\{\mathcal{H}_0^{(1)} \otimes \mathcal{H}_1^{(2)}\right\} \oplus \left\{\mathcal{H}_1^{(1)} \otimes \mathcal{H}_1^{(2)}\right\} \\
&\triangleq \mathcal{H}^0 \oplus \mathcal{H}^1 \oplus \mathcal{H}^2 \oplus \mathcal{H}^3,
\end{aligned}
$$

where $\mathcal{H}_0^{(k)} = \{1\}$ and $\mathcal{H}_1^{(k)} = \{f \in W_2^1[0, 1] : \int_0^1 f dx_k = 0\}$ for $k = 1, 2$, $\mathcal{H}^0 = \mathcal{H}_0^{(1)} \otimes \mathcal{H}_0^{(2)}$, $\mathcal{H}^1 = \mathcal{H}_1^{(1)} \otimes \mathcal{H}_0^{(2)}$, $\mathcal{H}^2 = \mathcal{H}_0^{(1)} \otimes \mathcal{H}_1^{(2)}$, and $\mathcal{H}^3 = \mathcal{H}_1^{(1)} \otimes \mathcal{H}_1^{(2)}$. $\mathcal{H}^0$ is an one-dimensional space with basis $\phi(\boldsymbol{x}) = 1$. The RKs of $\mathcal{H}^1$, $\mathcal{H}^2$, and $\mathcal{H}^3$ can be calculated from RKs of $\mathcal{H}_0^{(k)}$ and $\mathcal{H}_1^{(k)}$ given in Table 2.2.

Now suppose we want to model both x_1 and x_2 using cubic splines. That is, we consider the tensor product space $W_2^2[0, 1] \otimes W_2^2[0, 1]$. Define four averaging operators

$$
\begin{aligned}
\mathcal{A}_1^{(k)}f &= \int_0^1 f dx_k, \\
\mathcal{A}_2^{(k)}f &= \left(\int_0^1 f' dx_k\right)(x_k - 0.5), \quad k = 1, 2,
\end{aligned}
$$

where $\mathcal{A}_1^{(1)}$ and $\mathcal{A}_1^{(2)}$ extract the constant function out of all possible functions for each variable, and $\mathcal{A}_2^{(1)}$ and $\mathcal{A}_2^{(2)}$ extract the linear function for each variable. Let $\mathcal{A}_3^{(k)} = I - \mathcal{A}_1^{(k)} - \mathcal{A}_2^{(k)}$ for $k = 1, 2$. Then

$$
\begin{aligned}
f &= \{\mathcal{A}_1^{(1)} + \mathcal{A}_2^{(1)} + \mathcal{A}_3^{(1)}\}\{\mathcal{A}_1^{(2)} + \mathcal{A}_2^{(2)} + \mathcal{A}_3^{(2)}\}f \\
&= \mathcal{A}_1^{(1)}\mathcal{A}_1^{(2)}f + \mathcal{A}_1^{(1)}\mathcal{A}_2^{(2)}f + \mathcal{A}_1^{(1)}\mathcal{A}_3^{(2)}f \\
&\quad + \mathcal{A}_2^{(1)}\mathcal{A}_1^{(2)}f + \mathcal{A}_2^{(1)}\mathcal{A}_2^{(2)}f + \mathcal{A}_2^{(1)}\mathcal{A}_3^{(2)}f \\
&\quad + \mathcal{A}_3^{(1)}\mathcal{A}_1^{(2)}f + \mathcal{A}_3^{(1)}\mathcal{A}_2^{(2)}f + \mathcal{A}_3^{(1)}\mathcal{A}_3^{(2)}f \\
&\triangleq \mu + \beta_2 \times (x_2 - 0.5) + f_2^s(x_2) \\
&\quad + \beta_1 \times (x_1 - 0.5) + \beta_3 \times (x_1 - 0.5) \times (x_2 - 0.5) + f_{12}^{ls}(x_1, x_2) \\
&\quad + f_1^s(x_1) + f_{12}^{sl}(x_1, x_2) + f_{12}^{ss}(x_1, x_2), \quad\quad\quad (4.21)
\end{aligned}
$$

where μ represents the overall mean; $\beta_1 \times (x_1 - 0.5)$ and $\beta_2 \times (x_2 - 0.5)$ represent the linear main effects of x_1 and x_2; $f_1^s(x_1)$ and $f_2^s(x_2)$ represent the smooth main effect of x_1 and x_2; $\beta_3 \times (x_1 - 0.5) \times (x_2 - 0.5)$, $f_{12}^{ls}(x_1, x_2)$, $f_{12}^{sl}(x_1, x_2)$ and $f_{12}^{ss}(x_1, x_2)$ represent the linear–linear, linear–smooth, smooth–linear, and smooth–smooth interactions between x_1 and x_2. The overall main effect of x_k

$$
f_k(x_k) = \beta_k \times (x_k - 0.5) + f_k^s(x_k), \quad k = 1, 2,
$$

and the overall interaction between x_1 and x_2

$$
\begin{aligned}
f_{12}(x_1, x_2) &= \beta_3 \times (x_1 - 0.5) \times (x_2 - 0.5) + f_{12}^{ls}(x_1, x_2) + f_{12}^{sl}(x_1, x_2) \\
&\quad + f_{12}^{ss}(x_1, x_2).
\end{aligned}
$$

The simple regression model with both x_1 and x_2 being modeled by straight lines is a special case of (4.21) with $f_1^s = f_2^s = f_{12}^{ls} = f_{12}^{sl} = f_{12}^{ss} = 0$.

In terms of the model space, (4.21) decomposes

$$
\begin{aligned}
& W_2^2[0, 1] \otimes W_2^2[0, 1] \\
&= \left\{\mathcal{H}_0^{(1)} \oplus \mathcal{H}_1^{(1)} \oplus \mathcal{H}_2^{(1)}\right\} \otimes \left\{\mathcal{H}_0^{(2)} \oplus \mathcal{H}_1^{(2)} \oplus \mathcal{H}_2^{(2)}\right\} \\
&= \left\{\mathcal{H}_0^{(1)} \otimes \mathcal{H}_0^{(2)}\right\} \oplus \left\{\mathcal{H}_1^{(1)} \otimes \mathcal{H}_0^{(2)}\right\} \oplus \left\{\mathcal{H}_2^{(1)} \otimes \mathcal{H}_0^{(2)}\right\} \\
&\quad \oplus \left\{\mathcal{H}_0^{(1)} \otimes \mathcal{H}_1^{(2)}\right\} \oplus \left\{\mathcal{H}_1^{(1)} \otimes \mathcal{H}_1^{(2)}\right\} \oplus \left\{\mathcal{H}_2^{(1)} \otimes \mathcal{H}_1^{(2)}\right\} \\
&\quad \oplus \left\{\mathcal{H}_0^{(1)} \otimes \mathcal{H}_2^{(2)}\right\} \oplus \left\{\mathcal{H}_1^{(1)} \otimes \mathcal{H}_2^{(2)}\right\} \oplus \left\{\mathcal{H}_2^{(1)} \otimes \mathcal{H}_2^{(2)}\right\} \\
&\triangleq \mathcal{H}^0 \oplus \mathcal{H}^1 \oplus \mathcal{H}^2 \oplus \mathcal{H}^3 \oplus \mathcal{H}^4 \oplus \mathcal{H}^5,
\end{aligned}
$$

where $\mathcal{H}_0^{(k)} = \{1\}$, $\mathcal{H}_1^{(k)} = \{x_k - 0.5\}$, and $\mathcal{H}_2^{(k)} = \{f \in W_2^2[0, 1] : \int_0^1 f \, dx_k = \int_0^1 f' dx_k = 0\}$ for $k = 1, 2$, $\mathcal{H}^0 = \{\mathcal{H}_0^{(1)} \otimes \mathcal{H}_0^{(2)}\} \oplus \{\mathcal{H}_1^{(1)} \otimes \mathcal{H}_0^{(2)}\} \oplus$

$\{\mathcal{H}_0^{(1)} \otimes \mathcal{H}_1^{(2)}\} \oplus \{\mathcal{H}_1^{(1)} \otimes \mathcal{H}_1^{(2)}\}$, $\mathcal{H}^1 = \mathcal{H}_2^{(1)} \otimes \mathcal{H}_0^{(2)}$, $\mathcal{H}^2 = \mathcal{H}_2^{(0)} \otimes \mathcal{H}_1^{(2)}$, $\mathcal{H}^3 = \mathcal{H}_0^{(1)} \otimes \mathcal{H}_2^{(2)}$, $\mathcal{H}^4 = \mathcal{H}_1^{(1)} \otimes \mathcal{H}_2^{(2)}$, and $\mathcal{H}^5 = \mathcal{H}_2^{(2)} \otimes \mathcal{H}_2^{(2)}$. $\mathcal{H}^0$ is a four-dimensional space with basis functions $\phi_1(\boldsymbol{x}) = 1$, $\phi_2(\boldsymbol{x}) = x_1 - 0.5$, $\phi_3(\boldsymbol{x}) = x_2 - 0.5$, and $\phi_4(\boldsymbol{x}) = (x_1 - 0.5) \times (x_2 - 0.5)$. The RKs of $\mathcal{H}^1$, $\mathcal{H}^2$, $\mathcal{H}^3$, $\mathcal{H}^4$ and $\mathcal{H}^5$ can be calculated from RKs of $\mathcal{H}_0^{(k)}$, $\mathcal{H}_1^{(k)}$, and $\mathcal{H}_2^{(k)}$ given in Table 2.2.

4.4.4 Decomposition of $\mathbb{R}^a \otimes W_2^m(per)$

Suppose x_1 is a discrete variable with a levels, and x_2 is a continuous variable in $[0, 1]$. In addition, suppose that f is a periodic function of x_2. A natural model space for x_1 is $\mathbb{R}^a$, and a natural model space for x_2 is $W_2^m(per)$. Therefore, we consider the tensor product space $\mathbb{R}^a \otimes W_2^m(per)$.

Define two averaging operators $\mathcal{A}_1^{(1)}$ and $\mathcal{A}_1^{(2)}$ as

$$\mathcal{A}_1^{(1)} f = \frac{1}{a} \sum_{x_1=1}^{a} f,$$

$$\mathcal{A}_1^{(2)} f = \int_0^1 f \, dx_2.$$

Let $\mathcal{A}_2^{(1)} = I - \mathcal{A}_1^{(1)}$ and $\mathcal{A}_2^{(2)} = I - \mathcal{A}_1^{(2)}$. Then

$$\begin{aligned}
f &= \{\mathcal{A}_1^{(1)} + \mathcal{A}_2^{(1)}\}\{\mathcal{A}_1^{(2)} + \mathcal{A}_2^{(2)}\} f \\
&= \mathcal{A}_1^{(1)} \mathcal{A}_1^{(2)} f + \mathcal{A}_2^{(1)} \mathcal{A}_1^{(2)} f + \mathcal{A}_1^{(1)} \mathcal{A}_2^{(2)} f + \mathcal{A}_2^{(1)} \mathcal{A}_2^{(2)} f \\
&\triangleq \mu + f_1(x_1) + f_2(x_2) + f_{12}(x_1, x_2),
\end{aligned} \tag{4.22}$$

where μ represents the overall mean, $f_1(x_1)$ represents the main effect of x_1, $f_2(x_2)$ represents the main effect of x_2, and $f_{12}(x_1, x_2)$ represents the interaction between x_1 and x_2. Write $\mathbb{R}^a = \mathcal{H}_0^{(1)} \oplus \mathcal{H}_1^{(1)}$ and $W_2^m(per) = \mathcal{H}_0^{(2)} \oplus \mathcal{H}_1^{(2)}$, where $\mathcal{H}_0^{(1)}$ and $\mathcal{H}_1^{(1)}$ are given in (4.5), $\mathcal{H}_0^{(2)} = \{1\}$, and $\mathcal{H}_1^{(2)} = \{f \in W_2^m(per) : \int_0^1 f \, du = 0\}$. Then, in terms of the model space, (4.22) decomposes

$$\begin{aligned}
&\mathbb{R}^a \otimes W_2^m(per) \\
&= \left\{\mathcal{H}_0^{(1)} \oplus \mathcal{H}_1^{(1)}\right\} \otimes \left\{\mathcal{H}_0^{(2)} \oplus \mathcal{H}_1^{(2)}\right\} \\
&= \left\{\mathcal{H}_0^{(1)} \otimes \mathcal{H}_0^{(2)}\right\} \oplus \left\{\mathcal{H}_1^{(1)} \otimes \mathcal{H}_0^{(2)}\right\} \oplus \left\{\mathcal{H}_0^{(1)} \otimes \mathcal{H}_1^{(2)}\right\} \oplus \left\{\mathcal{H}_1^{(1)} \otimes \mathcal{H}_1^{(2)}\right\} \\
&\triangleq \mathcal{H}^0 \oplus \mathcal{H}^1 \oplus \mathcal{H}^2 \oplus \mathcal{H}^3,
\end{aligned}$$

where $\mathcal{H}^0 = \mathcal{H}_0^{(1)} \otimes \mathcal{H}_0^{(2)}$, $\mathcal{H}^1 = \mathcal{H}_1^{(1)} \otimes \mathcal{H}_0^{(2)}$, $\mathcal{H}^2 = \mathcal{H}_0^{(1)} \otimes \mathcal{H}_1^{(2)}$, and $\mathcal{H}^3 = \mathcal{H}_1^{(1)} \otimes \mathcal{H}_1^{(2)}$. $\mathcal{H}^0$ is an one-dimensional space with basis function $\phi(\boldsymbol{x}) = 1$. The RKs of $\mathcal{H}^1$, $\mathcal{H}^2$, and $\mathcal{H}^3$ can be calculated from RKs of $\mathcal{H}_0^{(1)}$ and $\mathcal{H}_1^{(1)}$ given in (4.6) and RKs of $\mathcal{H}_0^{(2)}$ and $\mathcal{H}_1^{(2)}$ given in (2.33).

4.4.5 Decomposition of $W_2^{m_1}(per) \otimes W_2^{m_2}[0,1]$

Suppose both x_1 and x_2 are continuous variables in $[0,1]$. In addition, suppose f is a periodic function of x_1. A natural model space for x_1 is $W_2^{m_1}(per)$, and a natural model space for x_2 is $W_2^{m_2}[0,1]$. Therefore, we consider the tensor product space $W_2^{m_1}(per) \otimes W_2^{m_2}[0,1]$. For simplicity, we derive the SS ANOVA decomposition for $m_2 = 2$ only.

Define three averaging operators

$$\mathcal{A}_1^{(1)} f = \int_0^1 f \, dx_1,$$

$$\mathcal{A}_1^{(2)} f = \int_0^1 f \, dx_2,$$

$$\mathcal{A}_2^{(2)} f = \left(\int_0^1 f' dx_2 \right) (x_2 - 0.5).$$

Let $\mathcal{A}_2^{(1)} = I - \mathcal{A}_1^{(1)}$ and $\mathcal{A}_3^{(2)} = I - \mathcal{A}_1^{(2)} - \mathcal{A}_2^{(2)}$. Then

$$\begin{aligned} f &= \left\{ \mathcal{A}_1^{(1)} + \mathcal{A}_2^{(1)} \right\} \left\{ \mathcal{A}_1^{(2)} + \mathcal{A}_2^{(2)} + \mathcal{A}_3^{(2)} \right\} f \\ &= \mathcal{A}_1^{(1)} \mathcal{A}_1^{(2)} f + \mathcal{A}_1^{(1)} \mathcal{A}_2^{(2)} f + \mathcal{A}_1^{(1)} \mathcal{A}_3^{(2)} f \\ &\quad + \mathcal{A}_2^{(1)} \mathcal{A}_1^{(2)} f + \mathcal{A}_2^{(1)} \mathcal{A}_2^{(2)} f + \mathcal{A}_2^{(1)} \mathcal{A}_3^{(2)} f \\ &\triangleq \mu + \beta \times (x_2 - 0.5) + f_2^s(x_2) \\ &\quad + f_1(x_1) + f_{12}^{sl}(x_1, x_2) + f_{12}^{ss}(x_1, x_2), \end{aligned} \quad (4.23)$$

where μ represents the overall mean, $f_1(x_1)$ represents the main effect of x_1, $\beta \times (x_2 - 0.5)$ and $f_2^s(x_2)$ represent the linear and smooth main effects of x_2, $f_{12}^{sl}(x_1, x_2)$ and $f_{12}^{ss}(x_1, x_2)$ represent the smooth–linear and smooth–smooth interactions. The overall main effect of x_2

$$f_2(x_2) = \beta \times (x_2 - 0.5) + f_2^s(x_2),$$

and the overall interaction between x_1 and x_2

$$f_{12}(x_1, x_2) = f_{12}^{sl}(x_1, x_2) + f_{12}^{ss}(x_1, x_2).$$

Write $W_2^{m_1}(per) = \mathcal{H}_0^{(1)} \oplus \mathcal{H}_1^{(1)}$ and $W_2^2[0,1] = \mathcal{H}_0^{(2)} \oplus \mathcal{H}_1^{(2)} \oplus \mathcal{H}_2^{(2)}$, where $\mathcal{H}_0^{(1)} = \{1\}$, $\mathcal{H}_1^{(1)} = \{f \in W_2^{m_1}(per) : \int_0^1 f du = 0\}$, $\mathcal{H}_0^{(2)} = \{1\}$, $\mathcal{H}_1^{(2)} = \{x_2 - 0.5\}$, and $\mathcal{H}_2^{(2)} = \{f \in W_2^2[0,1] : \int_0^1 f du = \int_0^1 f' du = 0\}$. Then, in terms of the model space, (4.23) decomposes

$$W_2^{m_1}(per) \otimes W_2^2[0,1]$$
$$= \left\{\mathcal{H}_0^{(1)} \oplus \mathcal{H}_1^{(1)}\right\} \otimes \left\{\mathcal{H}_0^{(2)} \oplus \mathcal{H}_1^{(2)} \oplus \mathcal{H}_2^{(2)}\right\}$$
$$= \left\{\mathcal{H}_0^{(1)} \otimes \mathcal{H}_0^{(2)}\right\} \oplus \left\{\mathcal{H}_0^{(1)} \otimes \mathcal{H}_1^{(2)}\right\} \oplus \left\{\mathcal{H}_0^{(1)} \otimes \mathcal{H}_2^{(2)}\right\}$$
$$\oplus \left\{\mathcal{H}_1^{(1)} \otimes \mathcal{H}_0^{(2)}\right\} \oplus \left\{\mathcal{H}_1^{(1)} \otimes \mathcal{H}_1^{(2)}\right\} \oplus \left\{\mathcal{H}_1^{(1)} \otimes \mathcal{H}_2^{(2)}\right\}$$
$$\triangleq \mathcal{H}^0 \oplus \mathcal{H}^1 \oplus \mathcal{H}^2 \oplus \mathcal{H}^3 \oplus \mathcal{H}^4,$$

where $\mathcal{H}^0 = \{\mathcal{H}_0^{(1)} \otimes \mathcal{H}_0^{(2)}\} \oplus \{\mathcal{H}_0^{(1)} \otimes \mathcal{H}_1^{(2)}\}$, $\mathcal{H}^1 = \mathcal{H}_0^{(1)} \otimes \mathcal{H}_2^{(2)}$, $\mathcal{H}^2 = \mathcal{H}_1^{(1)} \otimes \mathcal{H}_0^{(2)}$, $\mathcal{H}^3 = \mathcal{H}_1^{(1)} \otimes \mathcal{H}_1^{(2)}$, and $\mathcal{H}^4 = \mathcal{H}_1^{(1)} \otimes \mathcal{H}_2^{(2)}$. $\mathcal{H}^0$ is a two-dimensional space with basis functions $\phi(\boldsymbol{x}) = 1$ and $\phi(\boldsymbol{x}) = x_2 - 0.5$. The RKs of $\mathcal{H}^1$, $\mathcal{H}^2$, $\mathcal{H}^3$, and $\mathcal{H}^4$ can be calculated from the RKs $\mathcal{H}_0^{(1)}$ and $\mathcal{H}_1^{(1)}$ given in (2.33) and the RKs of $\mathcal{H}_0^{(2)}$, $\mathcal{H}_1^{(2)}$, and $\mathcal{H}_2^{(2)}$ given in Table 2.2.

4.4.6   Decomposition of $W_2^2(\mathbb{R}^2) \otimes W_2^m(per)$

Suppose $\boldsymbol{x}_1 = (x_{11}, x_{12})$ is a bivariate continuous variable in $\mathbb{R}^2$, and x_2 is a continuous variable in $[0, 1]$. In addition, suppose that f is a periodic function of x_2. We consider the tensor product space $W_2^2(\mathbb{R}^2) \otimes W_2^2(per)$ for the joint regression function $f(\boldsymbol{x}_1, x_2)$.

Let $\phi_1(\boldsymbol{x}_1) = 1$, $\phi_2(\boldsymbol{x}_1) = x_{11}$, and $\phi_3(\boldsymbol{x}_1) = x_{12}$ be polynomials of total degree less than 2. Define three averaging operators

$$\mathcal{A}_1^{(1)} f = \sum_{j=1}^J w_j f(\boldsymbol{u}_j),$$

$$\mathcal{A}_2^{(1)} f = \sum_{j=1}^J w_j f(\boldsymbol{u}_j)\{\tilde{\phi}_2(\boldsymbol{u}_j)\tilde{\phi}_2 + \tilde{\phi}_3(\boldsymbol{u}_j)\tilde{\phi}_3\},$$

$$\mathcal{A}_1^{(2)} f = \int_0^1 f dx_2,$$

where $\boldsymbol{u}_j$ are fixed points in $\mathbb{R}^2$, w_j are fixed positive weights such that $\sum_{j=1}^J w_j = 1$, $\tilde{\phi}_1 = 1$, and $\tilde{\phi}_2$ and $\tilde{\phi}_3$ are orthonormal bases based on the norm (2.41). Let $\mathcal{A}_3^{(1)} = I - \mathcal{A}_1^{(1)} - \mathcal{A}_2^{(1)}$ and $\mathcal{A}_2^{(2)} = I - \mathcal{A}_1^{(2)}$. Then

$$
\begin{aligned}
f &= \left\{ \mathcal{A}_1^{(1)} + \mathcal{A}_2^{(1)} + \mathcal{A}_3^{(1)} \right\} \left\{ \mathcal{A}_1^{(2)} + \mathcal{A}_2^{(2)} \right\} f \\
&= \mathcal{A}_1^{(1)} \mathcal{A}_1^{(2)} f + \mathcal{A}_2^{(1)} \mathcal{A}_1^{(2)} f + \mathcal{A}_3^{(1)} \mathcal{A}_1^{(2)} f \\
&\quad + \mathcal{A}_1^{(1)} \mathcal{A}_2^{(2)} f + \mathcal{A}_2^{(1)} \mathcal{A}_2^{(2)} f + \mathcal{A}_3^{(1)} \mathcal{A}_2^{(2)} f \\
&= \mu + \beta_1 \tilde{\phi}_2(\boldsymbol{x}_1) + \beta_2 \tilde{\phi}_3(\boldsymbol{x}_1) \\
&\quad + f_1^s(\boldsymbol{x}_1) + f_2(x_2) + f_{12}^{ls}(\boldsymbol{x}_1, x_2) + f_{12}^{ss}(\boldsymbol{x}_1, x_2),
\end{aligned}
\tag{4.24}
$$

where μ is the overall mean, $\beta_1 \tilde{\phi}_2(\boldsymbol{x}_1) + \beta_2 \tilde{\phi}_3(\boldsymbol{x}_1)$ is the linear main effect of $\boldsymbol{x}_1$, $f_1^s(\boldsymbol{x}_1)$ is the smooth main effect of $\boldsymbol{x}_1$, $f_2(x_2)$ is the main effect of x_2, $f_{12}^{ls}(\boldsymbol{x}_1, x_2)$ is the linear–smooth interaction, and $f_{12}^{ss}(\boldsymbol{x}_1, x_2)$ is smooth–smooth interaction. The overall main effect of $\boldsymbol{x}_1$

$$
f_1(\boldsymbol{x}_1) = \beta_1 \tilde{\phi}_2(\boldsymbol{x}_1) + \beta_2 \tilde{\phi}_3(\boldsymbol{x}_1) + f_1^s(\boldsymbol{x}_1),
$$

and the overall interaction

$$
f_{12}(\boldsymbol{x}_1, x_2) = f_{12}^{ls}(\boldsymbol{x}_1, x_2) + f_{12}^{ss}(\boldsymbol{x}_1, x_2).
$$

Write $W_2^2(\mathbb{R}^2) = \mathcal{H}_0^{(1)} \oplus \mathcal{H}_1^{(1)} \oplus \mathcal{H}_2^{(1)}$ and $W_2^m(per) = \mathcal{H}_0^{(2)} \oplus \mathcal{H}_1^{(2)}$, where $\mathcal{H}_0^{(1)} = \{1\}$, $\mathcal{H}_1^{(1)} = \{\tilde{\phi}_2, \tilde{\phi}_3\}$, $\mathcal{H}_2^{(1)} = \{f \in W_2^2(\mathbb{R}^2) : J_2^2(f) = 0\}$, $\mathcal{H}_0^{(2)} = \{1\}$, and $\mathcal{H}_1^{(2)} = \{f \in W_2^{m_1}(per) : \int_0^1 f du = 0\}$. Then, in terms of the model space, (4.24) decomposes

$$
\begin{aligned}
&W_2^2(\mathbb{R}^2) \otimes W_2^m(per) \\
&= \left\{ \mathcal{H}_0^{(1)} \oplus \mathcal{H}_1^{(1)} \oplus \mathcal{H}_2^{(1)} \right\} \otimes \left\{ \mathcal{H}_0^{(2)} \oplus \mathcal{H}_1^{(2)} \right\} \\
&= \left\{ \mathcal{H}_0^{(1)} \otimes \mathcal{H}_0^{(2)} \right\} \oplus \left\{ \mathcal{H}_1^{(1)} \otimes \mathcal{H}_0^{(2)} \right\} \oplus \left\{ \mathcal{H}_2^{(1)} \otimes \mathcal{H}_0^{(2)} \right\} \\
&\quad \oplus \left\{ \mathcal{H}_0^{(1)} \otimes \mathcal{H}_1^{(2)} \right\} \oplus \left\{ \mathcal{H}_1^{(1)} \otimes \mathcal{H}_1^{(2)} \right\} \oplus \left\{ \mathcal{H}_2^{(1)} \otimes \mathcal{H}_1^{(2)} \right\} \\
&\triangleq \mathcal{H}^0 \oplus \mathcal{H}^1 \oplus \mathcal{H}^2 \oplus \mathcal{H}^3 \oplus \mathcal{H}^4,
\end{aligned}
\tag{4.25}
$$

where $\mathcal{H}^0 = \{\mathcal{H}_0^{(1)} \otimes \mathcal{H}_0^{(2)}\} \oplus \{\mathcal{H}_1^{(1)} \otimes \mathcal{H}_0^{(2)}\}$, $\mathcal{H}^1 = \{\mathcal{H}_2^{(1)} \otimes \mathcal{H}_0^{(2)}\}$, $\mathcal{H}^2 = \{\mathcal{H}_0^{(1)} \otimes \mathcal{H}_1^{(2)}\}$, $\mathcal{H}^3 = \{\mathcal{H}_1^{(1)} \otimes \mathcal{H}_1^{(2)}\}$, and $\mathcal{H}^4 = \{\mathcal{H}_2^{(1)} \otimes \mathcal{H}_1^{(2)}\}$. The basis functions of $\mathcal{H}^0$ are 1, $\tilde{\phi}_2$, and $\tilde{\phi}_3$. The RKs of $\mathcal{H}_0^{(1)}$ and $\mathcal{H}_1^{(1)}$ are 1 and $\tilde{\phi}_2(\boldsymbol{x}_1)\tilde{\phi}_2(\boldsymbol{z}_1) + \tilde{\phi}_3(\boldsymbol{x}_1)\tilde{\phi}_3(\boldsymbol{z}_1)$, respectively. The RK of $\mathcal{H}_2^{(1)}$ is given in (2.42). The RKs of $\mathcal{H}_0^{(2)}$ and $\mathcal{H}_1^{(2)}$ are given in (2.33). The RKs of $\mathcal{H}^1$, $\mathcal{H}^2$, $\mathcal{H}^3$, and $\mathcal{H}^4$ can be calculated as products of the RKs of the involved marginal spaces.

4.5 General SS ANOVA Decomposition

Consider the general case with d independent variables $x_1 \in \mathcal{X}_1, x_2 \in \mathcal{X}_2, \ldots, x_d \in \mathcal{X}_d$, and the tensor product space $\mathcal{H}^{(1)} \otimes \mathcal{H}^{(2)} \otimes \cdots \otimes \mathcal{H}^{(d)}$ on $\mathcal{X} = \mathcal{X}_1 \times \mathcal{X}_2 \times \cdots \times \mathcal{X}_d$. For f as a function of x_k, assume the following one-way decomposition as in (4.3),

$$f = \mathcal{A}_1^{(k)} f + \cdots + \mathcal{A}_{r_k}^{(k)} f, \quad 1 \le k \le d, \tag{4.26}$$

where $\mathcal{A}_1^{(k)} + \cdots + \mathcal{A}_{r_k}^{(k)} = I$. Then, for the joint function,

$$\begin{aligned} f &= \left\{ \mathcal{A}_1^{(1)} + \cdots + \mathcal{A}_{r_1}^{(1)} \right\} \ldots \left\{ \mathcal{A}_1^{(d)} + \cdots + \mathcal{A}_{r_d}^{(d)} \right\} f \\ &= \sum_{j_1=1}^{r_1} \cdots \sum_{j_d=1}^{r_d} \mathcal{A}_{j_1}^{(1)} \ldots \mathcal{A}_{j_d}^{(d)} f. \end{aligned} \tag{4.27}$$

The above decomposition of the function f is referred to as the *SS ANOVA decomposition*.

Denote

$$\mathcal{H}^{(k)} = \mathcal{H}_{(1)}^{(k)} \oplus \cdots \oplus \mathcal{H}_{(r_k)}^{(k)}, \quad k = 1, 2, \ldots, d$$

as the one-way decomposition to $\mathcal{H}^{(k)}$ associated with (4.26). Then, (4.27) decomposes the tensor product space

$$\begin{aligned} & \mathcal{H}^{(1)} \otimes \mathcal{H}^{(2)} \otimes \cdots \otimes \mathcal{H}^{(d)} \\ &= \left\{ \mathcal{H}_{(1)}^{(1)} \oplus \cdots \oplus \mathcal{H}_{(r_1)}^{(1)} \right\} \otimes \cdots \otimes \left\{ \mathcal{H}_{(1)}^{(d)} \oplus \cdots \oplus \mathcal{H}_{(r_d)}^{(d)} \right\} \\ &= \sum_{j_1=1}^{r_1} \cdots \sum_{j_d=1}^{r_d} \mathcal{H}_{(j_1)}^{(1)} \otimes \cdots \otimes \mathcal{H}_{(j_d)}^{(d)}. \end{aligned} \tag{4.28}$$

The RK of $\mathcal{H}_{(j_1)}^{(1)} \otimes \cdots \otimes \mathcal{H}_{(j_d)}^{(d)}$ equals $\prod_{k=1}^{d} R_{(j_k)}^{(k)}$, where $R_{(j_k)}^{(k)}$ is the RK of $\mathcal{H}_{(j_k)}^{(k)}$ for $k = 1, \ldots, d$.

Consider the special case when $r_k = 2$ for all $k = 1, \ldots, d$. Assume that $\mathcal{A}_1^{(k)} f$ is independent of x_k, or equivalently, $\mathcal{H}_{(1)}^{(k)} = \{1\}$. Then the decomposition (4.27) can be written as

$$\begin{aligned} f &= \sum_{B \subseteq \{1, \ldots, d\}} \left\{ \prod_{k \in B} \mathcal{A}_2^{(k)} \prod_{k \in B^c} \mathcal{A}_1^{(k)} f \right\} \\ &= \mu + \sum_{k=1}^{d} f_k(x_k) + \sum_{k<l} f_{kl}(x_k, x_l) + \cdots + f_{1 \ldots d}(x_1, \ldots, x_d), \tag{4.29} \end{aligned}$$

where μ represents the grand mean, $f_k(x_k)$ represents the main effect of x_k, $f_{kl}(x_k, x_l)$ represents the two-way interaction between x_k and x_l, and the remaining terms represent higher-order interactions.

For general r_k, assuming $\mathcal{A}_1^{(k)} f$ is independent of x_k, the decomposition (4.29) can be derived by combining operators $\mathcal{A}_2^{(k)}, \ldots, \mathcal{A}_{r_k}^{(k)}$ into one averaging operator $\tilde{\mathcal{A}}_2^{(k)} = \mathcal{A}_2^{(k)} + \cdots + \mathcal{A}_{r_k}^{(k)}$. Therefore, decomposition (4.29) combines components in (4.27) and reorganizes them into overall main effects and interactions.

When all x_k are discrete variables, the SS ANOVA decomposition (4.27) leads to the classical d-way ANOVA model. Therefore, SS ANOVA decompositions are natural extensions of classical ANOVA decompositions from discrete domains to general domains and from finite dimensional spaces to infinite dimensional spaces. They decompose tensor product RKHS's into meaningful subspaces. As the classical ANOVA decompositions, SS ANOVA decompositions lead to hierarchical structures that are useful for model selection and interpretation.

Different SS ANOVA decompositions can be derived based on different averaging operators (or equivalently different decompositions of marginal spaces). Therefore, the SS ANOVA decomposition should be regarded as a general prescription for building multivariate nonparametric models rather than some fixed models. They can also be used to construct submodels for components in more complicated models.

4.6 SS ANOVA Models and Estimation

The *curse of dimensionality* is a major problem in dealing with multivariate functions. From the SS ANOVA decomposition in (4.27), it is clear that the number of components in the decomposition increases exponentially as the dimension d increases. To overcome this problem, as in classical ANOVA, high-order interactions are often deleted from the model space.

A model containing any subset of components in the SS ANOVA decomposition (4.27) is referred to as an *SS ANOVA model*. The well-known *additive model* is a special case that contains main effects only (Hastie and Tibshirani 1990). Given an SS ANOVA model, we can regroup and write the model space as

$$\mathcal{M} = \mathcal{H}^0 \oplus \mathcal{H}^1 \oplus \cdots \oplus \mathcal{H}^q, \tag{4.30}$$

where $\mathcal{H}^0$ is a finite dimensional space collecting all functions that are not going to be penalized, and $\mathcal{H}^1, \ldots, \mathcal{H}^q$ are orthogonal RKHS's with RKs R^j for $j = 1, \ldots, q$. The RK R^j equals the product of RKs of the subspaces involved in the tensor product space $\mathcal{H}^j$. The norms on the composite $\mathcal{H}^j$ are the tensor product norms induced by the norms on the component subspaces. Details about the induced norm can be found in Aronszajn (1950), and an illustrative example can be found in Chapter 10 of Wahba (1990). Note that $\|f\|^2 = \|P_0 f\|^2 + \sum_{j=1}^q \|P_j f\|^2$, where P_j is the orthogonal projector in $\mathcal{M}$ onto $\mathcal{H}^j$.

For generality, suppose observations are generated by

$$y_i = \mathcal{L}_i f + \epsilon_i, \quad i = 1, \ldots, n, \tag{4.31}$$

where f is a multivariate function in the model space $\mathcal{M}$ defined in (4.30), $\mathcal{L}_i$ are bounded linear functionals, and ϵ_i are zero-mean independent random errors with a common variance σ^2.

The PLS estimate of f is the solution to

$$\min_{f \in \mathcal{M}} \left\{ \frac{1}{n} \sum_{i=1}^n (y_i - \mathcal{L}_i f)^2 + \sum_{j=1}^q \lambda_j \|P_j f\|^2 \right\}. \tag{4.32}$$

Different smoothing parameters λ_j allow different penalties for each component. For fixed smoothing parameters, the following rescaling allows us to derive and compute the solution to (4.32) using results in Chapter 2. Let $\mathcal{H}_1^* = \oplus_{j=1}^q \mathcal{H}^j$. Then, for any $f \in \mathcal{H}_1^*$,

$$f(\boldsymbol{x}) = f_1(\boldsymbol{x}) + \cdots + f_q(\boldsymbol{x}), \quad f_j \in \mathcal{H}^j, \ j = 1, \ldots, q.$$

Write $\lambda_j \triangleq \lambda / \theta_j$. The set of parameters λ and θ_j are overparameterized. The penalty is controlled by the ratio λ / θ_j, that is, λ_j. Define the inner product in $\mathcal{H}_1^*$ as

$$(f, g)_* = \sum_{j=1}^q \theta_j^{-1} (f_j, g_j). \tag{4.33}$$

Then, $\|f\|_*^2 = \sum_{j=1}^q \theta_j^{-1} \|f_j\|^2$. Let $R_1^* = \sum_{j=1}^q \theta_j R^j$. Since

$$(R_1^*(\boldsymbol{x}, \cdot), f(\cdot))_* = \sum_{j=1}^q \theta_j^{-1} (\theta_j R^j(\boldsymbol{x}, \cdot), f_j(\cdot)) = \sum_{j=1}^q f_j(\boldsymbol{x}) = f(\boldsymbol{x}),$$

then R_1^* is the RK of $\mathcal{H}_1^*$ with the inner product (4.33).

Let $P_1^* = \sum_{j=1}^q P_j$ be the orthogonal projection in $\mathcal{M}$ onto $\mathcal{H}_1^*$. Then the minimization problem (4.32) is reduced to

$$\min_{f \in \mathcal{M}} \left\{ \frac{1}{n} \sum_{i=1}^n (y_i - \mathcal{L}_i f)^2 + \lambda \|P_1^* f\|_*^2 \right\}. \tag{4.34}$$

The PLS (4.34) has the same form as (2.11), with $\mathcal{H}_1$ and P_1 being replaced by $\mathcal{H}_1^*$ and P_1^*, respectively. Therefore, results in Section 2.4 apply. Specifically, let $\boldsymbol{\theta} = (\theta_1, \ldots, \theta_q)$, $\phi_1, \ldots, \phi_p$ be basis functions of $\mathcal{H}^0$, and

$$T = \{\mathcal{L}_i \phi_\nu\}_{i=1\ \nu=1}^{n\ \ \ p},$$
$$\Sigma_k = \{\mathcal{L}_i \mathcal{L}_j R^k\}_{i,j=1}^n, \quad k = 1, \ldots, q, \qquad (4.35)$$
$$\Sigma_{\boldsymbol{\theta}} = \theta_1 \Sigma_1 + \cdots + \theta_q \Sigma_q.$$

Applying the Kimeldorf–Wahba representer theorem and noting that

$$\xi_i(\boldsymbol{x}) = \mathcal{L}_{i(\boldsymbol{z})} R_1^*(\boldsymbol{x}, \boldsymbol{z}) = \sum_{j=1}^q \theta_j \mathcal{L}_{i(\boldsymbol{z})} R^j(\boldsymbol{x}, \boldsymbol{z}),$$

the solution can be represented as

$$\hat{f}(\boldsymbol{x}) = \sum_{\nu=1}^p d_\nu \phi_\nu(\boldsymbol{x}) + \sum_{i=1}^n c_i \sum_{j=1}^q \theta_j \mathcal{L}_{i(\boldsymbol{z})} R^j(\boldsymbol{x}, \boldsymbol{z}), \qquad (4.36)$$

where $\boldsymbol{d} = (d_1, \ldots, d_p)^T$ and $\boldsymbol{c} = (c_1, \ldots, c_n)^T$ are solutions to

$$(\Sigma_{\boldsymbol{\theta}} + n\lambda I)\boldsymbol{c} + T\boldsymbol{d} = \boldsymbol{y},$$
$$T^T \boldsymbol{c} = \boldsymbol{0}. \qquad (4.37)$$

Equations in (4.37) have the same form as those in (2.21), with Σ being replaced by $\Sigma_{\boldsymbol{\theta}}$. Therefore, the coefficients $\boldsymbol{c}$ and $\boldsymbol{d}$ can be computed similarly.

Let $\hat{\boldsymbol{f}} = (\mathcal{L}_1 \hat{f}, \ldots, \mathcal{L}_n \hat{f})^T$ be the vector of fitted values. Let the QR decomposition of T be

$$T = (Q_1\ Q_2) \begin{pmatrix} R \\ 0 \end{pmatrix}$$

and $M = \Sigma_{\boldsymbol{\theta}} + n\lambda I$. Then

$$\hat{\boldsymbol{f}} = H(\lambda, \boldsymbol{\theta})\boldsymbol{y}$$

where
$$H(\lambda, \boldsymbol{\theta}) = I - n\lambda Q_2 (Q_2^T M Q_2)^{-1} Q_2^T \qquad (4.38)$$

is the *hat* matrix.

The `ssr` function in the `assist` package can be used to fit the SS ANOVA model (4.31). As in Chapter 2, observations $\boldsymbol{y}$ and T matrix can be specified using the argument `formula`. Instead of a single matrix

Σ, we now have multiple matrices Σ_j for $j = 1, \ldots, q$. They are specified as elements of a *list* for the argument `rk`. Examples are given in Section 4.9.

Sometimes it may be desirable to use the same smoothing parameter for a subset of penalties in the PLS (4.32). For illustration, suppose we want to solve the PLS (4.32) with $\lambda_{q-1} = \lambda_q$. This can be achieved by combining $\mathcal{H}^{q-1}$ and $\mathcal{H}^q$ into one space, say $\tilde{\mathcal{H}}^{q-1}$, and fit the SS ANOVA model with model space $\mathcal{M} = \mathcal{H}^0 \oplus \mathcal{H}^1 \oplus \cdots \oplus \tilde{\mathcal{H}}^{q-1}$. The RK of $\tilde{\mathcal{H}}^{q-1}$ is $\tilde{R}^{q-1} = R^{q-1} + R^q$. Then the model can be fitted by a call to the `ssr` function with the combined RK. The same approach can be applied to multiple subsets such that penalties in each subset share the same smoothing parameter. When appropriate, this approach can greatly reduce the computation time when q is large. An example will be given in Section 4.9.1.

4.7 Selection of Smoothing Parameters

The set of parameters λ and $\boldsymbol{\theta}$ are overparameterized. Therefore, even though the criteria in this section are presented as functions of λ and $\boldsymbol{\theta}$, they should be understood as functions of $(\lambda_1, \ldots, \lambda_q)$, where $\lambda_j = \lambda/\theta_j$.

Define *mean squared error* (MSE) as

$$\text{MSE}(\lambda, \boldsymbol{\theta}) = \text{E}\left(\frac{1}{n}||\hat{\boldsymbol{f}} - \boldsymbol{f}||^2\right), \tag{4.39}$$

where $\boldsymbol{f} = (\mathcal{L}_1 f, \ldots, \mathcal{L}_n f)^T$. Following the same arguments as in Section 3.3, it is easy to check that the function

$$\text{UBR}(\lambda, \boldsymbol{\theta}) \triangleq \frac{1}{n}||(I - H(\lambda, \boldsymbol{\theta}))\boldsymbol{y}||^2 + \frac{2\sigma^2}{n}\text{tr}H(\lambda, \boldsymbol{\theta}) \tag{4.40}$$

is an unbiased estimate of $\text{MSE}(\lambda, \boldsymbol{\theta}) + \sigma^2$. The function $\text{UBR}(\lambda, \boldsymbol{\theta})$ is referred to as the *unbiased risk* (UBR) criterion and the minimizer of $\text{UBR}(\lambda, \boldsymbol{\theta})$ is referred to as the UBR estimate of $(\lambda, \boldsymbol{\theta})$. The UBR method requires an estimate of error variance σ^2. Few methods are available for estimating σ^2 in a multivariate nonparametric model without needing to estimate the function f first. When the product domain $\mathcal{X}$ is equipped with a norm, the method in Tong and Wang (2005) may be used to estimate σ^2.

A parallel derivation as in Section 3.4 leads to the following extension of the GCV criterion

$$\text{GCV}(\lambda, \boldsymbol{\theta}) \triangleq \frac{\frac{1}{n}\sum_{i=1}^{n}(\mathcal{L}_i\hat{f} - y_i)^2}{\left\{\frac{1}{n}\text{tr}(I - H(\lambda, \boldsymbol{\theta}))\right\}^2}. \tag{4.41}$$

The GCV estimate of $(\lambda, \boldsymbol{\theta})$ is the minimizer of $\text{GCV}(\lambda, \boldsymbol{\theta})$.

We now construct a Bayes model for the SS ANOVA model (4.31). Assume a prior for f as

$$F(\boldsymbol{x}) = \sum_{\nu=1}^{p} \zeta_\nu \phi_\nu(\boldsymbol{x}) + \delta^{\frac{1}{2}} \sum_{j=1}^{q} \sqrt{\theta_j} U_j(\boldsymbol{x}), \tag{4.42}$$

where $\zeta_\nu \overset{iid}{\sim} \text{N}(0, \kappa)$; $U_j(\boldsymbol{x})$ are independent, zero-mean Gaussian stochastic processes with covariance function $R^j(\boldsymbol{x}, \boldsymbol{z})$; ζ_ν and $U_j(\boldsymbol{x})$ are mutually independent; and κ and δ are positive constants. Suppose observations are generated from

$$y_i = \mathcal{L}_i F + \epsilon_i, \quad i = 1, \ldots, n, \tag{4.43}$$

where $\epsilon_i \overset{iid}{\sim} \text{N}(0, \sigma^2)$.

Let $\mathcal{L}_0$ be a bounded linear functional on $\mathcal{M}$. Let $\lambda = \sigma^2/n\delta$. The same arguments in Section 3.6 hold when $M = \Sigma + n\lambda I$ is replaced by $M = \Sigma_{\boldsymbol{\theta}} + n\lambda I$ in this chapter. Therefore,

$$\lim_{\kappa \to \infty} \text{E}(\mathcal{L}_0 F | \boldsymbol{y}) = \mathcal{L}_0 \hat{f}.$$

That is, the PLS estimate $\hat{f}$ is a Bayes estimate. Furthermore, we have the following extension of the GML criterion:

$$\text{GML}(\lambda, \boldsymbol{\theta}) \triangleq \frac{\boldsymbol{y}^T(I - H(\lambda, \boldsymbol{\theta}))\boldsymbol{y}}{[\det^+((I - H(\lambda, \boldsymbol{\theta})))]^{\frac{1}{n-p}}}. \tag{4.44}$$

All three forms of LME models for the SSR model in Section 3.5 can be extended for the SS ANOVA model. We present the extension of (3.35) only. Let $\Sigma_k = Z_k Z_k^T$, where Z_k is an $n \times m_k$ matrix with $m_k = \text{rank}(\Sigma_k)$. It is not difficult to see that the GML criterion is the REML criterion based on the following linear mixed-effects model

$$\boldsymbol{y} = T\boldsymbol{\zeta} + \sum_{k=1}^{q} Z_k \boldsymbol{u}_k + \boldsymbol{\epsilon}, \tag{4.45}$$

where $\boldsymbol{\zeta} = (\zeta_1, \ldots, \zeta_p)^T$ are deterministic parameters, $\boldsymbol{u}_k$ are mutually independent random effects, $\boldsymbol{u}_k \sim \text{N}(\boldsymbol{0}, \sigma^2 \theta_k I_{m_k}/n\lambda)$, $\boldsymbol{\epsilon} \sim \text{N}(\boldsymbol{0}, \sigma^2 I)$, and $\boldsymbol{u}_k$ are independent of $\boldsymbol{\epsilon}$. Details can be found in Chapter 9.

4.8 Confidence Intervals

Any function $f \in \mathcal{M}$ can be represented as

$$f = \sum_{\nu=1}^{p} f_{0\nu} + \sum_{j=1}^{q} f_{1j}, \qquad (4.46)$$

where $f_{0\nu} \in \text{span}\{\phi_\nu\}$ for $\nu = 1, \ldots, p$, and $f_{1j} \in \mathcal{H}_j$ for $j = 1, \ldots, q$. Our goal is to construct Bayesian confidence intervals for

$$\mathcal{L}_0 f \boldsymbol{\gamma} = \sum_{\nu=1}^{p} \gamma_\nu \mathcal{L}_0 f_{0\nu} + \sum_{j=1}^{q} \gamma_{p+j} \mathcal{L}_0 f_{1j} \qquad (4.47)$$

for any bounded linear functional $\mathcal{L}_0$ and any $\boldsymbol{\gamma} = (\gamma_1, \ldots, \gamma_{p+q})^T$, where $\gamma_k = 1$ when the corresponding component in (4.46) is to be included and 0 otherwise.

Let $F_{0\nu} = \zeta_\nu \phi_\nu$ for $\nu = 1, \ldots, p$, and $F_{1j} = \sqrt{\delta \theta_j} U_j$ for $j = 1, \ldots, q$. Let $\mathcal{L}_0$, $\mathcal{L}_{01}$, and $\mathcal{L}_{02}$ be bounded linear functionals.

Posterior means and covariances
For $\nu, \mu = 1, \ldots, p$ and $j, k = 1, \ldots, q$, the posterior means are

$$\begin{aligned} E(\mathcal{L}_0 F_{0\nu}|\boldsymbol{y}) &= (\mathcal{L}_0 \phi_\nu) \boldsymbol{e}_\nu^T \boldsymbol{d}, \\ E(\mathcal{L}_0 F_{1j}|\boldsymbol{y}) &= \theta_j (\mathcal{L}_0 \boldsymbol{\xi}_j)^T \boldsymbol{c}, \end{aligned} \qquad (4.48)$$

and the posterior covariances are

$$\begin{aligned} \delta^{-1} Cov(\mathcal{L}_{01} F_{0\nu}, \mathcal{L}_{02} F_{0\mu}|\boldsymbol{y}) &= (\mathcal{L}_{01}\phi_\nu)(\mathcal{L}_{02}\phi_\mu)\boldsymbol{e}_\nu^T A \boldsymbol{e}_\mu, \\ \delta^{-1} Cov(\mathcal{L}_{01} F_{0\nu}, \mathcal{L}_{02} F_{1j}|\boldsymbol{y}) &= -\theta_j(\mathcal{L}_{01}\phi_\nu)\boldsymbol{e}_\nu^T B(\mathcal{L}_{02}\boldsymbol{\xi}_j), \\ \delta^{-1} Cov(\mathcal{L}_{01} F_{1j}, \mathcal{L}_{02} F_{1k}|\boldsymbol{y}) &= \delta_{j,k}\theta_j \mathcal{L}_{01}\mathcal{L}_{02} R^j - \theta_j\theta_k(\mathcal{L}_{01}\boldsymbol{\xi}_j)^T C(\mathcal{L}_{02}\boldsymbol{\xi}_k), \end{aligned} \qquad (4.49)$$

where $\boldsymbol{e}_\nu$ is a vector of dimension p with the νth element being one and all other elements being zero, $\delta_{j,k}$ is the Kronecker delta, $\boldsymbol{c}$ and $\boldsymbol{d}$ are solutions to (4.37), $\mathcal{L}\boldsymbol{\xi}_j = (\mathcal{L}\mathcal{L}_1 R^j, \ldots, \mathcal{L}\mathcal{L}_n R^j)^T$ for any well-defined $\mathcal{L}$, $M = \Sigma_{\boldsymbol{\theta}} + n\lambda I$, $A = (T^T M^{-1} T)^{-1}$, $B = A T^T M^{-1}$, and $C = M^{-1}(I - B)$.

As in Section 3.8.1, even though not explicitly expressed, a diffuse prior is assumed for ζ with $\kappa \to \infty$. The first two equations in (4.48) state that the projections of $\hat{f}$ on subspaces are the posterior means of the corresponding components in the Bayes model (4.42). The next three equations in (4.49) can be used to compute posterior covariances of the spline estimates and their projections. Based on these posterior covariances, we construct Bayesian confidence intervals for the overall function f and its components in (4.46). Specifically, posterior mean and variance for $\mathcal{L}_0 f_{\boldsymbol{\gamma}}$ in (4.47) can be calculated using the formulae in (4.48) and (4.49). Then $100(1 - \alpha)\%$ Bayesian confidence interval for $\mathcal{L}_0 f_{\boldsymbol{\gamma}}$ is

$$\mathrm{E}\{\mathcal{L}_0 F_{\boldsymbol{\gamma}} | \boldsymbol{y}\} \pm z_{\frac{\alpha}{2}} \sqrt{\mathrm{Var}\{\mathcal{L}_0 F_{\boldsymbol{\gamma}} | \boldsymbol{y}\}}$$

where

$$F_{\boldsymbol{\gamma}}(\boldsymbol{x}) = \sum_{\nu=1}^{p} \gamma_{\nu} F_{0\nu}(\boldsymbol{x}) + \sum_{j=1}^{q} \gamma_{p+j} F_{1j}(\boldsymbol{x}).$$

Confidence intervals for a collection of points can be constructed similarly.

The same approach in Section 3.8.2 can be used to construct bootstrap confidence intervals. The extension is straightforward.

4.9 Examples

4.9.1 Tongue Shapes

Consider the ultrasound data. Let y be the response variable `height`; x_1 be the index of `environment` with $x_1 = 1, 2, 3$ corresponding to *2words*, *cluster*, and *schwa* respectively; x_2 be the variable `length` scaled into $[0, 1]$; and x_3 be the variable `time` scaled into $[0, 1]$.

We first investigate how tongue shapes for an articulation differ under different environments at a particular time, say, at time 60 ms. Observations are shown in Figure 4.2. Consider a bivariate regression function $f(x_1, x_2)$ where x_1 (`environment`) is a discrete variable with three levels, and x_2 (`length`) is a continuous variable in $[0, 1]$. Therefore, we model the joint function using the tensor product space $\mathbb{R}^3 \otimes W_2^m[0, 1]$. The SS ANOVA decompositions of $\mathbb{R}^a \otimes W_2^m[0, 1]$ are given in (4.16) for $m = 1$ and (4.18) for $m = 2$.

The following statements fit the SS ANOVA model (4.16):

```
> data(ultrasound)
> ultrasound$y <- ultrasound$height
> ultrasound$x1 <- ultrasound$env
> ultrasound$x2 <- ident(ultrasound$length)
> ssr(y~1, data=ultrasound, subset=ultrasound$time==60,
```

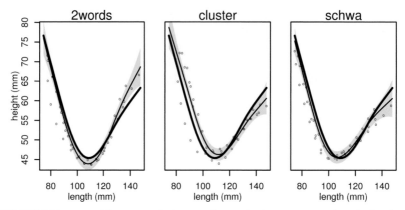

FIGURE 4.2 Ultrasound data, plots of observations (circles), fits (thin lines), and 95% Bayesian confidence intervals (shaded regions), and the mean curves among three environments (thicker lines). The tongue root is on the left, and the tongue tip is on the right in each plot.

```
rk=list(shrink1(x1),
        linear(x2),
        rk.prod(shrink1(x1),linear(x2))))
```

where the `ident` function transforms a variable into $[0, 1]$, the `shrink1` function computes the RK of $\mathcal{H}_1^{(1)} = \{f : \sum_{i=1}^{3} f(i) = 0\}$ using the formula in (4.6), and the `rk.prod` function computes the product of RKs.

The following statements fit the SS ANOVA model (4.18) and summarize the fit:

```
> ultra.el.c.fit <- ssr(y~x2, data=ultrasound,
    subset=ultrasound$time==60,
    rk=list(shrink1(x1),
            cubic(x2),
            rk.prod(shrink1(x1),kron(x2-.5)),
            rk.prod(shrink1(x1),cubic(x2))))
> summary(ultra.el.c.fit)
...
GCV estimate(s) of smoothing parameter(s) : 6.061108e-02
  2.650966e-05 1.557379e-03 3.177218e-05
Equivalent Degrees of Freedom (DF):  13.57631
Estimate of sigma:  2.654545
```

where the function `kron` computes the RK, $k_1(x_2)k_1(z_2)$, of the space $\mathcal{H}_1^{(2)} = \{k_1(x_2)\}$.

Estimates of the smoothing parameters λ_3 and λ_4 are small, which indicates that both interactions $\gamma_{x_1} \times (x_2 - 0.5)$ and $f_{12}^{ss}(x_1, x_2)$ may not be negligible. We compute the posterior mean and standard deviation for the overall interaction $f_{12}(x_1, x_2) = \gamma_{x_1} \times (x_2 - 0.5) + f_{12}^{ss}(x_1, x_2)$ on grid points as follows:

```
> grid <- seq(0,1,len=40)
> predict(ultra.el.c.fit, terms=c(0,0,0,0,1,1),
    newdata=expand.grid(x2=grid,x1=as.factor(1:3)))
```

The overall interaction and its 95% Bayesian confidence intervals are shown in Figure 4.3. It is clear that the interaction is nonzero.

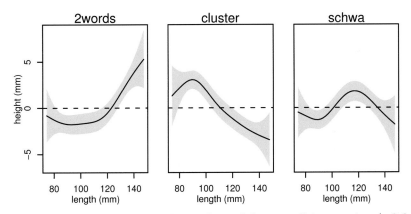

FIGURE 4.3 Ultrasound data, plots of the overall interaction (solid lines), and 95% Bayesian confidence intervals (shaded regions). Dashed line in each plot represents the constant function zero.

The posterior mean and standard deviation of the function $f(x_1, x_2)$ can be calculated as follows:

```
> predict(ultra.el.c.fit, terms=c(1,1,1,1,1,1),
    newdata=expand.grid(x2=grid,x1=as.factor(1:3)))
```

The default for the option `terms` is a vector of all 1's. Therefore, this option can be dropped in the above statement. The fits and 95% Bayesian confidence intervals are shown in Figure 4.2.

Note that in model (4.18) the first three terms represent the mean curve among three environments and the last three terms represent the departure of a particular environment from the mean curve. For comparison, we compute the estimate of the mean curve among three environments:

```
> predict(ultra.el.c.fit, terms=c(1,1,0,1,0,0),
    newdata=expand.grid(x2=grid,x1=as.factor(1)))
```

The estimate of the mean curve is also displayed in Figure 4.2. The difference between the tongue shape under a particular environment and the average tongue shape can be made by comparing two lines in each plot. To look at the effect of each environment more closely, we calculate the estimate of the departure from the mean curve for each environment:

```
> predict(ultra.el.c.fit, terms=c(0,0,1,0,1,1),
    newdata=expand.grid(x2=grid,x1=as.factor(1:3)))
```

The estimates of environment effects are shown in Figure 4.4. We can see that, comparing to the average shape, the tongue shape for *2words* is front-raising, and the tongue shape for *cluster* is back-raising. The tongue shape for *schwa* is close to the average shape.

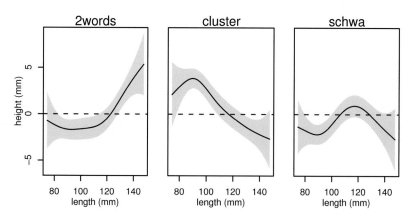

FIGURE 4.4 Ultrasound data, plots of effects of environment, and 95% Bayesian confidence intervals. The dashed line in each plot represents the constant function zero.

The model space of the SS ANOVA model (4.18) is $\mathcal{M} = \mathcal{H}^0 \oplus \mathcal{H}^1 \oplus \mathcal{H}^2 \oplus \mathcal{H}^3 \oplus \mathcal{H}^4$, where $\mathcal{H}^j$ for $j = 0, \ldots, 4$ are defined in (4.19). In particular, the spaces $\mathcal{H}^3$ and $\mathcal{H}^4$ contain smooth–linear and smooth–smooth interactions between x_1 and x_2. For illustration, suppose now that we want to fit model (4.18) with the same smoothing parameter for penalties to functions in $\mathcal{H}^3$ and $\mathcal{H}^4$. That is to set $\lambda_3 = \lambda_4$ in the PLS. As discussed in Section 4.6, this can be achieved by combining $\mathcal{H}^3$ and $\mathcal{H}^4$ into one space. The following statements fit the SS ANOVA model (4.18) with $\lambda_3 = \lambda_4$:

```
> ultra.el.c.fit1 <- ssr(y~x2, data=ultra,
    subset=ultra$time==60,
    rk=list(shrink1(x1),
            cubic(x2),
            rk.prod(shrink1(x1),kron(x2-.5))+
            rk.prod(shrink1(x1),cubic(x2))))
> summary(ultra.el.c.fit1)
...
GCV estimate(s) of smoothing parameter(s) : 5.648863e-02
    2.739598e-05 3.766634e-05
Equivalent Degrees of Freedom (DF):  13.52603
Estimate of sigma:  2.65806
```

Next we investigate how the tongue shapes change over time for each environment. Figure 4.1 shows 3-d plots of observations. For a fixed environment, consider a bivariate regression function $f(x_2, x_3)$ where both x_2 and x_3 are continuous variables. Therefore, we model the joint function using the tensor product space $W_2^{m_1}[0,1] \otimes W_2^{m_2}[0,1]$. The SS ANOVA decompositions of $W_2^{m_1}[0,1] \otimes W_2^{m_2}[0,1]$ were presented in Section 4.4.3. Note that variables x_2 and x_3 in this section correspond to x_1 and x_2 in Section 4.4.3.

The following statements fit the tensor product of linear splines with $m_1 = m_2 = 1$, that is, the SS ANOVA model (4.20), under environment *2words* ($x_1 = 1$):

```
> ultrasound$x3 <- ident(ultrasound$time)
> ssr(height~1, data=ultrasound, subset=ultrasound$env==1,
    rk=list(linear(x2),
            linear(x3),
            rk.prod(linear(x2),linear(x3))))
```

The following statements fit the tensor product of cubic splines with $m_1 = m_2 = 2$, that is, the SS ANOVA model (4.21), under environment *2words* and calculate estimates at grid points:

```
> ultra.lt.c.fit[[1]] <- ssr(height~x2+x3+x2*x3,
    data=ultrasound, subset=ultrasound$env==1,
```

```
rk=list(cubic(x2),
        cubic(x3),
        rk.prod(kron(x2),cubic(x3)),
        rk.prod(cubic(x2),kron(x3)),
        rk.prod(cubic(x2),cubic(x3))))
> grid <- seq(0,1,len=20)
> ultra.lt.c.pred <- predict(ultra.lt.c.fit[[1]],
  newdata=expand.grid(x2=grid,x3=grid))
```

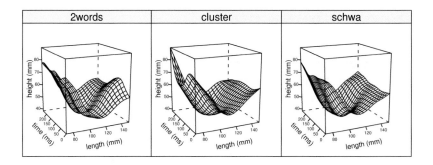

FIGURE 4.5 Ultrasound data, 3-d plots of the estimated tongue shapes as functions of `length` and `time` based on the SS ANOVA model (4.21).

The SS AONVA model (4.21) for environments *cluster* and *schwa* can be fitted similarly. The estimates of all three environments are shown in Figure 4.5. These surfaces show how the tongue shapes change over time. Note that $\mu + \beta_1 \times (x_2 - 0.5) + f_1^s(x_2)$ represents the mean tongue shape over the time period $[0, 140]$, and the rest in (4.21), $\beta_2 \times (x_3 - 0.5) + f_2^s(x_3) + \beta_3 \times (x_2 - 0.5) \times (x_3 - 0.5) + f_{12}^{ls}(x_2, x_3) + f_{12}^{sl}(x_2, x_3) + f_{12}^{ss}(x_2, x_3)$, represents the departure from the mean shape at time x_3. To look at the time effect, we compute posterior means and standard deviations of the departure on grid points:

```
> predict(ultra.lt.c.fit[[1]], term=c(0,0,1,1,0,1,1,1,1),
  newdata=expand.grid(x2=grid,x3=grid))
```

Figure 4.6 shows the contour plots of the estimated time effect for three environments. Regions where the lower bounds of the 95% Bayesian confidence intervals are positive are shaded in dark grey, while regions where the upper bounds of the 95% Bayesian confidence intervals are negative are shaded in light grey.

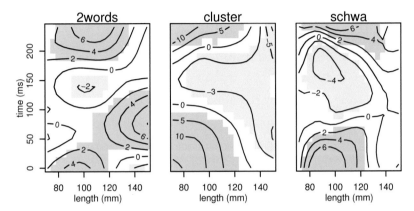

FIGURE 4.6 Ultrasound data, contour plots of the estimated time effect for three environments based on the SS ANOVA model (4.21). Regions where the lower bounds of the 95% Bayesian confidence intervals are positive are shaded in dark grey. Regions where the upper bounds of the 95% Bayesian confidence intervals are negative are shaded in light grey.

Finally we investigate how the changes of tongue shapes over time differ among different environments. Consider a trivariate regression function $f(x_1, x_2, x_3)$ in tensor product space $\mathbb{R}^3 \otimes W_2^{m_1}[0,1] \otimes W_2^{m_2}[0,1]$. For simplicity, we derive the SS ANOVA decomposition for $m_1 = m_2 = 2$ only. Define averaging operators:

$$\mathcal{A}_1^{(1)} f = \frac{1}{3} \sum_{x_1=1}^{3} f,$$

$$\mathcal{A}_1^{(k)} f = \int_0^1 f \, dx_k,$$

$$\mathcal{A}_2^{(k)} f = \left(\int_0^1 f' \, dx_k \right) (x_k - 0.5), \quad k = 2, 3,$$

where $\mathcal{A}_1^{(1)}$, $\mathcal{A}_1^{(2)}$, and $\mathcal{A}_1^{(3)}$ extract the constant function out of all possible functions for each variable, and $\mathcal{A}_2^{(2)}$ and $\mathcal{A}_2^{(3)}$ extract the linear function for x_2 and x_3. Let $\mathcal{A}_2^{(1)} = I - \mathcal{A}_1^{(1)}$ and $\mathcal{A}_3^{(k)} = I - \mathcal{A}_1^{(k)} - \mathcal{A}_2^{(k)}$ for $k = 2, 3$. Then

$$
\begin{aligned}
f &= \{\mathcal{A}_1^{(1)} + \mathcal{A}_2^{(1)}\}\{\mathcal{A}_1^{(2)} + \mathcal{A}_2^{(2)} + \mathcal{A}_3^{(2)}\}\{\mathcal{A}_1^{(3)} + \mathcal{A}_2^{(3)} + \mathcal{A}_3^{(3)}\}f \\
&= \mathcal{A}_1^{(1)}\mathcal{A}_1^{(2)}\mathcal{A}_1^{(3)}f + \mathcal{A}_1^{(1)}\mathcal{A}_1^{(2)}\mathcal{A}_2^{(3)}f + \mathcal{A}_1^{(1)}\mathcal{A}_1^{(2)}\mathcal{A}_3^{(3)}f \\
&\quad + \mathcal{A}_1^{(1)}\mathcal{A}_2^{(2)}\mathcal{A}_1^{(3)}f + \mathcal{A}_1^{(1)}\mathcal{A}_2^{(2)}\mathcal{A}_2^{(3)}f + \mathcal{A}_1^{(1)}\mathcal{A}_2^{(2)}\mathcal{A}_3^{(3)}f \\
&\quad + \mathcal{A}_1^{(1)}\mathcal{A}_3^{(2)}\mathcal{A}_1^{(3)}f + \mathcal{A}_1^{(1)}\mathcal{A}_3^{(2)}\mathcal{A}_2^{(3)}f + \mathcal{A}_1^{(1)}\mathcal{A}_3^{(2)}\mathcal{A}_3^{(3)}f \\
&\quad + \mathcal{A}_2^{(1)}\mathcal{A}_1^{(2)}\mathcal{A}_1^{(3)}f + \mathcal{A}_2^{(1)}\mathcal{A}_1^{(2)}\mathcal{A}_2^{(3)}f + \mathcal{A}_2^{(1)}\mathcal{A}_1^{(2)}\mathcal{A}_3^{(3)}f \\
&\quad + \mathcal{A}_2^{(1)}\mathcal{A}_2^{(2)}\mathcal{A}_1^{(3)}f + \mathcal{A}_2^{(1)}\mathcal{A}_2^{(2)}\mathcal{A}_2^{(3)}f + \mathcal{A}_2^{(1)}\mathcal{A}_2^{(2)}\mathcal{A}_3^{(3)}f \\
&\quad + \mathcal{A}_2^{(1)}\mathcal{A}_3^{(2)}\mathcal{A}_1^{(3)}f + \mathcal{A}_2^{(1)}\mathcal{A}_3^{(2)}\mathcal{A}_2^{(3)}f + \mathcal{A}_2^{(1)}\mathcal{A}_3^{(2)}\mathcal{A}_3^{(3)}f \\
&\triangleq \mu + \beta_2 \times (x_3 - 0.5) + f_3^s(x_3) \\
&\quad + \beta_1 \times (x_2 - 0.5) + \beta_3 \times (x_2 - 0.5)(x_3 - 0.5) + f_{23}^{ls}(x_2, x_3) \\
&\quad + f_2^s(x_2) + f_{23}^{sl}(x_2, x_3) + f_{23}^{ss}(x_2, x_3) \\
&\quad + f_1(x_1) + f_{13}^{sl}(x_1, x_3) + f_{13}^{ss}(x_1, x_3) \\
&\quad + f_{12}^{sl}(x_1, x_2) + f_{123}^{sll}(x_1, x_2, x_3) + f_{123}^{sls}(x_1, x_2, x_3) \\
&\quad + f_{12}^{ss}(x_1, x_2) + f_{123}^{ssl}(x_1, x_2, x_3) + f_{123}^{sss}(x_1, x_2, x_3), \qquad (4.50)
\end{aligned}
$$

where μ represents the overall mean; $f_1(x_1)$ represents the main effect of x_1; $\beta_1 \times (x_2 - 0.5)$ and $\beta_2 \times (x_3 - 0.5)$ represent the linear main effects of x_2 and x_3; $f_2^s(x_2)$ and $f_3^s(x_3)$ represent the smooth main effects of x_2 and x_3; $f_{12}^{sl}(x_1, x_2)$ ($f_{13}^{sl}(x_1, x_3)$) represents the smooth–linear interaction between x_1 and x_2 (x_3); $\beta_3 \times (x_2 - 0.5) \times (x_3 - 0.5)$, $f_{23}^{ls}(x_2, x_3)$, $f_{23}^{sl}(x_2, x_3)$ and $f_{23}^{ss}(x_2, x_3)$ represent linear–linear, linear–smooth, smooth–linear and smooth–smooth interactions between x_2 and x_3; and $f_{123}^{sll}(x_1, x_2, x_3)$, $f_{123}^{sls}(x_1, x_2, x_3)$, $f_{123}^{ssl}(x_1, x_2, x_3)$, and $f_{123}^{sss}(x_1, x_2, x_3)$ represent three-way interactions between x_1, x_2, and x_3. The overall main effect of x_k, $f_k(x_k)$, equals $\beta_{k-1} \times (x_k - 0.5) + f_k^s(x_k)$ for $k = 2, 3$. The overall interaction between x_1 and x_k, $f_{1k}(x_1, x_k)$, equals $f_{12}^{sl}(x_1, x_k) + f_{1k}^{ss}(x_1, x_k)$ for $k = 2, 3$. The overall interaction between x_2 and x_3, $f_{23}(x_2, x_3)$, equals $\beta_3 \times (x_2 - 0.5) \times (x_3 - 0.5) + f_{23}^{ls}(x_2, x_3) + f_{23}^{sl}(x_2, x_3) + f_{23}^{ss}(x_2, x_3)$. The overall three-way interaction, $f_{123}(x_1, x_2, x_3)$, equals $f_{123}^{sll}(x_1, x_2, x_3) + f_{123}^{sls}(x_1, x_2, x_3) + f_{123}^{ssl}(x_1, x_2, x_3) + f_{123}^{sss}(x_1, x_2, x_3)$.

We fit model (4.50) as follows:

```
> ssr(height~I(x2-.5)+I(x3-.5)+I((x2-.5)*(x3-.5)),
      data=ultrasound,
      rk=list(shrink1(x1),
```

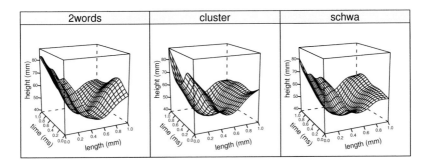

FIGURE 4.7 Ultrasound data, 3-d plots of the estimated tongue shape as a function of `environment`, `length` and `time` based on the SS ANOVA model (4.50).

```
cubic(x2),
cubic(x3),
rk.prod(shrink1(x1),kron(x2-.5)),
rk.prod(shrink1(x1),cubic(x2)),
rk.prod(shrink1(x1),kron(x3-.5)),
rk.prod(shrink1(x1),cubic(x3)),
rk.prod(cubic(x2),kron(x3-.5)),
rk.prod(kron(x2-.5),cubic(x3)),
rk.prod(cubic(x2),cubic(x3)),
rk.prod(shrink1(x1),kron(x2-.5),kron(x3-.5)),
rk.prod(shrink1(x1),kron(x2-.5),cubic(x3)),
rk.prod(shrink1(x1),cubic(x2),kron(x3-.5)),
rk.prod(shrink1(x1),cubic(x2),cubic(x3))))
```

The estimates of all three environments are shown in Figure 4.7. Note that the first nine terms in (4.50) represent the mean tongue shape surface over time, and the last nine terms in (4.50) represent the departure of an environment from this mean surface. To look at the environment effect on the tongue shape surface over time, we calculate the posterior mean and standard deviation of the departure for each environment:

```
> pred <- predict(ultra.elt.c.fit,
    newdata=expand.grid(x1=as.factor(1:3),x2=grid,x3=grid),
    terms=c(0,0,0,0,1,0,0,1,1,1,1,0,0,0,1,1,1,1))
```

The contour plots of the estimated departures for three environments are shown in Figure 4.8. Note that the significant regions at time 60 ms are similar to those in Figure 4.4.

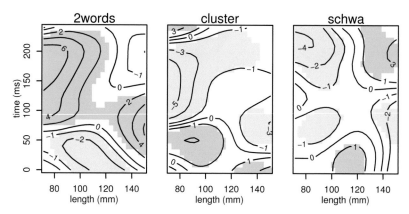

FIGURE 4.8 Ultrasound data, contour plots of estimated environment effect for three environments in the SS ANOVA model (4.50). Regions where the lower bounds of the 95% Bayesian confidence intervals are positive are shaded in dark grey. Regions where the upper bounds of the 95% Bayesian confidence intervals are negative are shaded in light grey.

4.9.2   Ozone in Arosa — Revisit

Suppose we want to investigate how ozone thickness changes over time by considering the effects of both `month` and `year`. In Section 2.10 we fitted an additive model (2.49) with the `month` effect modeled parametrically by a simple sinusoidal function, and the `year` effect modeled nonparametrically by a cubic spline. The following analyses show how to model both effects nonparametrically and investigate their interaction using SS ANOVA decompositions. We also show how to check the partial spline model (2.49).

Let y be the response variable `thick`, x_1 be the independent variable `month` scaled into $[0, 1]$, and x_2 be the independent variable `year` scaled into $[0, 1]$. It is reasonable to assume that the mean ozone thickness is a periodic function of x_1. We model the effect of x_1 using a periodic spline space $W_2^2(per)$ and the effect of x_2 using a cubic spline space $W_2^2[0, 1]$.

Therefore, we consider the SS ANOVA decomposition (4.23) for the tensor product space $W_2^{m_1}(per) \otimes W_2^2[0,1]$. The following statements fit model (4.23) with $m_1 = 2$:

```
> Arosa$x1 <- (Arosa$month-0.5)/12
> Arosa$x2 <- (Arosa$year-1)/45
> arosa.ssanova.fit1 <- ssr(thick~I(x2-0.5), data=Arosa,
    rk=list(periodic(x1),
            cubic(x2),
            rk.prod(periodic(x1),kron(x2-.5)),
            rk.prod(periodic(x1),cubic(x2))))
> summary(arosa.ssanova.fit1)
...
GCV estimate(s) of smoothing parameter(s) : 5.442106e-06
  2.154531e-09 3.387917e-06 2.961559e-02
Equivalent Degrees of Freedom (DF):  50.88469
Estimate of sigma:  14.7569
```

The mean function $f(x)$ in model (4.23) evaluated at design points $f = (f(x_1), \ldots, f(x_n))^T$ can be represented as

$$f = \mu 1 + f_1 + f_2 + f_{12},$$

where 1 is a vector of all ones, $f_1 = (f_1(x_1), \ldots, f_1(x_n))^T$, $f_2 = (f_2(x_1), \ldots, f_2(x_n))^T$, $f_{12} = (f_{12}(x_1), \ldots, f_{12}(x_n))^T$, and $f_1(x)$, $f_2(x)$ and $f_{12}(x_1)$ are the main effect of x_1, the main effect of x_2, and the interaction between x_1 and x_2. Eliminating the constant term, we have

$$f^* = f_1^* + f_2^* + f_{12}^*,$$

where $a^* = a - \bar{a}1$, and $\bar{a} = \sum_{i=1}^n a_i/n$. Let $\hat{f}^*$, $\hat{f}_1^*$, $\hat{f}_2^*$, and $\hat{f}_{12}^*$ be the estimates of f^*, f_1^*, f_2^*, and f_{12}^*, respectively. To check the contributions of the main effects and interaction, we compute the quantities $\pi_k = (\hat{f}_k^*)^T \hat{f}^* / \|\hat{f}^*\|^2$ for $k = 1, 2, 12$, and the Euclidean norms of $\hat{f}^*$, $\hat{f}_1^*$, $\hat{f}_2^*$, and $\hat{f}_{12}^*$:

```
> f1 <- predict(arosa.ssanova.fit1, terms=c(0,0,1,0,0,0))
> f2 <- predict(arosa.ssanova.fit1, terms=c(0,1,0,1,0,0))
> f12 <- predict(arosa.ssanova.fit1, terms=c(0,0,0,0,1,1))
> fs1 <- scale(f1$fit, scale=F)
> fs2 <- scale(f2$fit, scale=F)
> fs12 <- scale(f12$fit, scale=F)
> ys <- fs1+fs2+fs12
> pi1 <- sum(fs1*ys)/sum(ys**2)
> pi2 <- sum(fs2*ys)/sum(ys**2)
```

```
> pi12 <- sum(fs12*ys)/sum(ys**2)
> print(round(c(pi1,pi2,pi12),4))
   0.9375 0.0592 0.0033
> print(round(sqrt(c(sum(fs1**2),sum(fs2**2),
                     sum(fs12**2))),2))
   768.31 186.21  30.25
```

See Gu (2002) for more details about the cosine diagnostics. It is clear that the contribution from the interaction to the total variation is negligible. We also compute the posterior mean and standard deviation of the interaction f_{12}:

```
> grid <- seq(0,1,len=50)
> predict(arosa.ssanova.fit1, terms=c(0,0,0,0,1,1),
    expand.grid(x1=grid,x2=grid))
```

The estimate of the interaction is shown in Figure 4.9(a). Except for a narrow region, the zero function is contained in the 95% Bayesian confidence intervals. Thus the interaction is negligible.

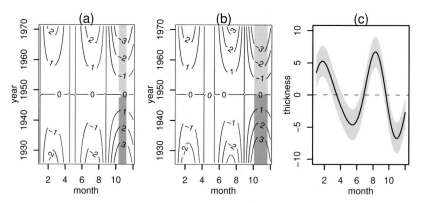

FIGURE 4.9 Arosa data, (a) plot of estimate of the interaction in model (4.23), (b) plot of estimate of the interaction in model (4.52), and (c) plot of estimate of the smooth component f_1^s in model (4.53). For plots (a) and (b), regions where the lower bounds of the confidence intervals are positive are shaded in dark grey, while regions where the upper bounds of the confidence intervals are negative are shaded in light grey. For plot (c), the solid line represents the estimate of f_1^s, the shaded region represents 95% Bayesian confidence intervals, and the dashed line represents the zero function.

We drop the interaction term and fit an additive model

$$f(x_1, x_2) = \mu + \beta \times (x_2 - 0.5) + f_2^s(x_2) + f_1(x_1) \tag{4.51}$$

as follows:

```
> update(arosa.ssanova.fit1,
          rk=list(periodic(x1),cubic(x2)))
```

The estimates of two main effects are shown in Figure 4.10.

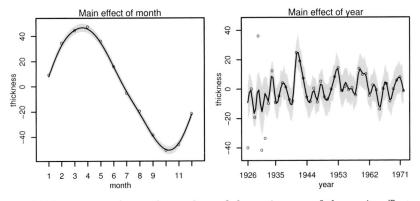

FIGURE 4.10   Arosa data, plots of the estimates of the main effects in the SS ANOVA model (4.51). Circles in the left panel represent monthly average thickness minus the overall mean. Circles in the right panel represent yearly average thickness minus the overall mean. Shaded regions are 95% Bayesian confidence intervals.

Now suppose we want to check if the partial spline model (2.49) is appropriate for the Arosa data. For the `month` effect, consider the trigonometric spline model $W_2^3(per)$ with $L = D\{D^2 + (2\pi)^2\}$ defined in Section 2.11.5. The null space span$\{1, \sin(2\pi x), \cos(2\pi x)\}$ corresponds to the sinusoidal model space assumed for the partial spline model (2.49). Consider the tensor product space $W_2^3(per) \otimes W_2^2[0, 1]$.   Define averaging operators:

$$A_1^{(1)} f = \int_0^1 f \, dx_1,$$

$$A_2^{(1)} f = \int_0^1 f \sin(2\pi x_1) \, dx_1,$$

$$A_3^{(1)} f = \int_0^1 f \cos(2\pi x_1) \, dx_1,$$

$$A_1^{(2)} f = \int_0^1 f \, dx_2,$$

$$A_2^{(2)} f = \left(\int_0^1 f' \, dx_2 \right) (x_2 - 0.5).$$

Let $A_4^{(1)} = I - A_1^{(1)} - A_2^{(1)} - A_3^{(1)}$ and $A_3^{(2)} = I - A_1^{(2)} - A_2^{(2)}$. Then

$$
\begin{aligned}
f &= \{A_1^{(1)} + A_2^{(1)} + A_3^{(1)} + A_4^{(1)}\}\{A_1^{(2)} + A_2^{(2)} + A_3^{(2)}\} f \\
&= A_1^{(1)} A_1^{(2)} f + A_2^{(1)} A_1^{(2)} f + A_3^{(1)} A_1^{(2)} f + A_4^{(1)} A_1^{(2)} f \\
&\quad + A_1^{(1)} A_2^{(2)} f + A_2^{(1)} A_2^{(2)} f + A_3^{(1)} A_2^{(2)} f + A_4^{(1)} A_2^{(2)} f \\
&\quad + A_1^{(1)} A_3^{(2)} f + A_2^{(1)} A_3^{(2)} f + A_3^{(1)} A_3^{(2)} f + A_4^{(1)} A_3^{(2)} f \\
&\triangleq \mu + \beta_1 \times \sin(2\pi x_1) + \beta_2 \times \cos(2\pi x_1) + f_1^s(x_1) \\
&\quad + \beta_3 \times (x_2 - 0.5) + \beta_4 \times \sin(2\pi x_1) \times (x_2 - 0.5) \\
&\quad + \beta_5 \times \cos(2\pi x_1) \times (x_2 - 0.5) + f_{12}^{sl}(x_1, x_2) \\
&\quad + f_2^s(x_2) + f_{12}^{1s}(x_1, x_2) + f_{12}^{2s}(x_1, x_2) + f_{12}^{ss}(x_1, x_2), \qquad (4.52)
\end{aligned}
$$

where μ represents the overall mean; $\beta_1 \times \sin(2\pi x_1)$ and $\beta_2 \times \cos(2\pi x_1)$ represent the parametric main effects of x_1; $f_1^s(x_1)$ represents the smooth main effect of x_1; $\beta_3 \times (x_2 - 0.5)$ represents the linear main effect of x_2; $f_2^s(x_2)$ represents the smooth main effects of x_2; and $\beta_4 \times \sin(2\pi x_1) \times (x_2 - 0.5)$, $\beta_5 \times \cos(2\pi x_1) \times (x_2 - 0.5)$, $f_{12}^{sl}(x_1, x_2)$, $f_{12}^{1s}(x_1, x_2)$, $f_{12}^{2s}(x_1, x_2)$, and $f_{12}^{ss}(x_1, x_2)$ represent interactions. We fit model (4.52) and compute the posterior mean and standard deviations for the overall interaction as follows:

```
> arosa.ssanova.fit3 <- ssr(thick~sin(2*pi*x1)+cos(2*pi*x1)
  +I(x2-0.5)+I(sin(2*pi*x1)*(x2-0.5))
  +I(cos(2*pi*x1)*(x2-0.5)), data=Arosa,
  rk=list(lspline(x1,type=''sine1''), cubic(x2),
   rk.prod(kron(sin(2*pi*x1)),cubic(x2)),
   rk.prod(kron(cos(2*pi*x1)),cubic(x2)),
```

```
      rk.prod(lspline(x1,type=''sine1''), kron(x2-.5)),
      rk.prod(lspline(x1,type=''sine1''), cubic(x2))))
```

```
> ngrid <- 50
> grid1 <- seq(.5/12,11.5/12,length=ngrid)
> grid2 <- seq(0,1,length=ngrid)
> predict(arosa.ssanova.fit3,
    expand.grid(x1=grid1,x2=grid2),
    terms=c(0,0,0,0,1,1,0,0,1,1,1,1))
```

The estimate of the interaction is shown in Figure 4.9(b). Except for a narrow region, the zero function is contained in the 95% Bayesian confidence intervals. Therefore, we drop interaction terms and consider the following additive model:

$$f(x_1, x_2) = \mu + \beta_1 \times \sin(2\pi x_1) + \beta_2 \times \cos(2\pi x_1) + f_1^s(x_1)$$
$$\beta_3 \times (x_2 - 0.5) + f_2^s(x_2). \tag{4.53}$$

Note that the partial spline model (2.49) is a special case of model (4.53) with $f_1^s(x_1) = 0$. We fit model (4.53) and compute posterior means and standard deviations for $f_1^s(x_1)$:

```
> arosa.ssanova.fit4 <- ssr(thick~sin(2*pi*x1)
    +cos(2*pi*x1)+I(x2-0.5), data=Aros,
    rk=list(lspline(x1,type=''sine1''), cubic(x2)))
> predict(arosa.ssanova.fit4, expand.grid(x1=grid1,x2=0),
    terms=c(0,0,0,0,1,0))
```

The estimate of $f_1^s(x_1)$ is shown in Figure 4.9(c) with 95% Bayesian confidence intervals. It is clear that $f_1^s(x_1)$ is nonzero, which indicates that the simple sinusoidal function is inadequate for modeling the month effect.

4.9.3 Canadian Weather — Revisit

Consider the Canadian weather data with annual temperature profiles from all 35 stations as functional data. To investigate how the weather patterns differ, Ramsay and Silverman (2005) divided Canada into four climatic regions: Atlantic, Continental, Pacific, and Arctic. Let y be the response variable temp, x_1 be the independent variable region, and x_2 be the independent variable month scaled into $[0, 1]$. The functional ANOVA (FANOVA) model (13.1) in Ramsay and Silverman (2005) assumes that

$$y_{k,x_1}(x_2) = \eta(x_2) + \alpha_{x_1}(x_2) + \epsilon_{k,x_1}(x_2), \tag{4.54}$$

where $y_{k,x_1}(x_2)$ is the temperature profile of station k in climate region x_1, $\eta(x_2)$ is the average temperature profile across all of Canada, $\alpha_{x_1}(x_2)$ is the departure of the region x_1 profile from the population average profile $\eta(x_2)$, and $\epsilon_{k,x_1}(x_2)$ are random errors. It is clear that the FANOVA can be derived from the SS ANOVA decomposition (4.22) by letting $\eta(x_2) = \mu + f_2(x_2)$ and $\alpha_{x_1}(x_2) = f_1(x_1) + f_{12}(x_1, x_2)$. The side condition for the FANOVA model (4.54), $\sum_{x_1=1}^{4} \alpha_{x_1}(x_2) = 0$ for all x_2, is satisfied from the construction of the SS ANOVA decomposition. Model (4.54) is an example of situation (ii) in Section 2.10 where the dependent variable involves functional data.

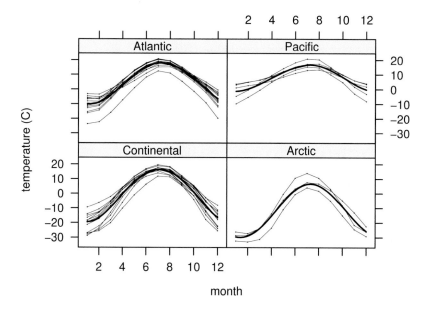

FIGURE 4.11 Canadian weather data, plots of temperature profiles of stations in four regions (thin lines), and the estimated profiles (thick lines).

Observed temperature profiles are shown in Figure 4.11. The following statements fit model (2.10) and compute posterior means and standard deviations for four regions:

```
> x1 <- rep(as.factor(region),rep(12,35))
```

```
> x2 <- (rep(1:12,35)-.5)/12
> y <- as.vector(monthlyTemp)
> canada.fit2 <- ssr(y~1,
    rk=list(shrink1(x1), periodic(x2),
            rk.prod(shrink1(x1),periodic(x2))))
> xgrid <- seq(.5/12,11.5/12,len=50)
> zone <- c(''Atlantic'',''Pacific'',
            ''Continental'',''Arctic'')
> grid <- data.frame(x1=rep(zone,rep(50,4)),
                     x2=rep(xgrid,4))
> canada.fit2.p1 <- predict(canada.fit2, newdata=grid)
```

Estimates of mean temperature functions for four regions are shown in Figure 4.11. To look at the region effects α_{x_1} more closely, we compute their posterior means and standard deviations:

```
> canada.fit2.p2 <- predict(canada.fit2, newdata=grid,
                     terms=c(0,1,0,1))
```

Estimates of region effects and 95% Bayesian confidence intervals are shown in Figure 4.12. These estimates are similar to those in Ramsay and Silverman (2005).

4.9.4 Texas Weather

Instead of dividing stations into geological regions as in model (4.54), suppose we want to investigate how the weather patterns depend on geographical locations in terms of latitude and longitude. For illustration, consider the Texas weather data consisting of average monthly temperatures during 1961–1990 from 48 weather stations in Texas. Denote $x_1 = $ (lat, long) as the geological location of a station, and x_2 as the month variable scaled into $[0, 1]$. We want to investigate how the expected temperature, $f(x_1, x_2)$, depends on both location and month variables. Average monthly temperatures are computed using monthly temperatures during 1986–1990 for all 48 stations and are used as observations of $f(x_1, x_2)$. For each fixed station, the annual temperature profile can be regarded as functional data on a continuous interval. Figure 4.13 shows these curves for all 48 stations. For each fixed month, the temperature surface as a function of latitude and longitude can be regarded as functional data on $\mathbb{R}^2$. Figure 4.14 shows contour plots of observed surfaces for January, April, July, and October.

A natural model space for the location variable is the thin-plate spline $W_2^2(\mathbb{R}^2)$, and a natural model space for the month variable is $W_2^2(per)$. Therefore, we fit the SS ANOVA model (4.24):

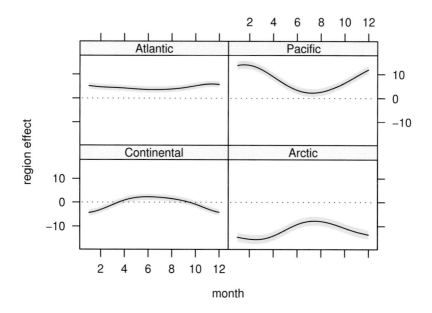

FIGURE 4.12 Canadian weather data, plots of the estimated region effects to temperature, and 95% Bayesian confidence intervals.

```
> data(TXtemp); TXtemp1 <- TXtemp[TXtemp$year>1985,]
> y <- gapply(TXtemp1, which=5,
    FUN=function(x) mean(x[x!=-99.99]),
    group=TXtemp1$stacod*TXtemp1$month)
> tx.dat <- data.frame(y=as.vector(t(matrix(y,48,12,
    byrow=F))))
> tx.dat$x2 <- rep((1:12-0.5)/12, 48)
> lat <- TXtemp$lat[seq(1, nrow(TXtemp),by=360)]
> long <- TXtemp$long[seq(1, nrow(TXtemp),by=360)]
> tx.dat$x11 <- rep(scale(lat), rep(12,48))
> tx.dat$x12 <- rep(scale(long), rep(12,48))
> tx.dat$stacod <- rep(TXtemp$stacod[seq(1,nrow(TXtemp),
    by=360)],rep(12,48))
> tx.ssanova <- ssr(y~x11+x12, data=tx.dat,
    rk=list(tp(list(x11,x12)),
            periodic(x2),
            rk.prod(tp.linear(list(x11,x12)),periodic(x2)),
            rk.prod(tp(list(x11,x12)), periodic(x2))))
```

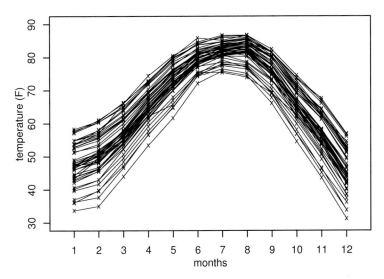

FIGURE 4.13 Texas weather data, plot of temperature profiles for all 48 stations.

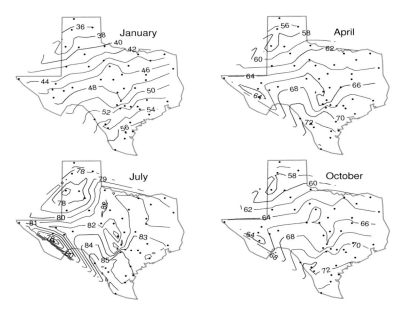

FIGURE 4.14 Texas weather data, contour plots of observations for January, April, July, and October. Dots represent station locations.

where `tp.linear` computes the RK, $\tilde{\phi}_2(\boldsymbol{x})\tilde{\phi}_2(\boldsymbol{z}) + \tilde{\phi}_3(\boldsymbol{x})\tilde{\phi}_3(\boldsymbol{z})$, of the subspace $\{\tilde{\phi}_2, \tilde{\phi}_3\}$.

The `location` effect equals $\beta_1\tilde{\phi}_1(\boldsymbol{x}_1)+\beta_2\tilde{\phi}_2(\boldsymbol{x}_1)+f_1^s(\boldsymbol{x}_1)+f_{12}^{ls}(\boldsymbol{x}_1,\boldsymbol{x}_2)+f_{12}^{ss}(\boldsymbol{x}_1,\boldsymbol{x}_2)$. We compute posterior means and standard deviations of the location effect for the southmost (Rio Grande City 3W), northmost (Stratford), westmost (El Paso WSO AP), and eastmost (Marshall) stations:

```
> selsta <- c(tx.dat[tx.dat$x11==min(tx.dat$x11),7][1],
              tx.dat[tx.dat$x11==max(tx.dat$x11),7][1],
              tx.dat[tx.dat$x12==min(tx.dat$x12),7][1],
              tx.dat[tx.dat$x12==max(tx.dat$x12),7][1])
> sellat <- sellong <- NULL
> for (i in 1:4) {
    sellat <- c(sellat,
                tx.dat$x11[tx.dat$stacod==selsta[i]][1])
    sellong <- c(sellong,
                 tx.dat$x12[tx.dat$stacod==selsta[i]][1])
  }
> grid <- data.frame(x11=rep(sellat,rep(40,4)),
                     x12=rep(sellong ,rep(40,4)),
                     x2=rep(seq(0,1,len=40), 4))
> tx.pred1 <- predict(tx.ssanova, grid,
                      terms=c(0,1,1,1,0,1,1))
```

The estimates of these effects and 95% Bayesian confidence intervals are shown in Figure 4.15. The curve in each plot shows how temperature profile of that particular station differ from the average profile among the 48 stations. It is clear that the temperature in Rio Grande City is higher than average, and the temperature in Stratford is lower than average, especially in the winter. The temperature in El Paso is close to average in the first half year, and lower than average in the second half year. The temperature in Marshall is slightly above average.

The `month` effect equals $f_2(x_2)+f_{12}^{ls}(\boldsymbol{x}_1,x_2)+f_{12}^{ss}(\boldsymbol{x}_1,x_2)$. We compute posterior means and standard deviations of the month effect:

```
> tx.pred2 <- predict(tx.ssanova, terms=c(0,0,0,0,1,1,1))
```

The estimates of these effects for January, April, July, and October are shown in Figure 4.16. Each plot in Figure 4.16 shows how temperature pattern for that particular month differ from the average pattern among all 12 months. As expected, the temperature in January is colder than average, while the temperature in July is warmer than average. In general, the difference becomes smaller from north to south. The temperatures in April and October are close to average.

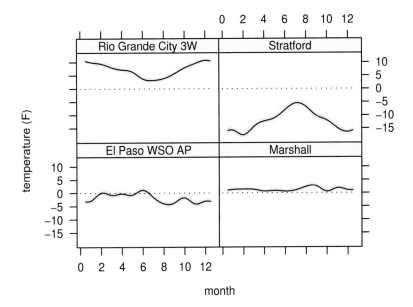

FIGURE 4.15 Texas weather data, plots of the `location` effect (solid lines) for four selected stations with 95% Bayesian confidence intervals (dashed lines).

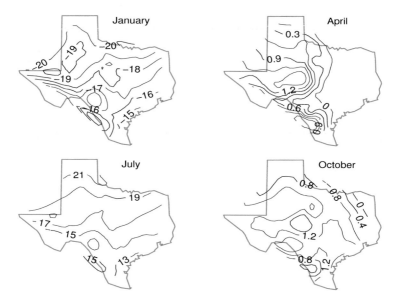

FIGURE 4.16 Texas weather data, plots of the `month` effect for January, April, July, and October.

Chapter 5

Spline Smoothing with Heteroscedastic and/or Correlated Errors

5.1 Problems with Heteroscedasticity and Correlation

In the previous chapters we have assumed that observations are independent with equal variances. These assumptions may not be appropriate for many applications. This chapter presents spline smoothing methods for heteroscedastic and/or correlated observations. Before introducing these methods, we first illustrate potential problems associated with the presence of heteroscedasticity and correlation in spline smoothing.

We use the following simulation to illustrate potential problems with heteroscedasticity. Observations are generated from model (1.1) with $f(x) = \sin(4\pi x^2)$, $x_i = i/n$ for $i = 1, \ldots, n$ and $n = 100$. Random errors are generated independently from the Gaussian distribution with mean zero and variance $\sigma^2 \exp\{\alpha|f(x)|\}$. Therefore, we have unequal variances when $\alpha \neq 0$. We set $\sigma = 0.05$ and $\alpha = 4$. For each simulated data, we first fit the cubic spline directly using PLS with GCV and GML choices of smoothing parameters. Note that these direct fits ignore heteroscedasticity. We then fit the cubic spline using the *penalized weighted LS* (PWLS) introduced in Section 5.2.1 with known weights $W = \text{diag}(\exp\{-4|f(x_1)|\}, \ldots, \exp\{-4|f(x_n)|\})$. For each fit, we compute weighted MSE (WMSE)

$$\text{WMSE} = \frac{1}{n} \sum_{i=1}^{n} w_i(\hat{f}(x_i) - f(x_i))^2,$$

where $w_i = \exp\{-4|f(x_i)|\}$. We also construct 95% Bayesian confidence intervals for each fit. The simulation is repeated 100 times. Figure 5.1 shows the performances of unweighted and weighted methods in terms of WMSE and coverages of Bayesian confidence intervals. Figure 5.1(a)

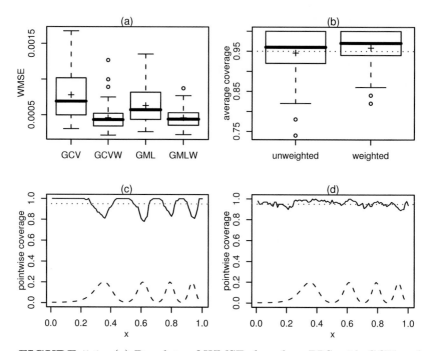

FIGURE 5.1 (a) Boxplots of WMSEs based on PLS with GCV and GML choices of the smoothing parameter and PWLS with GCV (labeled as GCVW) and GML (labeled as GMLW) choices of the smoothing parameter. Average WMSEs are marked as pluses; (b) boxplots of across-the-function coverages of 95% Bayesian confidence intervals for the PLS fits with GML choice of the smoothing parameter (labeled as unweighted) and PWLS fits with GML choice of the smoothing parameter (labeled as weighted). Average coverages are marked as pluses; (c) plot of pointwise coverages (solid line) of 95% Bayesian confidence intervals for the PLS fits with GML choice of the smoothing parameter; (d) plot of pointwise coverages (solid line) of 95% Bayesian confidence intervals for the PWLS fits with GML choice of the smoothing parameter. Dotted lines in (b), (c), and (d) represent the nominal value. Dashed lines in (c) and (d) represent a scaled version of the variance function $\exp\{\alpha|f(x)|\}$.

indicates that, even though ignoring heteroscedasticity, the unweighted methods provide good fits to the function. The weighted methods lead to better fits. Bayesian confidence intervals based on both methods provide the intended across-the-function coverages (Figure 5.1(b)). However, Figure 5.1(c) reveals the problem with heteroscedasticity: pointwise coverages in regions with larger variances are smaller than the nominal value, while pointwise coverages in other regions are larger than the nominal value. Obviously, this is caused by ignoring heteroscedasticity. Bayesian confidence intervals based on the PWLS method overcomes this problem (Figure 5.1(d)).

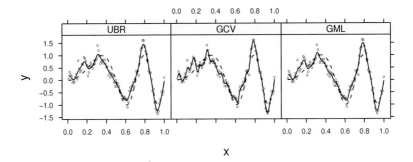

FIGURE 5.2 Plots of the true function (dashed lines), observations (circles), and the cubic spline fits (solid lines) with UBR, GCV, and GML choices of smoothing parameters. The true variance is used in the calculation of the UBR criterion.

Comparing with heteroscedasticity, the potential problems associated with correlation are more fundamental and difficult to deal with. For illustration, again we simulate data from model (1.1) with $f(x) = \sin(4\pi x^2)$, $x_i = i/n$ for $i = 1, \ldots, n$ and $n = 100$. Random errors ϵ_i are generated by a first-order autoregressive model (AR(1)) with mean zero, standard deviation 0.2, and first-order correlation 0.6. Figure 5.2 shows the simulated data, the true function, and three cubic spline fits with smoothing parameters chosen by the UBR, GCV, and GML methods, respectively. The true variance is used in the calculation of the UBR criterion. All fits are wiggly, which indicates that the estimated smoothing parameters are too small. This undersmoothing phenomenon is common in the presence of positive correlation. Ignoring correlation, the UBR, GCV, and GML methods perceive that all the trend (signal) in the data is due to the mean function f and attempt to incorporate that trend into the estimate. Correlated random errors may induce local trend and thus fool these methods to select smaller smoothing parameters such that the local trend can be picked up.

5.2   Extended SS ANOVA Models

Suppose observations are generated by

$$y_i = \mathcal{L}_i f + \epsilon_i, \quad i = 1, \ldots, n, \tag{5.1}$$

where f is a multivariate function in a model space $\mathcal{M}$, and $\mathcal{L}_i$ are bounded linear functionals. Let $\boldsymbol{\epsilon} = (\epsilon_1, \ldots, \epsilon_n)^T$. We assume that $\mathrm{E}(\boldsymbol{\epsilon}) = \mathbf{0}$ and $\mathrm{Cov}(\boldsymbol{\epsilon}) = \sigma^2 W^{-1}$. In this chapter we consider the SS ANOVA model space

$$\mathcal{M} = \mathcal{H}^0 \oplus \mathcal{H}^1 \oplus \cdots \oplus \mathcal{H}^q, \tag{5.2}$$

where $\mathcal{H}^0$ is a finite dimensional space collecting all functions that are not going to be penalized, and $\mathcal{H}^1, \ldots, \mathcal{H}^q$ are orthogonal RKHS's with RKs R^j for $j = 1, \ldots, q$. Model (5.1) is an extension of the SS ANOVA model (4.31) with non-iid random errors. See Section 8.2 for a more general model involving linear functionals and non-iid errors.

5.2.1   Penalized Weighted Least Squares

Our goal is to estimate f as well as W when it is unknown. We first assume that W is fixed and consider the estimation of the nonparametric function f. A direct generalization of the PLS (4.32) is the following PWLS:

$$\min_{f \in \mathcal{M}} \left\{ \frac{1}{n} (\boldsymbol{y} - \boldsymbol{f})^T W (\boldsymbol{y} - \boldsymbol{f}) + \lambda \sum_{j=1}^{q} \theta_j^{-1} \| P_j f \|^2 \right\}, \tag{5.3}$$

where $\boldsymbol{y} = (y_1, \ldots, y_n)^T$, $\boldsymbol{f} = (\mathcal{L}_1 f, \ldots, \mathcal{L}_n f)^T$, and P_j is the orthogonal projection in $\mathcal{M}$ onto $\mathcal{H}^j$.

Let $\boldsymbol{\theta} = (\theta_1, \ldots, \theta_q)$. Denote $\phi_1, \ldots, \phi_p$ as basis functions of $\mathcal{H}^0$, $T = \{\mathcal{L}_i \phi_\nu\}_{i=1 \ \nu=1}^{n \ \ p}$, $\Sigma_k = \{\mathcal{L}_i \mathcal{L}_j R^k\}_{i,j=1}^n$ for $k = 1, \ldots, q$, and $\Sigma_{\boldsymbol{\theta}} = \sum_{j=1}^{q} \theta_j \Sigma_j$. As in the previous chapters, we assume that T is of full column rank. The same arguments in Section 2.4 apply to the PWLS. Therefore, by the Kimeldorf–Wahba representer theorem, the solution to (5.3) exists and is unique, and the solution can be represented as

$$\hat{f}(\boldsymbol{x}) = \sum_{\nu=1}^{p} d_\nu \phi_\nu(\boldsymbol{x}) + \sum_{i=1}^{n} c_i \sum_{j=1}^{q} \theta_j \mathcal{L}_{i(\boldsymbol{z})} R^j(\boldsymbol{x}, \boldsymbol{z}).$$

Let $c = (c_1, \ldots, c_n)^T$ and $d = (d_1, \ldots, d_p)^T$. Let $\hat{f} = (\mathcal{L}_1 \hat{f}, \ldots, \mathcal{L}_n \hat{f})^T$ be the vector of fitted values. It is easy to check that $\hat{f} = Td + \Sigma_{\boldsymbol{\theta}} c$ and $\|P_1^* \hat{f}\|_*^2 = c^T \Sigma_{\boldsymbol{\theta}} c$. Then the PWLS (5.3) reduces to

$$\frac{1}{n}(y - Td - \Sigma_{\boldsymbol{\theta}} c)^T W (y - Td - \Sigma_{\boldsymbol{\theta}} c) + \lambda c^T \Sigma_{\boldsymbol{\theta}} c. \tag{5.4}$$

Taking the first derivatives leads to the following equations for c and d:

$$\begin{aligned}
(\Sigma_{\boldsymbol{\theta}} W \Sigma_{\boldsymbol{\theta}} + n\lambda \Sigma_{\boldsymbol{\theta}}) c + \Sigma_{\boldsymbol{\theta}} W Td &= \Sigma_{\boldsymbol{\theta}} W y, \\
T^T W \Sigma_{\boldsymbol{\theta}} c + T^T W Td &= T^T W y.
\end{aligned} \tag{5.5}$$

It is easy to check that a solution to

$$\begin{aligned}
(\Sigma_{\boldsymbol{\theta}} + n\lambda W^{-1}) c + Td &= y, \\
T^T c &= 0,
\end{aligned} \tag{5.6}$$

is also a solution to (5.5). Let $M = \Sigma_{\boldsymbol{\theta}} + n\lambda W^{-1}$ and

$$T = (Q_1 \ Q_2) \begin{pmatrix} R \\ 0 \end{pmatrix}$$

be the QR decomposition of T. Then the solutions to (5.6) are

$$\begin{aligned}
c &= Q_2 (Q_2^T M Q_2)^{-1} Q_2^T y, \\
d &= R^{-1} Q_1^T (y - Mc).
\end{aligned} \tag{5.7}$$

Based on the first equation in (5.6) and the fact that $\hat{f} = Td + \Sigma_{\boldsymbol{\theta}} c$, we have

$$\hat{f} = y - n\lambda W^{-1} c = H(\lambda, \boldsymbol{\theta}) y$$

where

$$H(\lambda, \boldsymbol{\theta}) = I - n\lambda W^{-1} Q_2 (Q_2^T M Q_2)^{-1} Q_2^T \tag{5.8}$$

is the *hat* matrix. The dependence of H on the smoothing parameters is expressed explicitly in (5.8). In the reminder of this chapter, for simplicity, the notation H will be used. Note that, different from the independent case, H may be asymmetric.

To solve (5.6), for fixed W and smoothing parameters, consider transformations $\tilde{y} = W^{1/2} y$, $\tilde{T} = W^{1/2} T$, $\tilde{\Sigma}_{\boldsymbol{\theta}} = W^{1/2} \Sigma_{\boldsymbol{\theta}} W^{1/2}$, $\tilde{c} = W^{-1/2} c$, and $\tilde{d} = d$. Then equations in (5.6) are equivalent to the following equations

$$\begin{aligned}
(\tilde{\Sigma}_{\boldsymbol{\theta}} + n\lambda I) \tilde{c} + \tilde{T} \tilde{d} &= \tilde{y}, \\
\tilde{T}^T \tilde{c} &= 0.
\end{aligned} \tag{5.9}$$

Note that equations in (5.9) have the same form as those in (2.21); thus methods in Section 2.4 can be used to compute $\tilde{c}$ and $\tilde{d}$. Transforming back, we have the solutions of c and d.

5.2.2   UBR, GCV and GML Criteria

We now extend the UBR, GCV, and GML criteria for the estimation of smoothing parameters λ and $\boldsymbol{\theta}$ as well as W when it is unknown.

Define the weighted version of the loss function

$$L(\lambda, \boldsymbol{\theta}) = \frac{1}{n}(\hat{\boldsymbol{f}} - \boldsymbol{f})^T W (\hat{\boldsymbol{f}} - \boldsymbol{f}) \qquad (5.10)$$

and weighted MSE as $\mathrm{WMSE}(\lambda, \boldsymbol{\theta}) \triangleq EL(\lambda, \boldsymbol{\theta})$. Then

$$\mathrm{WMSE}(\lambda, \boldsymbol{\theta}) = \frac{1}{n}\boldsymbol{f}^T(I - H^T)W(I - H)\boldsymbol{f} + \frac{\sigma^2}{n}\mathrm{tr}H^T W H W^{-1}. \quad (5.11)$$

It is easy to check that an unbiased estimate of $\mathrm{WMSE}(\lambda, \boldsymbol{\theta}) + \sigma^2$ is

$$\mathrm{UBR}(\lambda, \boldsymbol{\theta}) = \frac{1}{n}\boldsymbol{y}^T(I - H^T)W(I - H)\boldsymbol{y} + \frac{2\sigma^2}{n}\mathrm{tr}H. \qquad (5.12)$$

The UBR estimates of smoothing parameters and W when it is unknown are the minimizers of $\mathrm{UBR}(\lambda, \boldsymbol{\theta})$.

To extend the CV and GCV methods, we first consider the special case when W is diagonal: $W = \mathrm{diag}(w_1, \ldots, w_n)$. Denote $\hat{f}^{[i]}$ as the minimizer of the PWLS (5.3) based on all observations except y_i. It is easy to check that the leaving-out-one lemma in Section 3.4 still holds and $\mathcal{L}_i \hat{f}^{[i]} - y_i = (\mathcal{L}_i \hat{f} - y_i)/(1 - h_{ii})$, where h_{ii} are the diagonal elements of H. Then the cross-validation criterion is

$$\mathrm{CV}(\lambda, \boldsymbol{\theta}) \triangleq \frac{1}{n}\sum_{i=1}^n w_i \left(\mathcal{L}_i \hat{f}^{[i]} - y_i\right)^2 = \frac{1}{n}\sum_{i=1}^n w_i \frac{(\mathcal{L}_i \hat{f} - y_i)^2}{(1 - h_{ii})^2}. \qquad (5.13)$$

Replacing h_{ii} by the average of diagonal elements leads to the GCV criterion

$$\mathrm{GCV}(\lambda, \boldsymbol{\theta}) \triangleq \frac{\frac{1}{n}||W^{1/2}(I - H)\boldsymbol{y}||^2}{\left\{\frac{1}{n}\mathrm{tr}(I - H)\right\}^2}. \qquad (5.14)$$

Next consider the following model for clustered data

$$y_{ij} = \mathcal{L}_{ij} f + \epsilon_{ij}, \quad i = 1, \ldots, m; \quad j = 1, \ldots, n_i, \qquad (5.15)$$

where y_{ij} is the jth observation in cluster i, and $\mathcal{L}_{ij}$ are bounded linear functionals. Let $n = \sum_{i=1}^m n_i$, $\boldsymbol{y}_i = (y_{i1}, \ldots, y_{in_i})^T$, $\boldsymbol{y} = (\boldsymbol{y}_1^T, \ldots, \boldsymbol{y}_m^T)^T$, $\boldsymbol{\epsilon}_i = (\epsilon_{i1}, \ldots, \epsilon_{in_i})^T$, and $\boldsymbol{\epsilon} = (\boldsymbol{\epsilon}_1^T, \ldots, \boldsymbol{\epsilon}_m^T)^T$. We assume that $\mathrm{Cov}(\boldsymbol{\epsilon}_i) = \sigma^2 W_i^{-1}$ and observations between clusters are independent. Consequently, $\mathrm{Cov}(\boldsymbol{\epsilon}) = \sigma^2 W^{-1}$, where $W = \mathrm{diag}(W_1, \ldots, W_m)$ is a block diagonal matrix. Assume that $f \in \mathcal{M}$, where $\mathcal{M}$ is the model space in

(5.2). We estimate f using the PWLS (5.3). Let $\hat{f}$ be the PWLS estimate based on all observations, and $\hat{f}^{[i]}$ be the PWLS estimate based on all observations except those from the ith cluster $\boldsymbol{y}_i$. Let $\boldsymbol{f}_i = (\mathcal{L}_{i1}f, \ldots, \mathcal{L}_{in_i}f)^T$, $\boldsymbol{f} = (\boldsymbol{f}_1^T, \ldots, \boldsymbol{f}_m^T)^T$, $\hat{\boldsymbol{f}}_i = (\mathcal{L}_{i1}\hat{f}, \ldots, \mathcal{L}_{in_i}\hat{f})^T$, $\hat{\boldsymbol{f}} = (\hat{\boldsymbol{f}}_1^T, \ldots, \hat{\boldsymbol{f}}_m^T)^T$, $\hat{\boldsymbol{f}}_j^{[i]} = (\mathcal{L}_{j1}\hat{f}^{[i]}, \ldots, \mathcal{L}_{jn_j}\hat{f}^{[i]})^T$, and $\hat{\boldsymbol{f}}^{[i]} = ((\hat{\boldsymbol{f}}_1^{[i]})^T, \ldots, (\hat{\boldsymbol{f}}_m^{[i]})^T)^T$.

Leaving-out-one-cluster Lemma
For any fixed i, $\hat{f}^{[i]}$ is the minimizer of

$$\frac{1}{n}\left(\hat{\boldsymbol{f}}_i^{[i]} - \boldsymbol{f}_i\right)^T W_i \left(\hat{\boldsymbol{f}}_i^{[i]} - \boldsymbol{f}_i\right) + \frac{1}{n}\sum_{j\neq i}(\boldsymbol{y}_j - \boldsymbol{f}_j)^T W_j (\boldsymbol{y}_j - \boldsymbol{f}_j)$$

$$+\lambda \sum_{j=1}^{q}\theta_j^{-1}\|P_jf\|^2. \tag{5.16}$$

[Proof] For any function $f \in \mathcal{M}$, we have

$$\frac{1}{n}\left(\hat{\boldsymbol{f}}_i^{[i]} - \boldsymbol{f}_i\right)^T W_i \left(\hat{\boldsymbol{f}}_i^{[i]} - \boldsymbol{f}_i\right) + \frac{1}{n}\sum_{j\neq i}(\boldsymbol{y}_j - \boldsymbol{f}_j)^T W_j (\boldsymbol{y}_j - \boldsymbol{f}_j)$$

$$+\lambda \sum_{j=1}^{q}\theta_j^{-1}\|P_jf\|^2$$

$$\geq \frac{1}{n}\sum_{j\neq i}(\boldsymbol{y}_j - \boldsymbol{f}_j)^T W_j (\boldsymbol{y}_j - \boldsymbol{f}_j) + \lambda\sum_{j=1}^{q}\theta_j^{-1}\|P_jf\|^2$$

$$\geq \frac{1}{n}\sum_{j\neq i}\left(\boldsymbol{y}_j - \hat{\boldsymbol{f}}_j^{[i]}\right)^T W_j \left(\boldsymbol{y}_j - \hat{\boldsymbol{f}}_j^{[i]}\right) + \lambda\sum_{j=1}^{q}\theta_j^{-1}\|P_j\hat{f}^{[i]}\|^2$$

$$= \frac{1}{n}\left(\hat{\boldsymbol{f}}_i^{[i]} - \hat{\boldsymbol{f}}_i^{[i]}\right)^T W_i \left(\hat{\boldsymbol{f}}_i^{[i]} - \hat{\boldsymbol{f}}_i^{[i]}\right) + \frac{1}{n}\sum_{j\neq i}\left(\boldsymbol{y}_j - \hat{\boldsymbol{f}}_j^{[i]}\right)^T W_j \left(\boldsymbol{y}_j - \hat{\boldsymbol{f}}_j^{[i]}\right)$$

$$+\lambda\sum_{j=1}^{q}\theta_j^{-1}\|P_j\hat{f}^{[i]}\|^2.$$

The leaving-out-one-cluster lemma implies that $\hat{\boldsymbol{f}} = H\boldsymbol{y}$ and $\hat{\boldsymbol{f}}^{[i]} = H\boldsymbol{y}^{[i]}$, where H is the hat matrix and $\boldsymbol{y}^{[i]}$ is the same as $\boldsymbol{y}$ except that $\boldsymbol{y}_i$ is replaced by $\hat{\boldsymbol{f}}_i^{[i]}$. Divide the hat matrix H according to clusters such that $H = \{H_{ik}\}_{i,k=1}^m$, where H_{ik} is an $n_i \times n_k$ matrix. Then we have

$$\hat{\boldsymbol{f}}_i = \sum_{j=1}^m H_{ij}\boldsymbol{y}_j,$$

$$\hat{\boldsymbol{f}}_i^{[i]} = \sum_{j \neq i} H_{ij}\boldsymbol{y}_j + H_{ii}\hat{\boldsymbol{f}}_i^{[i]}.$$

Assume that $I - H_{ii}$ is invertible. Then

$$\hat{\boldsymbol{f}}_i^{[i]} - \boldsymbol{y}_i = (I - H_{ii})^{-1}(\hat{\boldsymbol{f}}_i - \boldsymbol{y}_i).$$

The cross-validation criterion

$$\mathrm{CV}(\lambda, \boldsymbol{\theta})$$

$$\triangleq \frac{1}{n} \sum_{i=1}^m \left(\hat{\boldsymbol{f}}_i^{[i]} - \boldsymbol{y}_i\right)^T W_i \left(\hat{\boldsymbol{f}}_i^{[i]} - \boldsymbol{y}_i\right)$$

$$= \frac{1}{n} \sum_{i=1}^m \left(\hat{\boldsymbol{f}}_i - \boldsymbol{y}_i\right)^T (I - H_{ii})^{-T} W_i (I - H_{ii})^{-1} \left(\hat{\boldsymbol{f}}_i - \boldsymbol{y}_i\right). \quad (5.17)$$

Replacing $I - H_{ii}$ by its generalized average G_i (Ma, Dai, Klein, Klein, Lee and Wahba 2010), we have

$$\mathrm{GCV}(\lambda, \boldsymbol{\theta}) \triangleq \frac{1}{n} \sum_{i=1}^m \left(\hat{\boldsymbol{f}}_i - \boldsymbol{y}_i\right)^T G_i W_i G_i \left(\hat{\boldsymbol{f}}_i - \boldsymbol{y}_i\right), \quad (5.18)$$

where $G_i = a_i I_{n_i} - b_i \mathbf{1}_{n_i} \mathbf{1}_{n_i}^T$, $a_i = 1/(\delta_i - \gamma_i)$, $b_i = \gamma_i/[(\delta_i - \gamma_i)\{\delta_i + (n_i - 1)\gamma_i\}]$, $\delta_i = (n - \mathrm{tr}H)/mn_i$, $\gamma_i = 0$ when $n_i = 1$ and $\gamma_i = -\sum_{i=1}^m \sum_{s \neq t} h_{st}^i/mn_i(n_i - 1)$ when $n_i > 1$, h_{st}^i is the element of the sth row and tth column of the matrix H_{ii}, I_{n_i} is an identity matrix of size n_i, and $\mathbf{1}_{n_i}$ is an n_i-vector of all ones. The CV and GCV estimates of smoothing parameters and W when it is unknown are the minimizers of the CV and GCV criteria.

To derive the GML criterion, we first construct a Bayes model for the extended SS ANOVA model (5.1). Assume the same prior for f as in (4.42):

$$F(\boldsymbol{x}) = \sum_{\nu=1}^p \zeta_\nu \phi_\nu(\boldsymbol{x}) + \delta^{\frac{1}{2}} \sum_{j=1}^q \sqrt{\theta_j} U_j(\boldsymbol{x}), \quad (5.19)$$

where $\zeta_\nu \overset{iid}{\sim} \mathrm{N}(0, \kappa)$, $U_j(\boldsymbol{x})$ are independent, zero-mean Gaussian stochastic processes with covariance function $R^j(\boldsymbol{x}, \boldsymbol{z})$, ζ_ν and $U_j(\boldsymbol{x})$ are mutually independent, and κ and δ are positive constants. Suppose observations are generated from

$$y_i = \mathcal{L}_i F + \epsilon_i, \quad i = 1, \ldots, n, \quad (5.20)$$

where $\boldsymbol{\epsilon} = (\epsilon_1, \ldots, \epsilon_n)^T \sim N(\boldsymbol{0}, \sigma^2 W^{-1})$.

Let $\mathcal{L}_0$ be a bounded linear functional on $\mathcal{M}$. Let $\lambda = \sigma^2/n\delta$. The same arguments in Section 3.6 hold when $M = \Sigma + n\lambda I$ is replaced by $M = \Sigma_{\boldsymbol{\theta}} + n\lambda W^{-1}$ in this chapter (Wang 1998b). Therefore,

$$\lim_{\kappa \to \infty} \mathrm{E}(\mathcal{L}_0 F | \boldsymbol{y}) = \mathcal{L}_0 \hat{f},$$

and an extension of the GML criterion is

$$\mathrm{GML}(\lambda, \boldsymbol{\theta}) = \frac{\boldsymbol{y}^T W (I - H) \boldsymbol{y}}{\{\det^+(W(I - H))\}^{\frac{1}{n-p}}}, \qquad (5.21)$$

where $\det^+$ is the product of the nonzero eigenvalues. The GML estimates of smoothing parameters and W when it is unknown are the minimizers of $\mathrm{GML}(\lambda, \boldsymbol{\theta})$.

The GML estimator of the variance σ^2 is (Wang 1998b)

$$\hat{\sigma}^2 = \frac{\boldsymbol{y}^T W (I - H) \boldsymbol{y}}{n - p}. \qquad (5.22)$$

5.2.3   Known Covariance

In this section we discuss the implementation of the UBR, GCV, and GML criteria in Section 5.2.2 when W is known. In this situation we only need to estimate the smoothing parameters λ and $\boldsymbol{\theta}$. Consider transformations discussed at the end of Section 5.2.1. Let $\tilde{\boldsymbol{f}} = W^{1/2} \hat{\boldsymbol{f}}$ be the fits to the transformed data, and $\tilde{H}$ be the hat matrix associated with the transformed data. Then $H\boldsymbol{y} = \hat{\boldsymbol{f}} = W^{-1/2} \tilde{\boldsymbol{f}} = W^{-1/2} \tilde{H} \tilde{\boldsymbol{y}} = W^{-1/2} \tilde{H} W^{1/2} \boldsymbol{y}$ for any $\boldsymbol{y}$. Therefore,

$$H = W^{-1/2} \tilde{H} W^{1/2}. \qquad (5.23)$$

From (5.23), the UBR, GCV, and GML in (5.12), (5.14), and (5.21) can be rewritten based on the transformed data as

$$\mathrm{UBR}(\lambda, \boldsymbol{\theta}) = \frac{1}{n} \|(I - \tilde{H}) \tilde{\boldsymbol{y}}\|^2 + \frac{2\sigma^2}{n} \mathrm{tr} \tilde{H}, \qquad (5.24)$$

$$\mathrm{GCV}(\lambda, \boldsymbol{\theta}) = \frac{\frac{1}{n} \sum_{i=1}^{n} \|(I - \tilde{H}) \tilde{\boldsymbol{y}}\|^2}{\left\{ \frac{1}{n} \mathrm{tr}(I - \tilde{H}) \right\}^2}, \qquad (5.25)$$

$$\mathrm{GML}(\lambda, \boldsymbol{\theta}) = \frac{C \tilde{\boldsymbol{y}}^T (I - \tilde{H}) \tilde{\boldsymbol{y}}}{\{\det^+(I - \tilde{H})\}^{\frac{1}{n-p}}}, \qquad (5.26)$$

where C in $\mathrm{GML}(\lambda, \boldsymbol{\theta})$ is a constant independent of λ and $\boldsymbol{\theta}$. Equations (5.24), (5.25), and (5.26) indicate that, when W is known, the UBR,

GCV, and GML estimates of smoothing parameters can be calculated based on the transformed data using the method described in Section 4.7.

5.2.4   Unknown Covariance

We now consider the case when W is unknown. When a separate method is available for estimating W, one approach is to estimate the function f and covariance W iteratively. For example, the following two-step procedure is simple and easy to implement: (1) Estimate the function with a "sensible" choice of the smoothing parameter; (2) Estimate the covariance using residuals; and (3) Estimate the function again using the estimated covariance. However, it may not work in certain situations due to the interplay between the smoothing parameter and correlation.

We use the following two simple simulations to illustrate the potential problem associated with the above iterative approach. In the first simulation, $n = 100$ observations are generated according to model (1.1) with $f(x) = \sin(4\pi x^2)$, $x_i = i/n$ for $i = 1, \ldots, n$, and $\epsilon_i \overset{iid}{\sim} N(0, 0.2^2)$. We fit a cubic spline with a fixed smoothing parameter λ such that $\log_{10}(n\lambda) = -3.5$. Figure 5.3(a) shows the fit, which is slightly oversmoothed. The estimated autocorrelation function (ACF) of residuals (Figure 5.3(b)) suggests an autoregressive structure even though the true random errors are independent. It is clear that the leftover signal due to oversmoothing shows up in the residuals. In the second simulation, $n = 100$ observations are generated according to model (1.1) with $f(x) \equiv 0$, $x_i = i/n$ for $i = 1, \ldots, n$, and ϵ_i are generated by an AR(1) model with mean zero, standard deviation 0.2, and first-order correlation 0.6. Figure 5.3(c) shows the cubic spline fit with GCV choice of the smoothing parameter. The fit picks up local trend in the AR(1) process, and the estimated ACF of residuals (Figure 5.3(d)) does not reveal any autoregressive structure. In both cases, the mean functions are incorrectly estimated, and the conclusions about the error structures are erroneous. These two simulations indicate that a wrong choice of the smoothing parameter in the first step will lead to a deceptive serial correlation in the second step.

In the most general setting where no parametric shape is assumed for the mean or the correlation function, the model is essentially unidentifiable. In the following we will model the correlation structure parametrically. Specifically we assume that W depends on an unknown vector of parameters $\boldsymbol{\tau}$. Models for W^{-1} will be discussed in Section 5.3.

When there is no strong connection between the smoothing and correlation parameters, an iterative procedure as described earlier may be

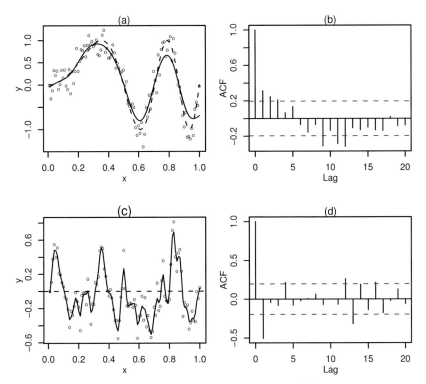

FIGURE 5.3 For the first simulation, plots of (a) the true function (dashed lines), observations (circles), and the cubic spline fits (solid line) with $\log_{10}(n\lambda) = -3.5$, and (b) estimated ACF of residuals. For the second simulation, plots of (c) the true function (dashed lines), observations (circles), and the cubic spline fits (solid line) with GCV choice of the smoothing parameter, and (d) estimated ACF of residuals.

used to estimate f and W. See Sections 5.4.1 and 6.4 for examples. When there is a strong connection, it is helpful to estimate the smoothing and correlation parameters simultaneously. One may use the UBR, GCV, or GML criterion to estimate the smoothing and correlation parameters simultaneously. It was found that the GML method performs better than the UBR and GCV methods (Wang 1998b). Therefore, in the following, we present the implementation of the GML method only.

To compute the minimizers of the GML criterion (5.21), we now construct a corresponding LME model for the extended SS ANOVA model (5.1). Let $\Sigma_k = Z_k Z_k^T$, where Z_k is an $n \times m_k$ matrix with

$m_k = \text{rank}(\Sigma_k)$. Consider the following LME model

$$y = T\zeta + \sum_{k=1}^{q} Z_k u_k + \epsilon, \qquad (5.27)$$

where ζ is a p-dimensional vector of deterministic parameters, u_k are mutually independent random effects, $u_k \sim \text{N}(0, \sigma^2 \theta_k I_{m_k}/n\lambda)$, I_{m_k} is the identity matrix of order m_k, $\epsilon \sim \text{N}(0, \sigma^2 W^{-1})$, and u_k are independent of ϵ. It is not difficult to show that the spline estimate based on the PWLS is the BLUP estimate, and the GML criterion (5.21) is the REML criterion based on the LME model (5.27) (Opsomer, Wang and Yang 2001). Details for the more complicated semiparametric mixed-effects models will be given in Chapter 9. In the `ssr` function, we first calculate Z_k through Cholesky decomposition. Then we calculate the REML (GML) estimates of λ, θ and τ using the function `lme` in the `nlme` library by Pinheiro and Bates (2000).

5.2.5   Confidence Intervals

Consider the Bayes model defined in (5.19) and (5.20). Conditional on W, it is not difficult to show that formulae for posterior means and covariances in Section 4.8 hold when $M = \Sigma_\theta + n\lambda I$ is replaced by $M = \Sigma_\theta + n\lambda W^{-1}$. Bayesian confidence intervals can be constructed similarly. When the matrix W involves unknown parameters τ, it is replaced by $W(\hat{\tau})$ in the construction of Bayesian confidence intervals. This simple plug-in approach does not account for the variation in estimating the covariance parameters τ. The bootstrap approach may also be used to construct confidence intervals. More research is necessary for inferences on the nonparametric function f as well as on parameters τ.

5.3   Variance and Correlation Structures

In this section we discuss commonly used models for the variance–covariance matrix W^{-1}. The matrix W^{-1} can be decomposed as

$$W^{-1} = VCV,$$

where V is a diagonal matrix with positive diagonal elements, and C is a correlation matrix with all diagonal elements equal to one. The matrices V and C describe the variance and correlation, respectively.

The above decomposition allows us to develop separate models for the *variance structure* and *correlation structure*. We now describe structures available for the `assist` package. Details about these structures can be found in Pinheiro and Bates (2000).

Consider the variance structure first. Define general variance function as

$$\text{Var}(\epsilon_i) = \sigma^2 v^2(\mu_i, z_i; \zeta), \tag{5.28}$$

where v is a known variance function of the mean $\mu_i = \text{E}(y_i)$ and a vector of covariates z_i associated with the variance, and ζ is a vector of variance parameters. As in Pinheiro and Bates (2000), when v depends on μ_i, the *pseudo-likelihood* method will be used to estimate all parameters and the function f.

In the `ssr` function, a variance matrix (vector) or a model for the variance function is specified using the `weights` argument. The input to the `weights` argument may be the matrix W when it is known. Furthermore, when W is a known diagonal matrix, the diagonal elements may be specified as a vector by the `weights` argument. The input to the `weights` argument may also be a `varFunc` structure specifying a model for variance function v in (5.28). All `varFunc` objects available in the `nlme` package are available for the `assist` package. The `varFunc` structure is defined through the function $v(s; \zeta)$, where s can be either a variance covariate or the fitted value. Standard `varFunc` classes and their corresponding variance functions v are listed in Table 5.1. Two or more variance models may be combined using the `varComb` constructor.

TABLE 5.1   Standard `varFunc` classes

Class	$v(s; \zeta)$
varFixed	$\sqrt{\|s\|}$
varIdent	ζ_s, s is a stratification variable
varPower	$\|s\|^\zeta$
varExp	$\exp(\zeta s)$
varConstPower	$\zeta_1 + \|s\|^{\zeta_2}$

Next consider the correlation structure. Assume the general *isotropic correlation* structure

$$\text{cor}(\epsilon_i, \epsilon_j) = h(d(p_i, p_j); \rho), \tag{5.29}$$

where p_i and p_j are position vectors associated with observations i and j respectively; h is a known correlation function of the distance $d(p_i, p_j)$

such that $h(0; \boldsymbol{\rho}) = 1$; and $\boldsymbol{\rho}$ is a vector of the correlation parameters. In the `ssr` function, correlation structures are specified as `corStruct` objects through the `correlation` argument. There are two common families of correlation structures: serial correlation structures for time series and spatial correlation structures for spatial data.

For time series data, observations are indexed by an one-dimensional position vector, and $d(p_i, p_j)$ represents lags that take nonnegative integers. Therefore, serial correlation is determined by the function $h(k; \boldsymbol{\rho})$ for $k = 1, 2, \ldots$. Standard `corStruct` classes for serial correlation structures and their corresponding correlation functions h are listed in Table 5.2.

TABLE 5.2 Standard `corStruct` classes for serial correlation structures

Class	$h(k; \boldsymbol{\rho})$
`corCompSymm`	ρ
`corSymm`	ρ_k
`corAR1`	ρ^k
`corARMA`	correlation function given in (5.30)

The classes `corCompSymm` and `corSymm` correspond to the compound symmetry and general correlation structures. An autoregressive-moving average (ARMA(p,q)) model assumes that

$$\epsilon_t = \sum_{i=1}^{p} \phi_i \epsilon_{t-i} + \sum_{j=1}^{q} \theta_j a_{t-j} + a_t,$$

where a_t are iid random variables with mean zero and a constant variance, $\phi_1, \ldots, \phi_p$ are autoregressive parameters, and $\theta_1, \ldots, \theta_q$ are moving average parameters. The correlation function is defined recursively as

$$h(k; \boldsymbol{\rho}) = \begin{cases} \phi_1 h(|k-1|; \boldsymbol{\rho}) + \cdots + \phi_p h(|k-p|; \boldsymbol{\rho}) + \\ \quad \theta_1 \psi(k-1; \boldsymbol{\rho}) + \cdots + \theta_q \psi(k-q; \boldsymbol{\rho}), & k \leq q, \quad (5.30) \\ \phi_1 h(|k-1|; \boldsymbol{\rho}) + \cdots + \phi_p h(|k-p|; \boldsymbol{\rho}), & k > q, \end{cases}$$

where $\boldsymbol{\rho} = (\phi_1, \ldots, \phi_p, \theta_1, \ldots, \theta_q)$ and $\psi(k; \boldsymbol{\rho}) = \mathrm{E}(\epsilon_{t-k} a_t) / \mathrm{Var}(\epsilon_t)$.

The continuous AR(1) correlation function is defined as

$$h(s; \rho) = \rho^s, \quad s \geq 0, \quad \rho \geq 0.$$

The `corCAR1` constructor specifies the continuous AR(1) structure. For spatial data, locations $\boldsymbol{p}_i \in \mathbb{R}^r$. Any well-defined distance such as

Euclidean $d(\boldsymbol{p}_i, \boldsymbol{p}_j) = \{\sum_{k=1}^{r}(p_{ik} - p_{jk})^2\}^{1/2}$, Manhattan $d(\boldsymbol{p}_i, \boldsymbol{p}_j) = \sum_{k=1}^{r}|p_{ik} - p_{jk}|$, and maximum distance $d(\boldsymbol{p}_i, \boldsymbol{p}_j) = \max_{1 \le k \le r}|p_{ik} - p_{jk}|$ may be used. The correlation function structure is defined through the function $h(s; \boldsymbol{\rho})$, where $s \ge 0$. Standard `corStruct` classes for spatial correlation structures and their corresponding correlation functions h are listed in Table 5.3. The function $I(s < \rho)$ equals 1 when $s < \rho$ and 0 otherwise. When desirable, the following correlation function allows a *nugget effect*:

$$h_{\text{nugg}}(s, c_0; \boldsymbol{\rho}) = \begin{cases} (1 - c_0)h_{\text{cont}}(s; \boldsymbol{\rho}), & s > 0, \\ 1, & s = 0, \end{cases}$$

where h_{cont} is any standard correlation function that is continuous in s and $0 < c_0 < 1$ is a `nugget effect`.

TABLE 5.3   Standard `corStruct` classes for spatial correlation structures

Class	$h(s; \boldsymbol{\rho})$
corExp	$\exp(-s/\rho)$
corGaus	$\exp\{-(s/\rho)^2\}$
corLin	$(1 - s/\rho)I(s < \rho)$
corRatio	$1/\{1 + (s/\rho)^2\}$
corSpher	$\{1 - 1.5(s/\rho) + 0.5(s/\rho)^3\}I(s < \rho)$

5.4   Examples

5.4.1   Simulated Motorcycle Accident — Revisit

Plot of observations in Figure 3.8 suggests that variances may change over time. For the partial spline fit to model (3.53), we compute squared residuals and plot the logarithm of squared residuals over time in Figure 5.4(a). Cubic spline fit to the logarithm of squared residuals in Figure 5.4(a) indicates that the constant variance assumption may be violated. We then fit model (3.53) again using PWLS with weights fixed as the estimated variances:

```
> r <- residuals(mcycle.ps.fit2); y <- log(r**2)
```

```
> mcycle.v <- ssr(y~x, cubic(x), spar=''m'')
> update(mcycle.ps.fit2, weights=exp(mcycle.v$fit))
> predict(mcycle.ps.fit3,
    newdata=data.frame(x=grid, s1=(grid-t1)*(grid>t1),
    s2=(grid-t2)*(grid>t2), s3=(grid-t3)*(grid>t3)))
```

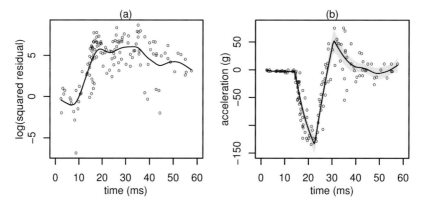

FIGURE 5.4   Motorcycle data, plots of (a) logarithm of the squared residuals (circles) based on model (3.53), and the cubic spline fit (line) to logarithm of the squared residuals; and (b) observations (circles), new PWLS fit (line), and 95% Bayesian confidence intervals (shaded region).

The PWLS fit and 95% Bayesian confidence intervals are shown in 5.4(b). The impact of the unequal variances is reflected in the widths of confidence intervals. The two-step approach adapted here is crude, and the variation in the estimation of the variance function has been ignored in the construction of the confidence intervals. Additional methods for estimating the mean and variance functions will be discussed in Section 6.4.

5.4.2   Ozone in Arosa — Revisit

Figure 2.2 suggests that the variances may not be a constant. Based on fit to the trigonometric spline model (2.75), we calculate residual variances for each month and plot them on the logarithm scale in Figure 5.5(a). It is obvious that variations depend on the time of the year. It seems that a simple sinusoidal function can be used to model the variance function.

```
> v <- sapply(split(arosa.ls.fit$resi,Arosa$month),var)
> a <- sort(unique(Arosa$x))
> b <- lm(log(v)~sin(2*pi*a)+cos(2*pi*a))
> summary(b)
...
Coefficients:
                  Estimate Std. Error t value Pr(>|t|)
(Intercept)        5.43715    0.05763  94.341 8.57e-15 ***
sin(2 * pi * a)    0.71786    0.08151   8.807 1.02e-05 ***
cos(2 * pi * a)    0.49854    0.08151   6.117 0.000176 ***
---
Residual standard error: 0.1996 on 9 degrees of freedom
Multiple R-squared: 0.9274,     Adjusted R-squared: 0.9113
F-statistic: 57.49 on 2 and 9 DF,  p-value: 7.48e-06
```

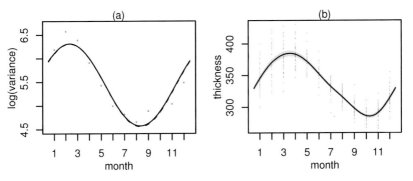

FIGURE 5.5   Arosa data, plots of (a) logarithm of residual variances (circles) based on the periodic spline fit in Section 2.7, the sinusoidal fit to logarithm of squared residuals (dashed line), and the fit from model (5.31) (solid line); and (b) observations (dots), the new PWLS fit (solid line), and 95% Bayesian confidence intervals (shaded region).

The fit of the simple sinusoidal model to the log variance is shown in Figure 5.5(a). We now assume the following variance function for the trigonometric spline model (2.75):

$$v(x) = \exp(\zeta_1 \sin 2\pi x + \zeta_2 \cos 2\pi x), \tag{5.31}$$

and fit the model as follows:

```
> arosa.ls.fit1 <- ssr(thick~sin(2*pi*x)+cos(2*pi*x),
    rk=lspline(x,type=''sine1''), spar=''m'', data=Arosa,
    weights=varComb(varExp(form=~sin(2*pi*x)),
                    varExp(form=~cos(2*pi*x))))
> summary(arosa.ls.fit1)
...
GML estimate(s) of smoothing parameter(s) : 3.675780e-09
Equivalent Degrees of Freedom (DF):   6.84728
Estimate of sigma:   15.22466
Combination of:
Variance function structure of class varExp representing
    expon
0.3555942
Variance function structure of class varExp representing
    expon
0.2497364
```

The estimated variance parameters, 0.3556 and 0.2497, are very close
to (up to a scale of 2 by definition) those in the sinusoidal model based
on residual variances — 0.7179 and 0.4985. The fitted variance function
is plotted in Figure 5.5(a) which is almost identical to the fit based
on residual variances. Figure 5.5(b) plots the trigonometric spline fit
to the mean function with 95% Bayesian confidence intervals. Note
that these confidence intervals are conditional on the estimated variance
parameters. Thus they may have smaller coverage than the nominal
value since variation in the estimation of variance parameters is not
counted. Nevertheless, we can see that unequal variances are reflected
in the widths of these confidence intervals.

Observations close in time may be correlated. We now consider a
first-order autoregressive structure for random errors. Since some observations are missing, we use the continuous AR(1) correlation structure
$h(s, \rho) = \rho^s$, where s represents distance in terms of calendar time between observations. We refit model (2.75) using the variance structure
(5.31) and continuous AR(1) correlation structure as follows:

```
> Arosa$time <- Arosa$month+12*(Arosa$year-1)
> arosa.ls.fit2 <- update(arosa.ls.fit1,
    corr=corCAR1(form=~time))
> summary(arosa.ls.fit2)
...
GML estimate(s) of smoothing parameter(s) : 3.575283e-09
Equivalent Degrees of Freedom (DF):  7.327834
Estimate of sigma:   15.31094
Correlation structure of class corCAR1 representing
```

```
       Phi
0.3411414
Combination of:
Variance function structure of class varExp representing
     expon
0.3602905
Variance function structure of class varExp representing
     expon
0.3009282
```

where the variable `time` represents the continuous calendar time in months.

5.4.3   Beveridge Wheat Price Index

The Beveridge data contain the time series of annual wheat price index from 1500 to 1869. The time series of price index on the logarithmic scale is shown in Figure 5.6.

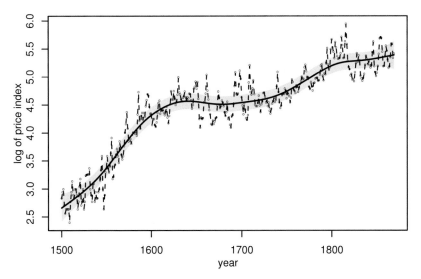

FIGURE 5.6 Beveridge data, observations (circles), the cubic spline fit under the assumption of independent random errors (dashed line), and the cubic spline fit under AR(1) correlation structure (solid line) with 95% Bayesian confidence intervals (shaded region).

Let x be the year scaled into the interval $[0, 1]$ and y be the logarithm of price index. Consider the following nonparametric regression model

$$y_i = f(x_i) + \epsilon_i, \quad i = 1, \ldots, n, \tag{5.32}$$

where $f \in W_2^2[0, 1]$ and ϵ_i are random errors. Under the assumption that ϵ_i are independent with a common variance, model (5.32) can be fitted as follows:

```
> library(tseries); data(bev)
> y <- log(bev); x <- seq(0,1,len=length(y))
> bev.fit1 <- ssr(y~x, rk=cubic(x))
> summary(bev.fit1)
...
GCV estimate(s) of smoothing parameter(s) : 3.032761e-13
Equivalent Degrees of Freedom (DF):   344.8450
Estimate of sigma:   0.02865570
```

where the GCV method was used to select the smoothing parameter. The estimate of the smoothing parameter is essentially zero, which indicates that the cubic spline fit interpolates observations (dashed line in Figure 5.6). The undersmoothing is likely caused by the autocorrelation in time series. Now consider an AR(1) correlation structure for random errors:

```
> bev.fit2 <- update(bev.fit1, spar=''m'',
                    correlation=corAR1(form=~1))
> summary(bev.fit2)
 GML estimate(s) of smoothing parameter(s) : 1.249805e-06
 Equivalent Degrees of Freedom (DF):   8.091024
 Estimate of sigma:   0.2243519
 Correlation structure of class corAR1 representing
       Phi
0.6936947
```

Estimate of f and 95% Bayesian confidence intervals under AR(1) correlation structure are shown in Figure 5.6 as the solid line and the shaded region.

5.4.4   Lake Acidity

The lake acidity data contain measurements of 112 lakes in the southern Blue Ridge mountains area. It is of interest to investigate the dependence of the water pH level on calcium concentration and geological location. To match notations in Chapter 4, we relabel calcium concentration `t1` as variable x_1, and geological location x1 (latitude) and x2 (longitude) as variables x_{21} and x_{22}, respectively. Let $\boldsymbol{x}_2 = (x_{21}, x_{22})$.

First, we fit an one-dimensional thin-plate spline to the response variable ph using one independent variable x_1 (calcium):

$$\text{ph}(x_{i1}) = f(x_{i1}) + \epsilon_i, \tag{5.33}$$

where $f \in W_2^2(\mathbb{R})$, and ϵ_i are zero-mean independent random errors with a common variance.

```
> data(acid)
> acid$x21 <- acid$x1; acid$x22 <- acid$x2
> acid$x1 <- acid$t1
> acid.tp.fit1 <- ssr(ph~x1, rk=tp(x1), data=acid)
> summary(acid.tp.fit1)
...
GCV estimate(s) of smoothing parameter(s) : 3.433083e-06
Equivalent Degrees of Freedom (DF):  8.20945
Estimate of sigma:  0.281299

Number of Observations:  112

> anova(acid.tp.fit1)


Testing H_0: f in the NULL space


      test.value simu.size simu.p-value
LMP 0.003250714       100         0.08
GCV 0.008239078       100         0.01
```

Both p-values from the LMP and GCV tests suggest that the departure from a straight line model is borderline significant. The estimate of the function f in model (5.33) is shown in the left panel of Figure 5.7.

Observations of the pH level close in geological locations are often correlated. Suppose that we want to model potential spatial correlation among random errors in model (5.33) using the exponential spatial correlation structure with nugget effect for the location variable $\boldsymbol{x}_2 = (x_{21}, x_{22})$. That is, we assume that

$$h_{\text{nugg}}(s, c_0, \rho) = \begin{cases} (1 - c_0) \exp(-s/\rho), & s > 0, \\ 1, & s = 0, \end{cases}$$

where c_0 is the nugget effect and s is the Euclidean distance between two geological locations. Model (5.33) with the above correlation structure can be fitted as follows:

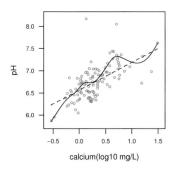
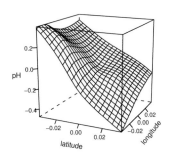

FIGURE 5.7 Lake acidity data, the left panel includes observations (circles), the fit from model (5.33) (solid line), the fit from model (5.33) with the exponential spatial correlation structure (dashed line), and estimate of the constant plus main effect of x_1 from model (5.36) (dotted line); the right panel includes estimate of the main effect of $\boldsymbol{x}_2$ from model (5.36).

```
> acid.tp.fit2 <- update(acid.tp.fit1, spar=''m'',
    corr=corExp(form=~x21+x22,nugget=T))
> summary(acid.tp.fit2)
...
GML estimate(s) of smoothing parameter(s) : 53310.63
Equivalent Degrees of Freedom (DF):  2
Estimate of sigma:  0.3131702
Correlation structure of class corExp representing
     range       nugget
0.02454532 0.62321744
```

The GML estimate of the smoothing parameter is large, and the spline estimate is essentially a straight line (left panel of Figure 5.7). The smaller smoothing parameter in the first fit with independence assumption might be caused by the spatial correlation. Equivalent degrees of freedom for f have been reduced from 8.2 to 2.

We can also model the effect of geological location directly in the mean function. That is to consider the mean pH level as a bivariate function of x_1 (calcium) and $\boldsymbol{x}_2$ (geological location):

$$\text{ph}(x_{i1}, \boldsymbol{x}_{i2}) = f(x_{i1}, \boldsymbol{x}_{i2}) + \epsilon_i, \tag{5.34}$$

where ϵ_i are zero-mean independent random errors with a common variance. One possible model space for x_1 is $W_2^2(\mathbb{R})$, and one possible model space for $\boldsymbol{x}_2$ is $W_2^2(\mathbb{R}^2)$. Therefore, we consider the tensor product space $W_2^2(\mathbb{R}) \otimes W_2^2(\mathbb{R}^2)$. Define three averaging operators

$$\mathcal{A}_1^{(1)} f = \sum_{j=1}^{J_1} w_{j1} f(u_{j1}),$$

$$\mathcal{A}_2^{(1)} f = \sum_{j=1}^{J_1} w_{j1} f(u_{j1}) \tilde{\phi}_{12}(u_{j1}) \tilde{\phi}_{12},$$

$$\mathcal{A}_1^{(2)} f = \sum_{j=1}^{J_2} w_{j2} f(\boldsymbol{u}_{j2}),$$

$$\mathcal{A}_2^{(2)} f = \sum_{j=1}^{J_2} w_{j2} f(\boldsymbol{u}_{j2}) \{ \tilde{\phi}_{22}(\boldsymbol{u}_{j2}) \tilde{\phi}_{22} + \tilde{\phi}_{23}(\boldsymbol{u}_{j2}) \tilde{\phi}_{23} \},$$

where u_{j1} and $\boldsymbol{u}_{j2}$ are fixed points in $\mathbb{R}$ and $\mathbb{R}^2$, and w_{j1} and w_{j2} are fixed positive weights such that $\sum_{j=1}^{J_1} w_{j1} = \sum_{j=1}^{J_2} w_{j2} = 1$; $\tilde{\phi}_{11} = 1$ and $\tilde{\phi}_{12}$ are orthonormal bases for the null space in $W_2^2(\mathbb{R})$ based on the norm (2.41); $\tilde{\phi}_{21} = 1$, and $\tilde{\phi}_{22}$ and $\tilde{\phi}_{23}$ are orthonormal bases for the null space in $W_2^2(\mathbb{R}^2)$ based on the norm (2.41). Let $\mathcal{A}_3^{(1)} = I - \mathcal{A}_1^{(1)} - \mathcal{A}_2^{(2)}$ and $\mathcal{A}_3^{(2)} = I - \mathcal{A}_1^{(2)} - \mathcal{A}_2^{(2)}$. Then we have the following SS ANOVA decomposition:

$$\begin{aligned} f &= \left\{ \mathcal{A}_1^{(1)} + \mathcal{A}_2^{(1)} + \mathcal{A}_3^{(1)} \right\} \left\{ \mathcal{A}_1^{(2)} + \mathcal{A}_2^{(2)} + \mathcal{A}_3^{(2)} \right\} f \\ &= \mathcal{A}_1^{(1)} \mathcal{A}_1^{(2)} f + \mathcal{A}_2^{(1)} \mathcal{A}_1^{(2)} f + \mathcal{A}_3^{(1)} \mathcal{A}_1^{(2)} f \\ &\quad + \mathcal{A}_1^{(1)} \mathcal{A}_2^{(2)} f + \mathcal{A}_2^{(1)} \mathcal{A}_2^{(2)} f + \mathcal{A}_3^{(1)} \mathcal{A}_2^{(2)} f \\ &\quad + \mathcal{A}_1^{(1)} \mathcal{A}_3^{(2)} f + \mathcal{A}_2^{(1)} \mathcal{A}_3^{(2)} f + \mathcal{A}_3^{(1)} \mathcal{A}_3^{(2)} f \\ &\triangleq \mu + \beta_1 \tilde{\phi}_{12}(x_1) + \beta_2 \tilde{\phi}_{22}(\boldsymbol{x}_2) + \beta_3 \tilde{\phi}_{23}(\boldsymbol{x}_2) + \beta_4 \tilde{\phi}_{12}(x_1) \tilde{\phi}_{22}(\boldsymbol{x}_2) \\ &\quad + \beta_5 \tilde{\phi}_{12}(x_1) \tilde{\phi}_{23}(\boldsymbol{x}_2) + f_1^s(x_1) + f_2^s(\boldsymbol{x}_2) + f_{12}^{ls}(x_1, \boldsymbol{x}_2) \\ &\quad + f_{12}^{sl}(x_1, \boldsymbol{x}_2) + f_{12}^{ss}(x_1, \boldsymbol{x}_2). \end{aligned} \tag{5.35}$$

Due to small sample size, we ignore all interactions and consider the following additive model

$$\begin{aligned} y_1 &= \mu + \beta_1 \tilde{\phi}_{12}(x_{i1}) + \beta_2 \tilde{\phi}_{22}(\boldsymbol{x}_{i2}) + \beta_3 \tilde{\phi}_{23}(\boldsymbol{x}_{i2}) \\ &\quad + f_1^s(x_{i1}) + f_2^s(\boldsymbol{x}_{i2}) + \epsilon_i, \end{aligned} \tag{5.36}$$

where ϵ_i are zero-mean independent random errors with a common variance. We fit model (5.36) as follows:

```
> acid.ssanova.fit <- ssr(ph~x1+x21+x22,
    rk=list(tp(x1), tp(list(x21,x22))),
    data=acid, spar=''m'')
> summary(acid.ssanova.fit)
...
GML estimate(s) of smoothing parameter(s) : 4.819636e-01
                                            2.870235e-07
Equivalent Degrees of Freedom (DF):  8.768487
Estimate of sigma:  0.2560684
```

The estimate of the main effect of x_1 plus the constant, $\mu + \beta_1 \tilde{\phi}_{12}(x_1) + f_1^s(x_1)$, is shown in Figure 5.7(a). This estimate is almost identical to that from model (5.33) with the exponential spatial correlation structure. The estimate of the main effect of x_2 is shown in the right panel of Figure 5.7.

An alternative approach to modeling the effect of geological location using mixed-effects will be discussed in Section 9.4.2.

Chapter 6

Generalized Smoothing Spline ANOVA

6.1 Generalized SS ANOVA Models

Generalized linear models (GLM) (McCullagh and Nelder 1989) provide a unified framework for analysis of data from exponential families. Denote $(\boldsymbol{x}_i, y_i)$ for $i = 1, \ldots, n$ as independent observations on independent variables $\boldsymbol{x} = (x_1, \ldots, x_d)$ and dependent variable y. Assume that y_i are generated from a distribution in the exponential family with the density function

$$g(y_i; f_i, \phi) = \exp\left\{ \frac{y_i h(f_i) - b(f_i)}{a_i(\phi)} + c(y_i, \phi) \right\}, \qquad (6.1)$$

where $f_i = f(\boldsymbol{x}_i)$, $h(f_i)$ is a monotone transformation of f_i known as the canonical parameter, and ϕ is a dispersion parameter. The function f models the effect of independent variables $\boldsymbol{x}$. Denote $\mu_i = \mathrm{E}(y_i)$, G_c as the canonical link such that $G_c(\mu_i) = h(f_i)$, and G as the link function such that $G(\mu_i) = f_i$. Then $h = G_c \circ G^{-1}$, and it reduces to the identity function when the canonical link is chosen for G. The last term $c(y_i, \phi)$ in (6.1) is independent of f.

A GLM assumes that $f(\boldsymbol{x}) = \boldsymbol{x}^T \boldsymbol{\beta}$. Similar to the linear models for Gaussian data, the parametric GLM may be too restrictive for some applications. We consider the nonparametric extension of the GLM in this chapter. In addition to providing more flexible models, the nonparametric extension also provides model building and diagnostic methods for GLMs.

Let the domain of each covariate x_k be an arbitrary set $\mathcal{X}_k$. Consider f as a multivariate function on the product domain $\mathcal{X} = \mathcal{X}_1 \times \mathcal{X}_2 \times \cdots \times \mathcal{X}_d$. The SS ANOVA decomposition introduced in Chapter 4 may be applied to construct candidate model spaces for f. In particular, we will assume that $f \in \mathcal{M}$, where

$$\mathcal{M} = \mathcal{H}^0 \oplus \mathcal{H}^1 \oplus \cdots \oplus \mathcal{H}^q$$

is an SS ANOVA model space defined in (4.30), $\mathcal{H}^0 = \text{span}\{\phi_1, \ldots, \phi_p\}$ is a finite dimensional space collecting all functions that are not penalized, and $\mathcal{H}^j$ for $j = 1, \ldots, q$ are orthogonal RKHS's with RKs R^j. The same notations in Chapter 4 will be used in this Chapter.

We assume the same density function (6.1) for y_i. However, for generality, we assume that f is observed through some linear functionals. Specifically, $f_i = \mathcal{L}_i f$, where $\mathcal{L}_i$ are known bounded linear functionals.

6.2 Estimation and Inference

6.2.1 Penalized Likelihood Estimation

Assume that $a_i(\phi) = a(\phi)/\varpi_i$ for $i = 1, \ldots, n$ where ϖ_i are known constants. Denote $\sigma^2 = a(\phi)$, $\boldsymbol{y} = (y_1, \ldots, y_n)^T$ and $\boldsymbol{f} = (f_1, \ldots, f_n)^T$. Let

$$l_i(f_i) = \varpi_i\{b(f_i) - y_i h(f_i)\}, \quad i = 1, \ldots, n, \tag{6.2}$$

and $l(\boldsymbol{f}) = \sum_{i=1}^n l_i(f_i)$. Then the log-likelihood

$$\sum_{i=1}^n \log g(y_i; f_i, \phi) = \sum_{i=1}^n \left\{ \frac{y_i h(f_i) - b(f_i)}{a_i(\phi)} + c(y_i, \phi) \right\}$$

$$= -\frac{1}{\sigma^2} l(\boldsymbol{f}) + C, \tag{6.3}$$

where C is independent of $\boldsymbol{f}$. Therefore, up to an additive and a multiplying constant, $l(\boldsymbol{f})$ is the negative log-likelihood. Note that $l(\boldsymbol{f})$ is independent of the dispersion parameter.

For a GLM, the MLE of parameters $\boldsymbol{\beta}$ are the maximizers of the log-likelihood. For a generalized SS ANOVA model, as in the Gaussian case, a penalty term is necessary to avoid overfitting. We will use the same form of penalty as in Chapter 4. Specifically, we estimate f as the solution to the following *penalized likelihood* (PL)

$$\min_{f \in \mathcal{M}} \left\{ l(\boldsymbol{f}) + \frac{n}{2} \sum_{j=1}^q \lambda_j \|P_j f\|^2 \right\}, \tag{6.4}$$

where P_j is the orthogonal projector in $\mathcal{M}$ onto $\mathcal{H}^j$, and λ_j are smoothing parameters. The multiplying term $1/\sigma^2$ is absorbed into smoothing parameters, and the constant C is dropped since it is independent of f. The multiplying constant $n/2$ is added such that the PL reduces to the

PLS (4.32) for Gaussian data. Under the new inner product defined in (4.33), the PL can be rewritten as

$$\min_{f \in \mathcal{M}} \left\{ l(\boldsymbol{f}) + \frac{n}{2} \lambda \|P_1^* f\|_*^2 \right\}, \tag{6.5}$$

where $\lambda_j = \lambda/\theta_j$, and $P_1^* = \sum_{j=1}^q P_j$ is the orthogonal projection in $\mathcal{M}$ onto $\mathcal{H}_1^* = \oplus_{j=1}^q \mathcal{H}^j$. Note that the RK of $\mathcal{H}_1^*$ under the new inner product is $R_1^* = \sum_{j=1}^q \theta_j R^j$.

It is easy to check that $l(\boldsymbol{f})$ is convex in f under the canonical link. In general, we assume that $l(\boldsymbol{f})$ is convex in f and has a unique minimizer in $\mathcal{H}^0$. Then the PL (6.5) has a unique minimizer (Theorem 2.9 in Gu (2002)). We now show the Kimeldorf–Wahba representer theorem holds. Let R^0 be the RK of $\mathcal{H}^0$ and $R = R^0 + R_1^*$. Let η_i be the representer associated with $\mathcal{L}_i$. Then, from (2.12),

$$\eta_i(x) = \mathcal{L}_{i(z)} R(x, z) = \mathcal{L}_{i(z)} R^0(x, z) + \mathcal{L}_{i(z)} R_1^*(x, z) \triangleq \delta_i(x) + \xi_i(x).$$

That is, the representers η_i for $\mathcal{L}_i$ belong to the finite dimensional subspace $\mathcal{S} = \mathcal{H}^0 \oplus \text{span}\{\xi_1, \dots, \xi_n\}$. Let $\mathcal{S}^c$ be the orthogonal complement of $\mathcal{S}$. Any $f \in \mathcal{H}$ can be decomposed into $f = \varsigma_1 + \varsigma_2$, where $\varsigma_1 \in \mathcal{S}$ and $\varsigma_2 \in \mathcal{S}^c$. Then we have

$$\mathcal{L}_i f = (\eta_i, f) = (\eta_i, \varsigma_1) + (\eta_i, \varsigma_2) = (\eta_i, \varsigma_1) = \mathcal{L}_i \varsigma_1.$$

Consequently, for any $f \in \mathcal{H}$, the PL (6.5) satisfies

$$\sum_{i=1}^n l_i(\mathcal{L}_i f) + \frac{n}{2} \lambda \|P_1^* f\|_*^2$$

$$= \sum_{i=1}^n l_i(\mathcal{L}_i \varsigma_1) + \frac{n}{2} \lambda (\|P_1^* \varsigma_1\|_*^2 + \|P_1^* \varsigma_2\|_*^2)$$

$$\geq \sum_{i=1}^n l_i(\mathcal{L}_i \varsigma_1) + \frac{n}{2} \lambda \|P_1^* \varsigma_1\|_*^2,$$

where equality holds iff $\|P_1^* \varsigma_2\|_* = \|\varsigma_2\|_* = 0$. Thus, the minimizer of the PL falls in the finite dimensional space $\mathcal{S}$, which can be represented as

$$\hat{f}(\boldsymbol{x}) = \sum_{\nu=1}^p d_\nu \phi_\nu(\boldsymbol{x}) + \sum_{i=1}^n c_i \xi_i$$

$$= \sum_{\nu=1}^p d_\nu \phi_\nu(\boldsymbol{x}) + \sum_{i=1}^n c_i \sum_{j=1}^q \theta_j \mathcal{L}_{i(z)} R^j(\boldsymbol{x}, z). \tag{6.6}$$

For simplicity of notation, the dependence of $\hat{f}$ on the smoothing parameters λ and $\boldsymbol{\theta} = (\theta_1, \dots, \theta_q)$ is not expressed explicitly. Let $T = \{\mathcal{L}_i \phi_\nu\}_{i=1}^n {}_{\nu=1}^p$, $\Sigma_k = \{\mathcal{L}_i \mathcal{L}_j R^k\}_{i,j=1}^n$ for $k = 1, \dots, q$ and $\Sigma_{\boldsymbol{\theta}} = \theta_1 \Sigma_1 + \dots + \theta_q \Sigma_q$. Let $\boldsymbol{d} = (d_1, \dots, d_p)^T$ and $\boldsymbol{c} = (c_1, \dots, c_n)^T$. Denote $\hat{\boldsymbol{f}} = (\mathcal{L}_1 \hat{f}, \dots, \mathcal{L}_n \hat{f})^T$. Then $\hat{\boldsymbol{f}} = T\boldsymbol{d} + \Sigma_{\boldsymbol{\theta}} \boldsymbol{c}$. Note that $\|P_1^* \hat{f}\|_*^2 = \boldsymbol{c}^T \Sigma_{\boldsymbol{\theta}} \boldsymbol{c}$. Substituting (6.6) into (6.5), we need to solve $\boldsymbol{c}$ and $\boldsymbol{d}$ by minimizing

$$I(\boldsymbol{c}, \boldsymbol{d}) = l(T\boldsymbol{d} + \Sigma_{\boldsymbol{\theta}} \boldsymbol{c}) + \frac{n}{2} \lambda \boldsymbol{c}^T \Sigma_{\boldsymbol{\theta}} \boldsymbol{c}. \tag{6.7}$$

Except for the Gaussian distribution, the function $l(\boldsymbol{f})$ is not quadratic and (6.7) cannot be solved directly. For fixed λ and $\boldsymbol{\theta}$, we will apply the Newton–Raphson procedure to compute $\boldsymbol{c}$ and $\boldsymbol{d}$. Let $u_i = dl_i/df_i$ and $w_i = d^2 l_i/df_i^2$, where $f_i = \mathcal{L}_i f$. Let $\boldsymbol{u}^T = (u_1, \dots, u_n)^T$ and $W = \operatorname{diag}(w_1, \dots, w_n)$. Then

$$\frac{\partial I}{\partial \boldsymbol{c}} = \Sigma_{\boldsymbol{\theta}} \boldsymbol{u} + n\lambda \Sigma_{\boldsymbol{\theta}} \boldsymbol{c},$$

$$\frac{\partial I}{\partial \boldsymbol{d}} = T^T \boldsymbol{u},$$

$$\frac{\partial^2 I}{\partial \boldsymbol{c} \partial \boldsymbol{c}^T} = \Sigma_{\boldsymbol{\theta}} W \Sigma_{\boldsymbol{\theta}} + n\lambda \Sigma_{\boldsymbol{\theta}},$$

$$\frac{\partial^2 I}{\partial \boldsymbol{c} \partial \boldsymbol{d}^T} = \Sigma_{\boldsymbol{\theta}} W T,$$

$$\frac{\partial^2 I}{\partial \boldsymbol{d} \partial \boldsymbol{d}^T} = T^T W T.$$

The Newton–Raphson procedure iteratively solves the linear system

$$\begin{pmatrix} \Sigma_{\boldsymbol{\theta}} W_- \Sigma_{\boldsymbol{\theta}} + n\lambda \Sigma_{\boldsymbol{\theta}} & \Sigma_{\boldsymbol{\theta}} W_- T \\ T^T W_- \Sigma_{\boldsymbol{\theta}} & T^T W_- T \end{pmatrix} \begin{pmatrix} \boldsymbol{c} - \boldsymbol{c}_- \\ \boldsymbol{d} - \boldsymbol{d}_- \end{pmatrix}$$
$$= \begin{pmatrix} -\Sigma_{\boldsymbol{\theta}} \boldsymbol{u}_- - n\lambda \Sigma_{\boldsymbol{\theta}} \boldsymbol{c}_- \\ -T^T \boldsymbol{u}_- \end{pmatrix}, \tag{6.8}$$

where the subscript minus indicates quantities evaluated at the previous Newton–Raphson iteration. Equations in (6.8) can be rewritten as

$$(\Sigma_{\boldsymbol{\theta}} W_- \Sigma_{\boldsymbol{\theta}} + n\lambda \Sigma_{\boldsymbol{\theta}}) \boldsymbol{c} + \Sigma_{\boldsymbol{\theta}} W_- T \boldsymbol{d} = \Sigma_{\boldsymbol{\theta}} W_- \hat{\boldsymbol{f}}_- - \Sigma_{\boldsymbol{\theta}} \boldsymbol{u}_-,$$
$$T^T W_- \Sigma_{\boldsymbol{\theta}} \boldsymbol{c} + T^T W_- T \boldsymbol{d} = T^T W_- \hat{\boldsymbol{f}}_- - T^T \boldsymbol{u}_-, \tag{6.9}$$

where $\hat{\boldsymbol{f}}_- = T\boldsymbol{d}_- + \Sigma_{\boldsymbol{\theta}} \boldsymbol{c}_-$. As discussed in Section 2.4, it is only necessary to derive one set of solutions. Let $\check{\boldsymbol{y}} = \hat{\boldsymbol{f}}_- - W_-^{-1} \boldsymbol{u}_-$. It is easy to see that a solution to

$$(\Sigma_{\boldsymbol{\theta}} + n\lambda W_-^{-1}) \boldsymbol{c} + T\boldsymbol{d} = \check{\boldsymbol{y}},$$
$$T^T \boldsymbol{c} = \boldsymbol{0}, \tag{6.10}$$

is also a solution to (6.9). Note that W_- is known at the current Newton–Raphson iteration. Since equations in (6.10) have the same form as those in (5.6), then methods in Section 5.2.1 can be used to solve (6.10). Furthermore, (6.10) corresponds to the minimizer of the following PWLS problem:

$$\frac{1}{n}\sum_{i=1}^{n} w_{i-}(\breve{y}_i - f_i)^2 + \sum_{j=1}^{q} \lambda_j \|P_j f\|^2, \tag{6.11}$$

where $\breve{y}_i$ is the ith element of $\breve{\boldsymbol{y}}$. Therefore, at each iteration, the Newton–Raphson procedure solves the PWLS criterion with working variables $\breve{y}_i$ and working weights w_{i-}. Consequently, the procedure can be regarded as iteratively reweighted PLS.

6.2.2 Selection of Smoothing Parameters

With canonical link such that $h(f) = f$, we have $\mathrm{E}(y_i) = b'(f_i)$, $\mathrm{Var}(y_i) = b''(f_i)a_i(\phi)$, $u_i = \varpi_i\{b'(f_i) - y_i\}$, and $w_i = \varpi_i b''(f_i)$. Therefore, $\mathrm{E}(u_i/w_i) = 0$ and $\mathrm{Var}(u_i/w_i) = \sigma^2 w_i^{-1}$. Consequently, when $\hat{f}_-$ is close to f and under some regularity conditions, it can be shown (Wang 1994, Gu 2002) that the working variables approximately have the same structure as in (5.1)

$$\breve{y}_i = \mathcal{L}_i f + \epsilon_i + o_p(1), \tag{6.12}$$

where ϵ_i has mean 0 and variance $\sigma^2 w_i^{-1}$. The Newton–Raphson procedure essentially reformulates the problem to model f on working variables at each iteration.

From the above discussion and note that W_- is known at the current Newton–Raphson iteration, we can use the UBR, GCV, and GML methods discussed in Section 5.2.3 to select smoothing parameters at each step of the Newton–Raphson procedure. Since working data are reformulated at each iteration, the target criteria of the above iterative smoothing parameter selection methods change throughout the iteration. Therefore, the overall target criteria of these iterative methods are not explicitly defined. A justification for the UBR criterion can be found in Gu (2002).

One practical problem with the iterative methods for selecting smoothing parameters is that they are not guaranteed to converge. Nevertheless, extensive simulations indicate that, in general, the above algorithm works reasonably well in practice (Wang 1994, Wang, Wahba, Chappell and Gu 1995).

Nonconvergence may become a serious problem for certain applications (Liu, Tong and Wang 2007). Some direct noniterative methods for selecting smoothing parameters have been proposed. For Poisson and gamma distributions, it is possible to derive unbiased estimates of the symmetrized Kullback–Leibler discrepancy (Wong 2006, Liu et al. 2007). Xiang and Wahba (1996) proposed a direct GCV method. Details about the direct GCV method can be found in Gu (2002). A direct GML method will be discussed in Section 6.2.4. These direct methods are usually more computationally intensive. They have not been implemented in the current version of the `assist` package. It is not difficult to write R functions to implement these direct methods. A simple implementation of the direct GML method for gamma distribution will be given in Sections 6.4 and 6.5.3.

6.2.3   Algorithm and Implementation

The whole procedure discussed in Sections 6.2.1 and 6.2.2 is summarized in the following algorithm.

Algorithm for generalized SS ANOVA models

1. Compute matrices T and Σ_k for $k = 1, \ldots, q$, and set an initial value for $\boldsymbol{f}$.

2. Compute $\boldsymbol{u}_{\_}$, $W_{\_}$, $\tilde{T} = W_{\_}^{1/2}T$, $\tilde{\Sigma}_k = W_{\_}^{1/2}\Sigma_k W_{\_}^{1/2}$ for $k = 1, \ldots, q$, and $\tilde{\boldsymbol{y}} = W_{\_}^{1/2}\check{\boldsymbol{y}}$, and fit the transformed data with smoothing parameters selected by the UBR, GCV, or GML method.

3. Iterate step 2 until convergence.

The above algorithm is easy to implement. All we need to do is to compute quantities u_i and w_i. We now compute these quantities for some special distributions.

First consider *logistic regression* with logit link function. Assume that $y \sim \text{Binomial}(m, p)$ with density function

$$g(y) = \binom{m}{y} p^y (1-p)^{m-y}, \quad y = 0, \ldots, m.$$

Then $\sigma^2 = 1$ and $l_i = -y_i f_i + m_i \log(1 + \exp(f_i))$, where $f_i = \log(p_i/(1 - p_i))$. Consequently, $u_i = -y_i + m_i p_i$ and $w_i = m_i p_i (1 - p_i)$. Note that binary data is a special case with $m_i = 1$.

Next consider *Poisson regression* with log-link function. Assume that $y \sim \text{Poisson}(\mu)$ with density function

$$g(y) = \frac{1}{y!} \mu^y \exp(-\mu), \quad y = 0, 1, \ldots .$$

Then $\sigma^2 = 1$ and $l_i = -y_i f_i + \exp(f_i)$, where $f_i = \log \mu_i$. Consequently, $u_i = -y_i + \mu_i$ and $w_i = \mu_i$.

Last consider *gamma regression* with log-link function. Assume that $y \sim \text{Gamma}(\alpha, \beta)$ with density function

$$g(y) = \frac{1}{\Gamma(\alpha)\beta^\alpha} y^{\alpha-1} \exp\left(-\frac{y}{\beta}\right), \quad y > 0,$$

where $\alpha > 0$ and $\beta > 0$ are shape and scale parameters. We are interested in modeling the mean $\mu \triangleq \text{E}(y)$ as a function of independent variables. We assume that the shape parameter does not depend on independent variables. Note that $\mu = \alpha\beta$. The density function may be reparametrized as

$$g(y) = \frac{\alpha^\alpha}{\Gamma(\alpha)\mu^\alpha} y^{\alpha-1} \exp\left(-\frac{\alpha y}{\mu}\right), \quad y > 0.$$

The canonical parameter $-\mu^{-1}$ is negative. To avoid this constraint, we adopt the log-link. Then $\sigma^2 = \alpha^{-1}$ and $l_i = y_i \exp(-f_i) + f_i$, where $f_i = \log \mu_i$. Consequently, $u_i = -y_i/\mu_i + 1$ and $w_i = y_i/\mu_i$. Since w_i are nonnegative, then $l(\boldsymbol{f})$ is a convex function of f.

For the binomial and the Poisson distributions, the dispersion parameter σ^2 is fixed as $\sigma^2 = 1$. For the gamma distribution, the dispersion parameter $\sigma^2 = \alpha^{-1}$. Since this constant has been separated from the definition of l_i, then we can set $\sigma^2 = 1$ in the UBR criterion. Therefore, for binomial, Poisson, and gamma distributions, the UBR criterion reduces to

$$\text{UBR}(\lambda, \boldsymbol{\theta}) = \frac{1}{n}\|(I - \tilde{H})\tilde{\boldsymbol{y}}\|^2 + \frac{2}{n}tr\tilde{H}, \tag{6.13}$$

where $\tilde{H}$ is the hat matrix associated with the transformed data.

In general, the weighted average of residuals

$$\hat{\sigma}_-^2 = \frac{1}{n}\sum_{i=1}^n w_{i-}\left(\frac{u_{i-}}{w_{i-}}\right)^2 = \frac{1}{n}\sum_{i=1}^n \frac{u_{i-}^2}{w_{i-}} \tag{6.14}$$

provides an estimate of σ^2 at the current iteration when it is unknown. Then the UBR criterion reduces to

$$\text{UBR}(\lambda, \boldsymbol{\theta}) = \frac{1}{n}\|(I - \tilde{H})\tilde{\boldsymbol{y}}\|^2 + \frac{2\hat{\sigma}_-^2}{n}tr\tilde{H}. \tag{6.15}$$

There are two versions of the UBR criterion given in equations (6.13) and (6.15). The first version is favorable when σ^2 is known.

The above algorithm is implemented in the `ssr` function for binomial, Poisson, and gamma distributions based on a collection of Fortran subroutines called `GRKPACK` (Wang 1997). The specific distribution is specified by the `family` argument. The method for selecting smoothing parameters is specified by the argument `spar` with "u~", "v", and "m" representing UBR, GCV, and GML criteria defined in (6.15), (5.25) and (5.26), respectively. The UBR method with fixed dispersion parameter (6.13) is specified as `spar=''u''` together with the option `varht` for specifying the fixed dispersion parameter. Specifically, for binomial, Poisson and gamma distributions with $\sigma^2 = 1$, the combination `spar=''u''` and `varht=1` is used.

6.2.4   Bayes Model, Direct GML and Approximate Bayesian Confidence Intervals

Suppose observations y_i are generated from (6.1) with $f_i = \mathcal{L}_i f$. Assume the same prior for f as in (4.42):

$$F(\boldsymbol{x}) = \sum_{\nu=1}^{p} \zeta_\nu \phi_\nu(\boldsymbol{x}) + \delta^{\frac{1}{2}} \sum_{j=1}^{q} \sqrt{\theta_j} U_j(\boldsymbol{x}), \qquad (6.16)$$

where $\zeta_\nu \overset{iid}{\sim} N(0, \kappa)$; $U_j(\boldsymbol{x})$ are independent, zero-mean Gaussian stochastic processes with covariance function $R^j(\boldsymbol{x}, \boldsymbol{z})$; ζ_ν and $U_j(\boldsymbol{x})$ are mutually independent; and κ and δ are positive constants. Conditional on $\boldsymbol{\zeta} = (\zeta_1, \ldots, \zeta_p)^T$, $\boldsymbol{f}|\boldsymbol{\zeta} \sim N(T\boldsymbol{\zeta}, \delta\Sigma_{\boldsymbol{\theta}})$. Letting $\kappa \to \infty$ and integrating out $\boldsymbol{\zeta}$, Gu (1992) showed that the marginal density of $\boldsymbol{f}$

$$p(\boldsymbol{f}) \propto \exp\left\{ -\frac{1}{2\delta} \boldsymbol{f}^T \left(\Sigma_{\boldsymbol{\theta}}^+ - \Sigma_{\boldsymbol{\theta}}^+ T (T^T \Sigma_{\boldsymbol{\theta}}^+ T)^{-1} T^T \Sigma_{\boldsymbol{\theta}}^+ \right) \boldsymbol{f} \right\},$$

where $\Sigma_{\boldsymbol{\theta}}^+$ is the Moore–Penrose inverse of $\Sigma_{\boldsymbol{\theta}}$. The marginal density of $\boldsymbol{y}$,

$$p(\boldsymbol{y}) = \int p(\boldsymbol{y}|\boldsymbol{f}) p(\boldsymbol{f}) d\boldsymbol{f}, \qquad (6.17)$$

usually does not have a closed form since, except for the Gaussian distribution, the log-likelihood $\log p(\boldsymbol{y}|\boldsymbol{f})$ is not quadratic in $\boldsymbol{f}$. Note that

$$\log p(\boldsymbol{y}|\boldsymbol{f}) = \sum_{i=1}^{n} \log g(y_i; f_i, \phi) = -\frac{1}{\sigma^2} l(\boldsymbol{f}) + C.$$

We now approximate $l(\boldsymbol{f})$ by a quadratic function.

Let u_{ic} and w_{ic} be u_i and w_i evaluated at $\hat{f}$. Let $\boldsymbol{u}_c = (u_{1c}, \ldots, u_{nc})^T$, $W_c = \text{diag}(w_{1c}, \ldots, w_{nc})$, and $\boldsymbol{y}_c = \hat{\boldsymbol{f}} - W_c^{-1} \boldsymbol{u}_c$. Note that $\partial l(\boldsymbol{f})/\partial \boldsymbol{f}|_{\hat{\boldsymbol{f}}} =$

$\boldsymbol{u}_c$, and $\partial^2 l(\boldsymbol{f})/\partial\boldsymbol{f}\partial\boldsymbol{f}^T|_{\hat{\boldsymbol{f}}} = W_c$. Expanding $l(\boldsymbol{f})$ as a function of $\boldsymbol{f}$ around the fitted values $\hat{\boldsymbol{f}}$ to the second order leads to

$$l(\boldsymbol{f}) \approx \frac{1}{2}(\boldsymbol{f} - \boldsymbol{y}_c)^T W_c(\boldsymbol{f} - \boldsymbol{y}_c) + l(\hat{\boldsymbol{f}}) - \frac{1}{2}\boldsymbol{u}_c^T W_c^{-1}\boldsymbol{u}_c. \qquad (6.18)$$

Note that $\log p(\boldsymbol{f})$ is quadratic in $\boldsymbol{f}$. Then it can be shown that, applying approximation (6.18), the marginal density of $\boldsymbol{y}$ in (6.17) is approximately proportional to (Liu, Meiring and Wang 2005)

$$p(\boldsymbol{y}) \propto |W_c|^{-\frac{1}{2}}|V|^{-\frac{1}{2}}|T^T V^{-1}T|^{-\frac{1}{2}}p(\boldsymbol{y}|\hat{\boldsymbol{f}})\exp\left\{\frac{1}{2\sigma^2}\boldsymbol{u}_c^T W_c^{-1}\boldsymbol{u}_c\right\}$$
$$\times \exp\left\{-\frac{1}{2}\boldsymbol{y}_c^T\left(V^{-1} - V^{-1}T(T^T V^{-1}T)^{-1}T^T V^{-1}\right)\boldsymbol{y}_c\right\}, (6.19)$$

where $V = \delta\Sigma_{\boldsymbol{\theta}} + \sigma^2 W_c^{-1}$. When $\Sigma_{\boldsymbol{\theta}}$ is nonsingular, $\hat{\boldsymbol{f}}$ is the maximizer of the integrand $p(\boldsymbol{y}|\boldsymbol{f})p(\boldsymbol{f})$ in (6.17) (Gu 2002). In this case the foregoing approximation is simply the Laplace approximation.

Let $\tilde{\boldsymbol{y}}_c = W_c^{1/2}\boldsymbol{y}_c$, $\tilde{\Sigma}_{\boldsymbol{\theta}} = W_c^{1/2}\Sigma_{\boldsymbol{\theta}}W_c^{1/2}$, $\tilde{T} = W_c^{1/2}T$, and the QR decomposition of $\tilde{T}$ be $(\tilde{Q}_1\ \tilde{Q}_2)(\tilde{R}^T\ 0)^T$. Let UEU^T be the eigendecomposition of $\tilde{Q}_2^T\tilde{\Sigma}_{\boldsymbol{\theta}}\tilde{Q}_2$, where $E = \mathrm{diag}(e_1,\ldots,e_{n-p})$, $e_1 \geq e_2 \geq \ldots \geq e_{n-p}$ are eigenvalues. Let $\boldsymbol{z} = (z_1,\ldots,z_{n-p})^T \triangleq U^T\tilde{Q}_2^T\tilde{\boldsymbol{y}}$. Then it can be shown (Liu et al. 2005) that (6.19) is equivalent to

$$p(\boldsymbol{y}) \propto |\tilde{R}|^{-1}p(\boldsymbol{y}|\hat{\boldsymbol{f}})\exp\left\{\frac{1}{2\sigma^2}\boldsymbol{u}_c^T W_c^{-1}\boldsymbol{u}_c\right\}$$
$$\times \prod_{\nu=1}^{n-p}(\delta e_\nu + \sigma^2)^{-\frac{1}{2}}\exp\left\{-\frac{1}{2}\sum_{\nu=1}^{n-p}\frac{z_\nu^2}{\delta e_\nu + \sigma^2}\right\}.$$

Let $\delta = \sigma^2/n\lambda$. Then an approximation of the negative log-marginal likelihood is

$$\mathrm{DGML}(\lambda, \boldsymbol{\theta}) = \log|\tilde{R}| + \frac{1}{\sigma^2}l(\hat{\boldsymbol{f}}) - \frac{1}{2\sigma^2}\boldsymbol{u}_c^T W_c^{-1}\boldsymbol{u}_c$$
$$+ \frac{1}{2}\sum_{i=1}^{n-p}\left\{\log(e_\nu/n\lambda + 1) + \frac{z_v^2}{\sigma^2(e_\nu/n\lambda + 1)}\right\}. \qquad (6.20)$$

Notice that $\hat{\boldsymbol{f}}$, $\boldsymbol{u}_c$, W_c, $\tilde{R}$, e_ν, and z_ν all depend on λ and $\boldsymbol{\theta}$ even though the dependencies are not expressed explicitly. The function $\mathrm{DGML}(\lambda, \boldsymbol{\theta})$ is referred to as the *direct generalized maximum likelihood* (DGML) criterion, and the minimizers of $\mathrm{DGML}(\lambda, \boldsymbol{\theta})$ are called the DGML estimate of the smoothing parameter. Section 6.4 shows a simple implementation of the DGML criterion for the gamma distribution.

Let $\boldsymbol{y}_c = (y_{1c}, \ldots, y_{nc})^T$. Based on (6.12), consider the approximation model at convergence

$$y_{ic} = \mathcal{L}_i f + \epsilon_i, \quad i = 1, \ldots, n, \tag{6.21}$$

where ϵ_i has mean 0 and variance $\sigma^2 w_{ic}^{-1}$. Assume prior (6.16) for f. Then, as in Section 5.2.5, Bayesian confidence intervals can be constructed for f in the approximation model (6.21). They provide approximate confidence intervals for f in the generalized SS ANOVA model. The bootstrap approach may also be used to construct confidence intervals, and the extension is straightforward.

Connections between smoothing spline models and LME models are presented in Sections 3.5, 4.7, and 5.2.4. We now extend this connection to data from exponential families. Consider the following generalized linear mixed-effects model (GLMM) (Breslow and Clayton 1993)

$$G\{\mathrm{E}(\boldsymbol{y}|\boldsymbol{u})\} = T\boldsymbol{d} + Z\boldsymbol{u}, \tag{6.22}$$

where G is the link function, $\boldsymbol{d}$ are fixed effects, $\boldsymbol{u} = (\boldsymbol{u}_1^T, \ldots, \boldsymbol{u}_q^T)^T$ are random effects, $Z = (I_n, \ldots, I_n)$, $\boldsymbol{u}_k \sim \mathrm{N}(\boldsymbol{0}, \sigma^2 \theta_k \Sigma_k / n\lambda)$ for $k = 1, \ldots, q$, and $\boldsymbol{u}_k$ are mutually independent. Then $\boldsymbol{u} \sim \mathrm{N}(\boldsymbol{0}, \sigma^2 D / n\lambda)$, where $D = \mathrm{diag}(\theta_1 \Sigma_1, \ldots, \theta_q \Sigma_q)$. Setting $\boldsymbol{u}_k = \theta_k \Sigma_k \boldsymbol{c}$ as in the Gaussian case (Opsomer et al. 2001) and noting that $ZDZ^T = \Sigma_\theta$, we have $\boldsymbol{u} = DZ^T \boldsymbol{c}$ and $\boldsymbol{u}^T \{\mathrm{Cov}(\boldsymbol{u})\}^+ \boldsymbol{u} = n\lambda \boldsymbol{c}^T ZDD^+ DZ^T \boldsymbol{c} / \sigma^2 = n\lambda \boldsymbol{c}^T ZDZ^T \boldsymbol{c} / \sigma^2 = n\lambda \boldsymbol{c}^T \Sigma_\theta \boldsymbol{c} / \sigma^2$. Note that $Z\boldsymbol{u} = \Sigma_\theta \boldsymbol{c}$. Therefore the PL (6.7) is the same as the penalized quasi-likelihood (PQL) of the GLMM (6.22) (equation (6) in Breslow and Clayton (1993)).

6.3 Wisconsin Epidemiological Study of Diabetic Retinopathy

We use the Wisconsin Epidemiological Study of Diabetic Retinopathy (WESDR) data to illustrated how to fit an SS ANOVA model to binary responses. Based on Wahba, Wang, Gu, Klein and Klein (1995), we investigate how probability of progression to diabetic retinopathy at the first follow-up (prg) depends on the following covariates at baseline: duration of diabetes (dur), glycosylated hemoglobin (gly), and body mass index (bmi).

Let y be the response variable prg where $y = 1$ represents progression of retinopathy and $y = 0$ otherwise. Let x_1, x_2, and x_3 be the covariates dur, gly, and bmi transformed into $[0, 1]$. Let $\boldsymbol{x} = (x_1, x_2, x_3)$ and

$f(\boldsymbol{x}) = \text{logit} P(y = 1|\boldsymbol{x})$. We model f using the tensor product space $W_2^2[0,1] \otimes W_2^2[0,1] \otimes W_2^2[0,1]$. The three-way SS ANOVA decomposition can be derived similarly using the method in Chapter 4. For simplicity, we will ignore three-way interactions and start with the following SS ANOVA model with all two-way interactions:

$$
\begin{aligned}
f(\boldsymbol{x}) = {} & \mu + \beta_1 \times (x_1 - 0.5) + \beta_2 \times (x_2 - 0.5) + \beta_3 \times (x_3 - 0.5) \\
& + \beta_4 \times (x_1 - 0.5)(x_2 - 0.5) + \beta_5 \times (x_1 - 0.5)(x_3 - 0.5) \\
& + \beta_6 \times (x_2 - 0.5)(x_3 - 0.5) + f_1^s(x_1) + f_2^s(x_2) + f_3^s(x_3) \\
& + f_{12}^{ls}(x_1, x_2) + f_{12}^{sl}(x_1, x_2) + f_{12}^{ss}(x_1, x_2) \\
& + f_{13}^{ls}(x_1, x_3) + f_{13}^{sl}(x_1, x_3) + f_{13}^{ss}(x_1, x_3) \\
& + f_{23}^{ls}(x_2, x_3) + f_{23}^{sl}(x_2, x_3) + f_{23}^{ss}(x_2, x_3).
\end{aligned} \tag{6.23}
$$

The following statements fit model (6.23) with smoothing parameter selected by the UBR method:

```
> data(wesdr); attach(wesdr)
> y <- prg
> x1 <- (dur-min(dur))/diff(range(dur))
> x2 <- (gly-min(gly))/diff(range(gly))
> x3 <- (bmi-min(bmi))/diff(range(bmi))
> wesdr.fit1 <- ssr(y~I(x1-.5)+I(x2-.5)+I(x3-.5)+
   I((x1-.5)*(x2-.5))+I((x1-.5)*(x3-.5))+
   I((x2-.5)*(x3-.5)),
   rk=list(cubic(x1), cubic(x2), cubic(x3),
           rk.prod(kron(x1-.5),cubic(x2)),
           rk.prod(kron(x2-.5),cubic(x1)),
           rk.prod(cubic(x1),cubic(x2)),
           rk.prod(kron(x1-.5),cubic(x3)),
           rk.prod(kron(x3-.5),cubic(x1)),
           rk.prod(cubic(x1),cubic(x3)),
           rk.prod(kron(x2-.5),cubic(x3)),
           rk.prod(kron(x3-.5),cubic(x2)),
           rk.prod(cubic(x2),cubic(x3))),
           family=''binary'', spar=''u'', varht=1)
> summary(wesdr.fit1)
...
UBR estimate(s) of smoothing parameter(s) :
  6.913248e-06 1.920409e+04 9.516636e-01 2.966542e+03
  6.005694e+02 6.345814e+01 5.602521e+02 2.472658e+02
  1.816387e-07 9.820496e-07 1.481754e+03 2.789458e-07
Equivalent Degrees of Freedom (DF):  12.16
```

Components corresponding to large values of smoothing parameters are small. As in Section 4.9.2, we compute the Euclidean norm of the estimate for each component centered around zero:

```
> norm.cen <- function(x) sqrt(sum((x-mean(x))**2))
> comp.est1 <- predict(wesdr.fit1, pstd=F,
                        terms=diag(rep(1,19)))[-1,])
> comp.norm1 <- apply(comp.est1$fit, 2, norm.cen)
> print(round(comp.norm1,2))
  9.13 44.02 15.25  5.40 14.29 21.88  8.51  0.00  0.00
  0.00 0.00 0.00 0.00  0.00  5.17  5.39  0.00  2.41
```

Both the estimates of smoothing parameters and the norms indicate that the interaction between dur (x_1) and gly (x_2) can be dropped. Therefore, we fit the SS ANOVA model (6.23) with components f_{12}^{ls}, f_{12}^{sl}, and f_{12}^{ss} being eliminated:

```
> wesdr.fit2 <- update(wesdr.fit1,
    rk=list(cubic(x1), cubic(x2), cubic(x3),
            rk.prod(kron(x1-.5),cubic(x3)),
            rk.prod(kron(x3-.5),cubic(x1)),
            rk.prod(cubic(x1),cubic(x3)),
            rk.prod(kron(x2-.5),cubic(x3)),
            rk.prod(kron(x3-.5),cubic(x2)),
            rk.prod(cubic(x2),cubic(x3))))
> comp.est2 <- predict(wesdr.fit2,
                       terms=diag(rep(1,16))[-1,], pstd=F)
> comp.norm2 <- apply(comp.est2$fit, 2, norm.cen)
> print(round(comp.norm2,2))
  9.13 44.02 15.25  5.40 14.29 21.88  8.51  0.00  0.00
  0.00  0.00  5.17  5.39  0.00  2.41
```

Compared to other components, the norms of the nonparametric interactions f_{13}^{ls}, f_{13}^{sl}, f_{13}^{ss}, f_{23}^{ls}, f_{23}^{sl}, and f_{23}^{ss} are relatively small. We further compute 95% Bayesian confidence intervals for the overall nonparametric interaction between x_1 and x_3, $f_{13}^{ls} + f_{13}^{sl} + f_{13}^{ss}$, the overall nonparametric interaction between x_2 and x_3, $f_{23}^{ls} + f_{23}^{sl} + f_{23}^{ss}$, and the proportion of design points for which zero is outside these confidence intervals:

```
> int.dur.bmi <- predict(wesdr.fit2,
    terms=c(rep(0,10),rep(1,3),rep(0,3)))
> mean(((int.dur.bmi$fit-1.96*int.dur.bmi$pstd>0)|
       (int.dur.bmi$fit+1.96*int.dur.bmi$pstd<0))
  0
```

```
> int.gly.bmi <- predict(wesdr.fit2,
    terms=c(rep(0,13),rep(1,3)))
> mean((int.gly.bmi$fit-1.96*int.gly.bmi$pstd>0)|
        (int.gly.bmi$fit+1.96*int.gly.bmi$pstd<0))
0
```

Therefore, for both overall nonparametric interactions, the 95% confidence intervals contain zero at all design points. This suggests that the nonparametric interactions may be dropped. We fit the SS ANOVA model (6.23) with all nonparametric interactions being eliminated:

```
> wesdr.fit3 <- update(wesdr.fit2,
    rk=list(cubic(x1), cubic(x2), cubic(x3)))
> summary(wesdr.fit3)
UBR estimate(s) of smoothing parameter(s) :
 4.902745e-06 5.474122e+00 1.108322e-05
Equivalent Degrees of Freedom (DF):  11.78733

> comp.est3 <- predict(wesdr.fit3, pstd=F,
                    terms=diag(rep(1,10))[-1,])
> comp.norm3 <- apply(comp.est3$fit, 2, norm.cen)
> print(round(comp.norm3))
  7.22 33.71 12.32  4.02  8.99 11.95 10.15  0.00  6.79
```

Based on the estimates of smoothing parameters and the norms, the nonparametric main effect of x_2, f_2^s, can be dropped. Therefore, we fit the final model:

```
> wesdr.fit4 <- update(wesdr.fit3,
    rk=list(cubic(x1), cubic(x3)))
> summary(wesdr.fit4)
...
UBR estimate(s) of smoothing parameter(s) :
 4.902693e-06 1.108310e-05
Equivalent Degrees of Freedom (DF):  11.78733
Estimate of sigma:  1
```

To look at the effect of dur, with gly and bmi being fixed at the medians of their observed values, we compute estimates of the probabilities and posterior standard deviations at a grid point of dur. The estimated probability function and the approximate 95% Bayesian confidence intervals are shown in Figure 6.1(a). The risk of progression of retinopathy increases up to a duration of about 10 years and then decreases, possibly caused by censoring due to death in patients with longer durations. Similar plots for the effects of gly and bmi are shown in Figures 6.1(b) and

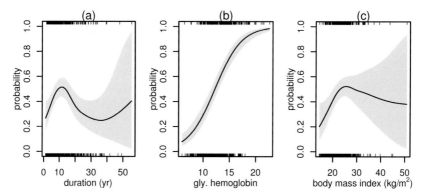

FIGURE 6.1   WESDR data, plots of (a) the estimated probability as a function of `dur` with `gly` and `bmi` being fixed at the medians of their observed values, (b) the estimated probability as a function of `gly` with `dur` and `bmi` being fixed at the medians of their observed values, and (c) the estimated probability as a function of `bmi` with `dur` and `gly` being fixed at the medians of their observed values. Shaded regions are approximate 95% Bayesian confidence intervals. Rugs on the bottom and the top of each plot are observations of `prg`.

6.1(c). The risk of progression of retinopathy increases with increasing glycosylated hemoglobin, and the risk increases with increasing body mass index until a value of about 25 kg/m^2, after which the trend is uncertain due to wide confidence intervals. As expected, the confidence intervals are wider in areas where observations are sparse.

6.4   Smoothing Spline Estimation of Variance Functions

Consider the following heteroscedastic SS ANOVA model

$$y_i = \mathcal{L}_{1i}f_1 + \exp\{\mathcal{L}_{2i}f_2/2\}\epsilon_i, \quad i = 1, \ldots, n, \qquad (6.24)$$

where f_k is a function on $\mathcal{X}_k = \mathcal{X}_{k1} \times \mathcal{X}_{k2} \times \cdots \times \mathcal{X}_{kd_k}$ with model space $\mathcal{M}_k = \mathcal{H}^{k0} \oplus \mathcal{H}^{k1} \oplus \cdots \oplus \mathcal{H}^{kq_k}$ for $k = 1, 2$; $\mathcal{L}_{1i}$ and $\mathcal{L}_{2i}$ are bounded linear functionals; and $\epsilon_i \overset{iid}{\sim} \mathrm{N}(0, 1)$. The goal is to estimate both the mean function f_1 and the variance function f_2. Note that both the mean and variance functions are modeled nonparametrically in this section. This is in contrast to the parametric model (5.28) for variance functions in Chapter 5.

One simple approach to estimating functions f_1 and f_2 is to use the following two-step procedure:

1. Estimate the mean function f_1 as if random errors are homoscedastic.

2. Estimate the variance function f_2 based on squared residuals from the first step.

3. Estimate the mean function again using the estimated variance function.

The PLS estimation method in Chapter 4 can be used in the first step. Denote the estimate at the first step as $\tilde{f}_1$ and $r_i = y_i - \mathcal{L}_{1i}\tilde{f}_1$ as residuals. Let $z_i = r_i^2$. Under suitable conditions, $z_i \approx \exp\{\mathcal{L}_{2i}f_2\}\chi_{i,1}^2$, where $\chi_{i,1}^2$ are iid Chi-square random variables with degree of freedom 1. Regarding Chi-square distribution as a special case of the gamma distribution, the PL method described in this chapter can be used to estimate the variance function f_2 at the second step. Denote the estimate at the second step as $\tilde{f}_2$. Then the PWLS method in Chapter 5 can be used in the third step with known covariance $W^{-1} = \text{diag}(\mathcal{L}_{21}\tilde{f}_2, \ldots, \mathcal{L}_{2n}\tilde{f}_2)$.

We now use the motorcycle data to illustrate the foregoing two-step procedure. We have fitted a cubic spline to the logarithm of squared residuals in Section 5.4.1. Based on model (3.53), consider the heteroscedastic partial spline model

$$y_i = f_1(x_i) + \exp\{f_2(x_i)/2\}\epsilon_i, \quad i = 1, \ldots, n, \tag{6.25}$$

where $f_1(x) = \sum_{j=1}^{3} \beta_j (x - t_j)_+ + g_1(x)$, $t_1 = 0.2128$, $t_2 = 0.3666$ and $t_3 = 0.5113$ are the change-points in the first derivative, and $\epsilon_i \overset{iid}{\sim} N(0,1)$. We model both g_1 and f_2 using the cubic spline model space $W_2^2[0,1]$. The following statements implement the two-step procedure for model (6.25):

```
# step 1
> t1 <- .2128; t2 <- .3666; t3 <- .5113;
> s1 <- (x-t1)*(x>t1); s2 <- (x-t2)*(x>t2)
> s3 <- (x-t3)*(x>t3)
> mcycle.ps.fit4 <- ssr(accel~x+s1+s2+s3, rk=cubic(x))

# step 2
> z1 <- residuals(mcycle.ps.fit4)**2
> mcycle.v.1 <- ssr(z1~x, cubic(x), limnla=c(-6,-1),
```

```
family=''gamma'', spar=''u'', varht=1)
```

```
# step 3
> mcycle.ps.fit5 <- update(mcycle.ps.fit4,
    weights=mcycle.v.1$fit)
```

In the second step, the search range for $\log_{10}(n\lambda)$ is set as $[-6, -1]$ to avoid numerical problems. The actual estimate of the smoothing parameter $\log_{10}(n\hat{\lambda}) = -6$, which leads to a rough estimate of f_2 (Figure 6.2(a)).

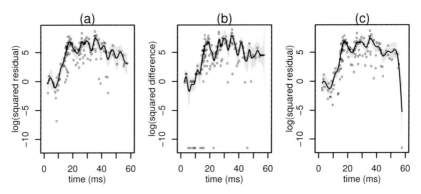

FIGURE 6.2   Motorcycle data, plots of the PL cubic spline estimates of f_2 (lines), and 95% approximate Bayesian confidence intervals (shaded regions) based on (a) the two-step procedure with squared residuals, (b) the two-step procedure with squared differences, and (c) the backfitting procedure. Circles are (a) logarithm of the squared residuals based on model (3.53), (b) logarithm of the squared differences, and (c) logarithm of the squared residuals based on the final fit to the mean function in the backfitting procedure.

The first step in the two-step procedure may be replaced by a difference-based method. Note that $x_1 \leq x_2 \leq \cdots \leq x_n$ in the motorcycle data. Let $z_i = (y_{i+1} - y_i)^2/2$ for $i = 1,\ldots,n-1$. When $x_{i+1} - x_i$ is small, $y_{i+1} - y_i = f_1(x_{i+1}) - f_1(x_i) + \exp\{f_2(x_{i+1})/2\}\epsilon_{i+1} - \exp\{f_2(x_i)/2\}\epsilon_i \approx \exp\{f_2(\tilde{x}_i)/2\}(\epsilon_{i+1} - \epsilon_i)$, where $\tilde{x}_i = (x_{i+1} + x_i)/2$. Then, $z_i \approx \exp\{f_2(\tilde{x}_i)\}\chi_{i,1}^2$, where $\chi_{i,1}^2$ are Chi-square random variables with degree of freedom 1. Ignoring correlations between neighboring observations, the following statements implement this difference-based method:

```
> z2 <- diff(accel)**2/2; z2[z2<.00001] <- .00001
```

```
> n <- length(x); xx <- (x[1:(n-1)]+x[2:n])/2
> mcycle.v.g <- ssr(z2~xx, cubic(xx), limnla=c(-6,-1),
    family=''gamma'', spar=''u'', varht=1)
> w <- predict(mcycle.v.g, pstd=F,
                newdata=data.frame(xx=x))
> mcycle.ps.fit6 <- update(mcycle.ps.fit4,
    weights=exp(w$fit))
```

As in Yuan and Wahba (2004), we set $z_i = \max\{.00001, (y_{i+1} - y_i)^2/2\}$ to avoid overfitting. Again, the actual estimate of the smoothing parameters reaches the lower bound such that $\log_{10}(n\hat{\lambda}) = -6$. The estimate of f_2 is rough (Figure 6.2(b)).

A more formal procedure is to estimate f_1 and f_2 in model (6.24) jointly as the minimizers of the following doubly penalized likelihood (DPL) (Yuan and Wahba 2004):

$$\frac{1}{n}\sum_{i=1}^{n}\left\{(y_i - \mathcal{L}_{1i}f_1)^2 \exp(-\mathcal{L}_{2i}f_2) + \mathcal{L}_{2i}f_2\right\} + \sum_{k=1}^{2}\sum_{j=1}^{q_k}\lambda_{kj}\|P_{kj}f_k\|^2,$$

(6.26)

where the first part is the negative log-likelihood, P_{kj} is the orthogonal projector in $\mathcal{M}_k$ onto $\mathcal{H}^{kj}$ for $k = 1, 2$, and λ_{kj} are smoothing parameters. Following the same arguments in Section 6.2.1, the minimizers of the DPL can be represented as

$$\hat{f}_k(\boldsymbol{x}) = \sum_{\nu=1}^{p_k}d_{k,\nu}\phi_{k,\nu}(\boldsymbol{x}) + \sum_{i=1}^{n}c_{k,i}\sum_{j=1}^{q_k}\theta_{kj}\mathcal{L}_{k,i(\boldsymbol{z})}R^{kj}(\boldsymbol{x},\boldsymbol{z}),$$

$$\boldsymbol{x}, \boldsymbol{z} \in \mathcal{X}_k, \; k = 1, 2,$$

(6.27)

where $\lambda_{kj} = \lambda_k/\theta_{kj}$, $\phi_{k,\nu}$ for $\nu = 1, \ldots, p_k$ are basis functions of $\mathcal{H}^{k0}$, and R^{kj} for $j = 1, \ldots, q_k$ are RKs of $\mathcal{H}^{kj}$. It is difficult to solve coefficients $d_{k,\nu}$ and $c_{k,\nu}$ directly. However, it is easy to implement a back-fitting procedure by iterating the following two steps until convergence: (a) Conditional on current estimates of $d_{2,\nu}$ and $c_{2,\nu}$, update $d_{1,\nu}$ and $c_{1,\nu}$; (b) Conditional on current estimates of $d_{1,\nu}$ and $c_{1,\nu}$, update $d_{2,\nu}$ and $c_{2,\nu}$. Note that, when $d_{2,\nu}$ and $c_{2,\nu}$ are fixed, f_2 in (6.26) is fixed, and the DPL reduces to the PWLS (5.3) with known weights. When $d_{1,\nu}$ and $c_{1,\nu}$ are fixed, f_1 in (6.27) is fixed and the DPL reduces to the PL (6.4) for observations $z_i = \exp\{\mathcal{L}_{2i}f_2\}\chi_{i,1}^2$. Therefore, steps (a) and (b) correspond to steps 2 and 3 in the two-step procedure, and the backfitting procedure extends the two-step procedure by iterating steps 2 and 3 until convergence. The above backfitting procedure is essentially the same as the iterative procedure in Yuan and Wahba (2004) where different methods were used to select smoothing parameters. The following is a simple R function that implements the backfitting procedure for motorcycle data:

```
> jemv <- function(x, y, prec=1e-6, maxit=30) {
    t1 <- .2128; t2 <- .3666; t3 <- .5113
    s1 <- (x-t1)*(x>t1); s2 <- (x-t2)*(x>t2)
    s3 <- (x-t3)*(x>t3)
    err <- 1e20; eta <- rep(1,length(x))
    while (err>prec&maxit>0) {
    fitf <- ssr(y~x+s1+s2+s3, cubic(x), weights=exp(eta))
    z <- fitf$resi**2; z[z<.00001] <- .00001
    fitv <- ssr(z~x, cubic(x), limnla=c(-5,-1),
     family=''gamma'', spar=''u'', varht=1)
    oldeta <- eta
    eta <- fitv$rkpk$eta
    err <- sqrt(mean(((eta-oldeta)/(1+abs(oldeta)))**2))
    maxit <- maxit-1
    }
    return(list(fitf=fitf,fitv=fitv))
    }
> mcycle.mv <- jemv(x, accel)
```

For estimation of the variance function, a new search range for $\log_{10}(n\lambda)$ has to be set as $[-5, -1]$ to avoid numerical problems. The backfitting algorithm converged in 13 iterations. The estimate of f_2 is shown in Figure 6.2(c).

The smoothing parameters in the above procedures are selected by the iterative UBR method. For Chi-square distributed response variables with small degrees of freedom, nonconvergence of the iterative approach may become a serious problem (Liu et al. 2007). Note that the degrees of freedom of Chi-square random variables in the above procedures equal to 1. The direct methods such as the DGML criterion guarantee convergence. We now show a simple implementation of the DGML method for gamma regression. Instead of the Newton–Raphson method, we use the Fisher scoring method, which leads to $u_i = 1 - y_i/\mu_i$ and $w_i = 1$. Consequently, $W_c = I$, and there is no need for the transformation. Furthermore, $|\tilde{R}|$ will be dropped since it is independent of the smoothing parameter. The following R function computes the DGML in (6.20) for gamma regression:

```
DGML <- function(th, nlaht, y, S, Q) {
  if (length(th)==1) { nlaht <- th; Qt <- Q }
  if (length(th)>1) {
```

```
  theta <- 10^th; Qt <- 0
  for (i in 1:dim(Q)[3]) Qt <- Qt + theta[i]*Q[,,i]
}
fit <- try(ssr(y~S-1, Qt, limnla=nlaht, family=''gamma'',
               spar=''u'', varht=1))
if(class(fit)==''ssr'') {
  fht <- fit$rkpk$eta
  tmp <- y*exp(-fht)
  uc <- 1-tmp
  yt <- fht-uc
  qrq <- qr.Q(qr(S), complete=T)
  q2 <- qrq[ ,(ncol(S)+1):nrow(S)]
  V <- t(q2)%*%Qt%*%q2
  l <- eigen((V + t(V))/2)
  U <- l$vectors
  e <- l$values
  z <- t(U)%*%t(q2)%*%yt
  delta <- 10^{-nlaht}
  GML <- sum(tmp+fit)-sum(uc^2)/2+sum(log(delta*e+1))/2
   +sum(z^2/(delta*e+1))/2
  return(GML)
}
else return(1e10)
}
```

where `fht`, `uc`, `yt`, `V`, `U`, `e`, `z`, and `delta` correspond to $\hat{f}$, u_c, $\tilde{y}$, $\tilde{Q}_2^T \tilde{\Sigma}_\theta \tilde{Q}_2$, U, $(e_1, \ldots, e_{n-p})^T$, z, and δ, respectively, in the definition of the DGML in (6.20). Note that $\sigma^2 = 1$. The R functions `qr` and `eigen` are used to compute the QR decomposition and eigendecomposition. For an SS ANOVA model with $q = 1$, the input `th` corresponds to $\log_{10}(n\lambda)$, `nlaht` is not used, `y` are observations, `S` corresponds to the matrix T, and `Q` corresponds to the matrix Σ. For an SS ANOVA model with $q > 1$, the input `th` corresponds to $\log_{10} \boldsymbol{\theta}$, `nlaht` corresponds to $\log_{10}(n\lambda)$, `y` are observations, `S` corresponds to the matrix T, and `Q` is an (n, n, q) array corresponding to the matrices Σ_k for $k = 1, \ldots, q$.

We now apply the above DGML method to the second step in the two-step procedure. We compute the DGML criterion on a grid of $\log_{10}(n\lambda)$, find the DGML estimate of the smoothing parameter as the minimizer of the DGML criterion, and fit again with the smoothing parameter being fixed as the DGML estimate:

```
> S <- cbind(1,x); Q <- cubic(x)
> lgrid <- seq(-6,-2,by=.1)
> gml <- sapply(lgrid, DGML, nlaht=0, y=z1, S=S, Q=Q)
```

```
> nlaht <- lgrid[order(gml)[1]]
> mcycle.v.g4 <- ssr(z1~x, cubic(x), limnla=nlaht,
    family=''gamma'', spar=''u'', varht=1)
```

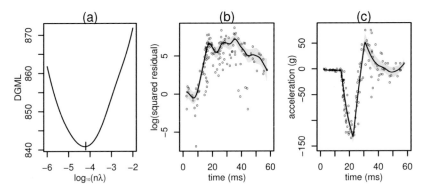

FIGURE 6.3   Motorcycle data, plots of (a) the DGML function where the minimum point is marked with a short bar at the bottom; (b) logarithm of squared residuals (circles) based on model (3.53), PL cubic spline estimate of f_2 (line) to squared residuals with the DGML estimate of the smoothing parameter, and 95% approximate Bayesian confidence intervals (shaded region); and (c) observations (circles) and PWLS partial spline estimate of f_1 (line) with 95% Bayesian confidence intervals (shaded region).

The DGML function is shown in Figure 6.3(a). It reaches the minimum at $\log_{10}(n\hat{\lambda}) = -4.2$. The PL cubic spline estimate of f_2 with the DGML choice of the smoothing parameter is shown in Figure 6.3(b). The PWLS partial spline estimate of f_1 and 95% Bayesian confidence intervals are shown in Figure 6.3(c). The effect of unequal variances is reflected in the widths of the confidence intervals.

6.5   Smoothing Spline Spectral Analysis

6.5.1   Spectrum Estimation of a Stationary Process

The spectrum is often used to describe the power distribution of a time series. Consider a zero-mean stationary time series $X_t, t = 0, \pm1, \pm2, \ldots$.

The spectrum is defined as

$$S(\omega) = \sum_{u=-\infty}^{\infty} \gamma(u)\exp(-i2\pi\omega u), \quad \omega \in [0,1], \tag{6.28}$$

where $\gamma(u) = \mathrm{E}(X_t X_{t+u})$ is the covariance function, and $i = \sqrt{-1}$. Let $X_0, X_1, \ldots, X_{T-1}$ be a finite sample of the stationary process and define the periodogram at frequency $\omega_k = k/T$ as

$$y_k = T^{-1}\Big|\sum_{t=0}^{T-1} X_t \exp(i2\pi\omega_k t)\Big|^2, \quad k = 0, \ldots, T-1. \tag{6.29}$$

The periodogram is an asymptotically unbiased but inconsistent estimator of the underlying true spectrum. Many different smoothing techniques have been proposed to overcome this problem. We now show how to estimate the spectrum using smoothing spline based on observations $\{(\omega_k, y_k), \; k = 0, \ldots, T-1\}$.

Under standard mixing conditions, the periodograms are asymptotically independent and distributed as

$$y_k \sim \begin{cases} S(\omega_k)\chi_1^2, & \omega_k = 0, 1/2, \\ S(\omega_k)\chi_2^2/2, & \omega_k \neq 0, 1/2. \end{cases}$$

Regarding Chi-square distribution as a special case of the gamma distribution, the method described in this chapter can be used to estimate the spectrum. Consider the logarithmic link function and let $f = \log(S)$ be the log spectrum. We model function f using the periodic spline space $W_2^m(per)$. Note that $f(\omega)$ is symmetric about $\omega = 0.5$. Therefore, it suffices to estimate $f(\omega)$ for $\omega \in [0, 0.5]$. Nevertheless, to estimate f as a periodic function, we will use periodograms at all frequencies in Section 6.5.3.

6.5.2 Time-Varying Spectrum Estimation of a Locally Stationary Process

Many time series are nonstationary. Locally stationary processes have been proposed to approximate the nonstationary time series. The time-varying spectrum of a locally stationary time series characterizes changes of stochastic variation over time.

A zero-mean stochastic process $\{X_t, \; t = 0, \ldots, T-1\}$ is a locally stationary process if

$$X_t = \int_0^1 A(\omega, t/T)\exp(i2\pi\omega t)dZ(\omega), \tag{6.30}$$

where $Z(\omega)$ is a zero-mean stochastic process on $[0,1]$, and $A(\omega, u)$ denotes a transfer function with continuous second-order derivatives for $(\omega, u) \in [0,1] \times [0,1]$. Define the time-varying spectrum as

$$S(\omega, u) = \|A(\omega, u)\|^2. \tag{6.31}$$

Consider the logarithmic link function and let $f(\omega, u) = \log S(\omega, u)$. Since it is a periodic function of ω, we model the log spectrum f using the tensor product space $W_2^2(per) \otimes W_2^2[0,1]$, where the SS ANOVA decomposition was derived in Section 4.4.5. The SS ANOVA model using notations in this section can be represented as

$$f(\omega, u) = \mu + \beta \times (u - 0.5) + f_2^s(u) + f_1(\omega) + f_{12}^{sl}(\omega, u) + f_{12}^{ss}(\omega, u), \tag{6.32}$$

where $\beta \times (u-0.5)$ and $f_2^s(u)$ are linear and smooth main effects of time, $f_1(\omega)$ is the smooth main effect of frequency, and $f_{12}^{sl}(\omega, u)$ and $f_{12}^{ss}(\omega, u)$ are linear–smooth and smooth–smooth interactions between frequency and time.

To estimate the bivariate function f, we compute local periodograms on some time-frequency grids. Specifically, divide the time domain into J disjoint blocks $[b_j, b_{j+1})$, where $0 = b_1 < b_2 < \cdots < b_J < b_{J+1} = 1$. Let $u_j = (b_j + b_{j+1})/2$ be the middle points of these J blocks. Let ω_k for $k = 1, \ldots, K$ be K frequencies in $[0,1]$. Define the local periodograms as

$$y_{(k-1)J+j} = \frac{|\sum_{t=b_j}^{b_{j+1}-1} X_t \exp(i2\pi\omega_k t)|^2}{|b_{j+1} - b_j|}, \quad k = 1, \ldots, K; \quad j = 1, \ldots, J. \tag{6.33}$$

Again, under regularity conditions, the local periodograms are asymptotically independent and distributed as (Guo, Dai, Ombao and von Sachs 2003)

$$y_{(k-1)J+j} \sim \begin{cases} \exp\{f(\omega_k, u_j)\}\chi_1^2, & \omega_k = 0, 1/2, \\ \exp\{f(\omega_k, u_j)\}\chi_2^2/2, & \omega_k \neq 0, 1/2. \end{cases}$$

Let $\boldsymbol{x} = (\omega, u)$ and $\boldsymbol{x}_{(k-1)J+j} = (\omega_k, u_j)$ for $k = 1, \ldots, K$, and $j = 1, \ldots, J$. Then the time-varying spectrum can be estimated based on observations $\{(\boldsymbol{x}_i, y_i), i = 1, \ldots, KJ\}$.

The SS ANOVA decomposition (6.32) may also be used to test if a locally stationary process is stationary. The locally stationary process $X(t)$ is stationary if $f(\omega, u)$ is independent of u. Let $h(u) = \beta \times (u - 0.5) + f_2^s(u) + f_{12}^{sl}(\omega, u) + f_{12}^{ss}(\omega, u)$, which collects all terms involving time in (6.32). Then the hypothesis is

$$H_0 : h(u) = 0 \text{ for all } u, \qquad H_1 : h(u) \neq 0 \text{ for some } u.$$

The full SS ANOVA model (6.32) is reduced to $f(\omega, u) = \mu + f_1(\omega)$ under H_0. Denote $\hat{f}^F$ and $\hat{f}^R$ as the estimates of f under the full and reduced models, respectively. Let

$$D_F = \sum_{i=1}^{n}\{\hat{f}_i^F + y_i \exp(-\hat{f}_i^F) - \log y_i - 1\},$$

$$D_R = \sum_{i=1}^{n}\{\hat{f}_i^R + y_i \exp(-\hat{f}_i^R) - \log y_i - 1\}$$

be deviances under the full and reduced models. We construct two test statistics

$$T_1 = D_R - D_F,$$

$$T_2 = \int_0^1 \int_0^1 \{\hat{f}^F(\omega, u) - \hat{f}^R(\omega, u)\}^2 d\omega du,$$

where T_1 corresponds to the Chi-square statistics commonly used for generalized linear models, and T_2 computes the L_2 distance between $\hat{f}^F$ and $\hat{f}^R$. We reject H_0 when these statistics are large. It is difficult to derive the null distributions of these statistics. Note that f does not depend on u under H_0. Therefore, the null distribution can be approximated by permutation. Specifically, permutation samples are generated by shuffling time grid, and statistics T_1 and T_2 are computed for each permutation sample. Then the p-values are approximated as the proportion of statistics based on permutation samples that are larger than those based on the original data.

6.5.3 Epileptic EEG

We now illustrate how to estimate the spectrum of a stationary process and the time-varying spectrum of a locally stationary process using the seizure data. The data contain two 5-minute intracranial electroencephalograms (IEEG) segments from a patient: one at baseline extracted at least 4 hours before the seizure's onset (labeled as baseline), and one right before a seizure's clinical onset (labeled as preseizure). Observations are shown in Figure 6.4.

First assume that both the baseline and preseizure series are stationary. Then the following statements compute the periodograms and fit a periodic spline model for the baseline series:

```
> data(seizure); attach(seizure)
> n <- nrow(seizure)
> x <- seq(1,n-1, by=120)/n
```

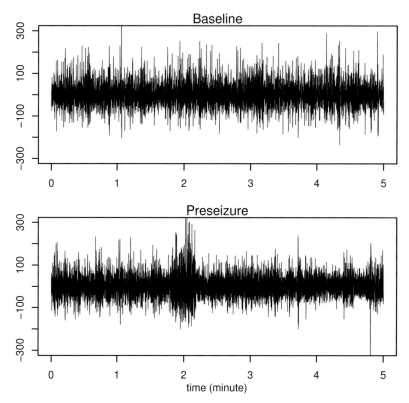

FIGURE 6.4 Seizure data, plots of 5-minute baseline segments collected hours away from the seizure (above), and 5-minute preseizure segments with the seizure's onset at the 5th minute (below). The sampling rate is 200 Hertz. The total number of time points is 60,000 for each segment.

```
> y <- (abs(fft(seizure$base))^2/n)[-1][seq(1,n-1, by=120)]
> seizure.s.base <- ssr(y~x, periodic(x), spar=``u'',
    varht=1, family=``gamma'',limnla=c(-5,1))
> grid <- data.frame(x=seq(0,.5,len=100))
> seizure.s.base.p <- predict(seizure.s.base,grid)
```

where the function `fft` computes an unscaled discrete Fourier transformation. A subset of periodograms is used. The UBR criterion is used to select the smoothing parameter with the dispersion parameter set to 1. We limit the search range for $\log_{10}(n\lambda)$ to avoid undersmoothing. Log periodograms, periodic spline estimate of the log spectrum, and 95% Bayesian confidence intervals of the baseline series are shown in the left panel of Figure 6.5. The log spectrum of the preseizure series is estimated similarly. Log periodograms, periodic spline estimate of the log spectrum, and 95% Bayesian confidence intervals of the preseizure series are shown in the right panel of Figure 6.5. Because the sampling rate is 200 HZ and the spectrum is symmetric around 100 HZ, we only show the estimated spectra in frequency bands 0–100 HZ.

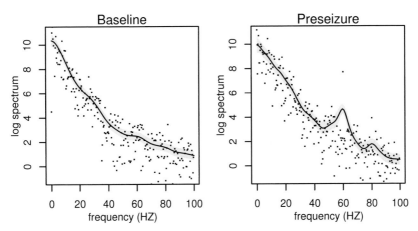

FIGURE 6.5 Seizure data, plots log periodograms (dots), estimate of the log spectra (line) based on the iterative UBR method, and 95% Bayesian confidence intervals (shaded region) of the baseline series (left) and preseizure series (right).

The IEEG series may become nonstationary before seizure events. Assume that both the baseline and preseizure series are locally stationary. We create time-frequency grids data with $\omega_k = k/(K+1)$ and $u_j = (j - .5)/J + 1/2n$ for $k = 1, \ldots, K$ and $j = 1, \ldots, J$ where $n = 60000$ is the length of a time series, and $K = J = 32$. We compute local periodograms for the baseline series:

```
> pgram <- function(x, freqs) {
    y <- numeric(length(freqs))
    tser <- seq(length(x))-1
    for(i in seq(length(freqs))) y[i] <-
    Mod(sum(x*complex(mod=1, arg=-2*pi*freqs[i]*tser)))^2
    y/length(x)
  }

> lpgram <- function(x,times,freqs) {
```

```
  nsub <- floor(length(x)/length(times))
  ymat <- matrix(0, length(freqs), length(times))
  for (j in seq(length(times)))
    ymat[,j] <- pgram(x[((j-1)*nsub+1):(j*nsub)], freqs)
  as.vector(ymat)
  }
```

```
> nf <- nt <- 32; freqs <- 1:nf/(nf+1)
> nsub <- floor(n/nt)
> times <- (seq(from=(1+nsub)/2, by=nsub, length=nt))/n
> y <- lpgram(seizure$base, times, freqs)
```

where the functions pgram and lpgram, respectively, compute periodograms
for a stationary process and local periodograms for a locally stationary
process. We now fit the SS ANOVA model (6.32) for the baseline series:

```
> ftgrid <- expand.grid(freqs,times)
> x1 <- ftgrid[,1]
> x2 <- ftgrid[,2]
> seizure.ls.base <- ssr(y~x2,
    rk=list(periodic(x1), cubic(x2),
            rk.prod(periodic(x1),kron(x2-.5)),
            rk.prod(periodic(x1),cubic(x2))),
    spar=''u'', varht=1, family=''gamma'')
> grid <- expand.grid(x1=seq(0,.5,len=50),
                      x2=seq(0,1,len=50))
> seizure.ls.base.p <- predict(seizure.ls.base,
                              newdata=grid)
```

The SS ANOVA model (6.32) for the preseizure series is fitted similarly. Figure 6.6 shows estimates of time-varying spectra of the baseline
series and preseizure series. The iterative UBR method was used in
the above fits. The DGML method usually leads to a better estimate
of the spectrum or the time-varying spectrum (Qin and Wang 2008).
The following function spdest estimates a spectrum or a time-varying
spectrum using the DGML method.

```
spdest <- function(y, freq, time, process, control=list(
  optfactr=1e10, limnla=c(-6,1), prec=1e-6, maxit=30))
{
  if (process==''S'') {
   thhat <- rep(NA, 4)
   S <- as.matrix(rep(1,length(y)))
   Q <- periodic(freq)
   tmp <- try(optim(-2, DGML, y=y, S=S, Q=Q,
```

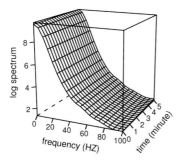 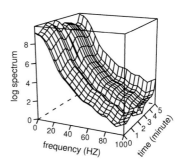

FIGURE 6.6 Seizure data, plots of estimates of the time-varying spectra of the baseline series (left), and preseizure series (right) based on the iterative UBR method.

```
method=''L-BFGS-B'',
lower=control$limnla[1], upper=control$limnla[2],
control=list(factr=control$optfactr)))
if (class(tmp)==''try-error'') {
info <- ''optim failure''
}
else {
  if (tmp$convergence==0) {
   nlaht <- tmp$par
   fit <- ssr(y~S-1, Q, limnla=tmp$par, family=''gamma'',
            spar=''u'', varht=1)
   info <- ''success''
  }
  else info <- paste(''optim failure'',
                  as.character(tmp$convergence))
 }
}
if (process==''LS'') {
 S <- cbind(1,time-.5)
 Q <- array(NA,c(length(freq),length(freq),4))
 Q[,,1] <- periodic(freq)
 Q[,,2] <- cubic(time)
 Q[,,3] <- rk.prod(periodic(freq),kron(time-.5))
```

```
Q[,,4] <- rk.prod(periodic(freq),cubic(time))

# compute initial values for optimization of theta
zz <- log(y)+.57721
tmp <- try(ssr(zz~S-1, Q, spar=''m''),T)
if (class(tmp)==''ssr'') {
 thini <- tmp$rkpk.obj$theta
 nlaht <- tmp$rkpk.obj$nlaht
}
else {
 thini <- rep(1,4)
 nlaht <- 1.e-8
}
tmp <- try(optim(thini, DGML, nlaht=nlaht,
           y=y, S=S, Q=Q, method=''L-BFGS-B'',
           control=list(factr=control$optfactr)))
if (class(tmp)==''try-error'') {info=''optim failure''}
else {
  if (tmp$convergence==0) {
   thhat <- tmp$par
   thetahat <- 10**thhat
   Qt <- 0
   for (i in 1:dim(Q)[3]) Qt <- Qt + thetahat[i]*Q[,,i]
   fit <- ssr(y~S-1, Qt, limnla=nlaht, family=''gamma'',
          spar=''u'', varht=1)
   info <- ''success''
  }
  else { info <- paste(''optim failure'',
    as.character(tmp$convergence)) }
 }
}
return(list(fit=fit, nlaht=nlaht, thhat=thhat, info=info))
}
```

The argument `process` specifies the type of process with "S" and "LS" corresponding to the stationary and locally stationary processes, respectively. For a stationary process, the argument y inputs periodograms at frequencies specified by `freq`. The argument `time` is irrelevant in this case. For a locally stationary process, the argument y inputs local periodograms at the time-frequency grid specified by `time` and `freq`, respectively. For locally stationary processes, we fit the logarithmic transformed periodograms to get initial values for smoothing parameters (Wahba 1980, Qin and Wang 2008). The `DGML` function was presented in Section 6.4.

The following statements estimate the spectrum of the baseline series as a stationary process using the DGML method and compute posterior means and standard deviations:

```
> x <- seq(1,n-1,by=120)/n
> y <- pgram(seizure$base, x)
> tmp <- spdest(y, x, 0, ''S'')
> seizure.s.base.dgml <- ssr(y~x, periodic(x), spar=''u'',
    varht=1, family=''gamma'',limnla=tmp$nlaht)
> grid <- data.frame(x=seq(0,.5,len=100))
> seizure.s.base.dgml.p <- predict(seizure.s.base.dgml,
                                    grid)
```

Note that, to use the **predict** function, we need to call **ssr** again using the DGML estimate of the smoothing parameter. The spectrum of the preseizure series as a stationary can be estimated similarly. The estimates of spectra are similar to those in Figure 6.5.

Next we estimate the time-varying spectrum of the baseline series as a locally stationary process using the DGML method and compute posterior means:

```
> y <- lpgram(seizure$base, times, freqs)
> tmp <- spdest(y, x1, x2, ''LS'')
> th <- 10**tmp$thhat
> S <- cbind(1, x2-.5)
> Q1 <- periodic(x1)
> Q2 <- cubic(x2)
> Q3 <- rk.prod(periodic(x1), kron(x2-.5))
> Q4 <- rk.prod(periodic(x1), cubic(x2))
> Qt <- th[1]*Q1+th[2]*Q2+th[3]*Q3+th[4]*Q4
> seizure.ls.base.dgml <- ssr(y~x2, rk=Qt, spar=''u'',
    varht=1, family=''gamma'', limnla=tmp$nlaht)
> grid <- expand.grid(x1=seq(0,.5,len=50),
                      x2=seq(0,1,len=50))
> Sg <- cbind(1,grid$x2-.5)
> Qg1 <- periodic(grid$x1,x1)
> Qg2 <- cubic(grid$x2,x2)
> Qg3 <- rk.prod(periodic(grid$x1,x1),
                kron(grid$x2-.5,x2-.5))
> Qg4 <- rk.prod(periodic(grid$x1,x1), cubic(grid$x2,x2))
> Qgt <- th[1]*Qg1+th[2]*Qg2+th[3]*Qg3+th[4]*Qg4
> seizure.ls.base.dglm.p <-
    Sg%*%seizure.ls.base.dgml$coef$d+
    Qgt%*%seizure.ls.base.dgml$coef$c
```

where the `spdest` function was used to find the DGML estimates of the smoothing parameters, and the `ssr` function was used to calculate the coefficients c and d. The time-varying spectrum of the preseizure series as a locally stationary process can be estimated similarly. The estimates of time-varying spectra are shown in Figure 6.7. The estimate of the preseizure series (right panel in Figure 6.7) is smoother than that based on the iterative UBR method (right panel in Figure 6.6).

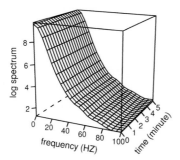 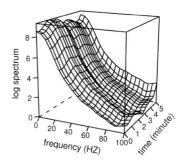

FIGURE 6.7 Seizure data, plots of estimates of the time-varying spectra based on the DGML method of the baseline series (left) and preseizure series (right).

It appears that the baseline spectrum does not vary much over time, while the preseizure spectrum varies over time. Therefore, the baseline series may be stationary, while the preseizure series may be nonstationary. The following statements perform the permutation test for the baseline series with 100 permutations.

```
z <- seizure$base
n <- length(z); nt <- 32; nf <- 32
freqs <- 1:nf/(nf+1); nsub <- floor(n/nt)
times <- (seq(from=(1+nsub)/2, by=nsub, length=nt))/n
y <- lpgram(z,times,freqs)
ftgrid <- expand.grid(freqs,times)
x1 <- ftgrid[,1]
x2 <- ftgrid[,2]
```

```
full <- spdest(y, x1, x2, ''LS'')
reduced <- spdest(y, x1, x2, ''S'')
fitf <- full$fit$rkpk$eta
fitr <- reduced$fit$rkpk$eta
devf <- sum(-1-log(y)+y*exp(-fitf)+fitf)
devr <- sum(-1-log(y)+y*exp(-fitr)+fitr)
cstat <- devr-devf
l2d <- mean((fitf-fitr)**2)

nperm <- 100; totperm <- 0
cstatp <- l2dp <- NULL
while (totperm < nperm) {
 x2p <- rep(sample(times),rep(nf,nt))
 full <- spdest(y, x1, x2p, ''LS'')
 if (full$info==''success'') {
  totperm <- totperm+1
  fitf <- full$fit$rkpk$eta
  devf <- sum(-1-log(y)+y*exp(-fitf)+fitf)
  cstatp <- c(cstatp, devr-devf)
  l2dp <- c(l2dp, mean((fitf-fitr)**2))
 }
}
print(c(mean(cstatp>cstat),mean(l2dp>l2d)))
```

Note that there is no need to fit the reduced model for the permuted data since the log spectrum does not depend on time under the null hypothesis. The permutation test for the preseizure series is performed similarly. The p-values for testing stationarity based on two statistics are 0.82 and 0.73 for the baseline series, and 0.03 and 0.06 for the preseizure series. Therefore, the processes far away from the seizure's clinical onset can be regarded as stationary, while the processes close to the seizure's clinical onset are nonstationary.

Chapter 7

Smoothing Spline Nonlinear Regression

7.1 Motivation

The general smoothing spline regression model (2.10) in Chapter 2 and the SS ANOVA model (4.31) in Chapter 4 assume that the unknown function is observed through some *linear* functionals. This chapter deals with situations when some unknown functions are observed indirectly through *nonlinear* functionals. We discuss some potential applications in this section. More examples can be found in Section 7.6.

In some applications, the theoretical models depend on the unknown functions nonlinearly. For example, in remote sensing, the satellite up-welling radiance measurements R_v are related to the underlying atmospheric temperature distribution f through the following equation

$$R_v(f) = B_v(f(x_s))\tau_v(x_s) - \int_{x_0}^{x_s} B_v(f(x))\tau_v'(x)dx,$$

where x is some monotone transformation of pressure p, for example, the kappa units $x(p) = p^{5/8}$; x_0 and x_s are x values at the surface and top of the atmosphere; $\tau_v(x)$ is the transmittance of the atmosphere above x at wavenumber v; and $B_v(t)$ is the Planck's function, $B_v(t) = c_1 v^3/\{exp(c_2 v/t) - 1\}$, with known constants c_1 and c_2. The goal is to estimate f as a function of x based on noisy observations of $R_v(f)$. Obviously, $R_v(f)$ is nonlinear in f. Other examples involving reservoir modeling and three-dimensional atmospheric temperature distribution from satellite-observed radiances can be found in Wahba (1987) and O'Sullivan (1986).

Very often there are certain constraints such as positivity and monotonicity on the function of interest, and sometimes nonlinear transformations may be used to relax those constraints. For example, consider the following nonparametric regression model

$$y_i = g(x_i) + \epsilon_i, \quad x_i \in [0,1], \ i = 1, \ldots, n. \tag{7.1}$$

Suppose g is known to be positive. The transformation $g = \exp(f)$ substitutes the original constrained estimation of g by the unconstrained estimation of f. The resulting transformed model,

$$y_i = \exp\{f(x_i)\} + \epsilon_i, \quad x_i \in [0,1], \ i = 1, \ldots, n, \tag{7.2}$$

depends on f nonlinearly. Monotonicity is another common constraint. Consider model (7.1) again and suppose g is known to be strictly increasing with $g'(x) > 0$. Write g' as $g'(x) = \exp\{f(x)\}$. Reexpressing g as $g(x) = f(0) + \int_0^x \exp\{f(s)\}ds$ leads to the following model

$$y_i = \beta + \int_0^{x_i} \exp\{f(s)\}ds + \epsilon_i, \quad x_i \in [0,1], \ i = 1, \ldots, n. \tag{7.3}$$

The function f is free of constraints and acts nonlinearly. Strictly speaking, model (7.3) is a semiparametric nonlinear regression model in Section 8.3 since it contains a parameter β.

Sometimes one may want to consider empirical models that depend on unknown functions nonlinearly. One such model, the *multiplicative model*, will be introduced in Section 7.6.4.

7.2 Nonparametric Nonlinear Regression Models

A general *nonparametric nonlinear regression* (NNR) model assumes that

$$y_i = \mathcal{N}_i(\boldsymbol{f}) + \epsilon_i, \quad i = 1, \ldots, n, \tag{7.4}$$

where $\boldsymbol{f} = (f_1, \ldots, f_r)$ are r unknown functions, f_k belongs to an RKHS $\mathcal{H}_k$ on an arbitrary domain $\mathcal{X}_k$ for $k = 1, \ldots, r$, $\mathcal{N}_i$ are known nonlinear functionals on $\mathcal{H}_1 \times \cdots \times \mathcal{H}_r$, and ϵ_i are zero-mean independent random errors with a common variance σ^2. Note that domains $\mathcal{X}_k$ for different functions f_k may be the same or different. It is obvious that the NNR model (7.4) is an extension of both the SSR model (2.10) and the SS ANOVA model (4.31). It may be considered as an extension of the parametric nonlinear regression model with functions in infinite dimensional spaces as parameters.

An interesting special case of the NNR model (7.4) is when $\mathcal{N}_i$ depends on $\boldsymbol{f}$ through a nonlinear function and some linear functionals. Specifically,

$$y_i = \psi(\mathcal{L}_{1i}f_1, \ldots, \mathcal{L}_{ri}f_r) + \epsilon_i, \quad i = 1, \ldots, n, \tag{7.5}$$

where ψ is a known nonlinear function, and $\mathcal{L}_{1i}, \ldots, \mathcal{L}_{ri}$ are bounded linear functionals. Model (7.2) for positive constraint is a special case of (7.5) with $r = 1$, $\psi(z) = \exp(z)$ and $\mathcal{L}_i f = f(x_i)$. However, model (7.3) for monotonicity constraint cannot be written in the form of (7.5).

7.3 Estimation with a Single Function

In this section we restrict our attention to the case when $r = 1$, and drop subscript k for simplicity of notation. Our goal is to estimate the nonparametric function f in $\mathcal{H}$.

Assume that $\mathcal{H} = \mathcal{H}_0 \oplus \mathcal{H}_1$, where $\mathcal{H}_0 = \operatorname{span}\{\phi_1, \ldots, \phi_p\}$ is an RKHS with RK R_0, and $\mathcal{H}_1$ is an RKHS with RK R_1. We estimate f as the minimizer of the following PLS:

$$\min_{f \in \mathcal{H}} \left\{ \frac{1}{n} \sum_{i=1}^n (y_i - \mathcal{N}_i f)^2 + \lambda \|P_1 f\|^2 \right\}, \tag{7.6}$$

where P_1 is a projection operator from $\mathcal{H}$ onto $\mathcal{H}_1$, and λ is a smoothing parameter. We assume that the solution to (7.6) exists and is unique (conditions can be found in the supplement document of Ke and Wang (2004)), and denote the solution as $\hat{f}$.

7.3.1 Gauss–Newton and Newton–Raphson Methods

We first consider the special NNR model (7.5) in this section. Let

$$\eta_i(x) = \mathcal{L}_{i(z)} R(x, z) = \mathcal{L}_{i(z)} R_0(x, z) + \mathcal{L}_{i(z)} R_1(x, z) \triangleq \delta_i(x) + \xi_i(x).$$

Then $\eta_i \in \mathcal{S}$, where $\mathcal{S} = \mathcal{H}_0 \oplus \operatorname{span}\{\xi_1, \ldots, \xi_n\}$. Any $f \in \mathcal{H}$ can be decomposed into $f = \varsigma_1 + \varsigma_2$, where $\varsigma_1 \in \mathcal{S}$ and $\varsigma_2 \in \mathcal{S}^c$. Furthermore, $\mathcal{L}_i f = \mathcal{L}_i \varsigma_1$. Denote

$$\mathrm{LS}(\psi(\mathcal{L}_1 f), \ldots, \psi(\mathcal{L}_n f)) = \frac{1}{n} \sum_{i=1}^n (y_i - \psi(\mathcal{L}_i f))^2$$

as the least squares. Then the penalized least squares

$$\begin{aligned}
& \mathrm{PLS}(\psi(\mathcal{L}_1 f), \ldots, \psi(\mathcal{L}_n f)) \\
& \triangleq \mathrm{LS}(\psi(\mathcal{L}_1 f), \ldots, \psi(\mathcal{L}_n f)) + \lambda \|P_1 f\|^2 \\
& = \mathrm{LS}(\psi(\mathcal{L}_1 \varsigma_1), \ldots, \psi(\mathcal{L}_n \varsigma_1)) + \lambda(\|P_1 \varsigma_1\|^2 + \|P_1 \varsigma_2\|^2) \\
& \geq \mathrm{LS}(\psi(\mathcal{L}_1 \varsigma_1), \ldots, \psi(\mathcal{L}_n \varsigma_1)) + \lambda \|P_1 \varsigma_1\|^2 \\
& = \mathrm{PLS}(\psi(\mathcal{L}_1 \varsigma_1), \ldots, \psi(\mathcal{L}_n \varsigma_1)).
\end{aligned}$$

Equality holds iff $||P_1\varsigma_2|| = ||\varsigma_2|| = 0$. Thus the minimizer of the PLS falls in the finite dimensional space $\mathcal{S}$, which can be represented as

$$\hat{f}(x) = \sum_{\nu=1}^{p} d_\nu \phi_\nu(x) + \sum_{i=1}^{n} c_i \xi_i(x). \tag{7.7}$$

The representation in (7.7) is an extension of the Kimeldorf–Wahba representer theorem. Let $\boldsymbol{c} = (c_1, \ldots, c_n)^T$ and $\boldsymbol{d} = (d_1, \ldots, d_p)^T$. Based on (7.7), the PLS (7.6) becomes

$$\frac{1}{n} \sum_{i=1}^{n} (y_i - \psi(\mathcal{L}_i \hat{f}))^2 + \lambda \boldsymbol{c}^T \Sigma \boldsymbol{c}, \tag{7.8}$$

where $\mathcal{L}_i \hat{f} = \sum_{\nu=1}^{p} d_\nu \mathcal{L}_i \phi_\nu + \sum_{j=1}^{n} c_j \mathcal{L}_i \xi_j$, and $\Sigma = \{\mathcal{L}_i \mathcal{L}_j R_1\}_{i,j=1}^{n}$. We need to find minimizers $\boldsymbol{c}$ and $\boldsymbol{d}$. Standard nonlinear optimization procedures can be employed to solve (7.8). We now describe the Gauss–Newton and Newton–Raphson methods.

We first describe the Gauss–Newton method. Let $\boldsymbol{c}_-$, $\boldsymbol{d}_-$, and $\hat{f}_-$ be the current approximations of $\boldsymbol{c}$, $\boldsymbol{d}$, and $\hat{f}$, respectively. Replacing $\psi(\mathcal{L}_i \hat{f})$ by its first-order expansion at $\mathcal{L}_i \hat{f}_-$,

$$\psi(\mathcal{L}_i \hat{f}) \approx \psi(\mathcal{L}_i \hat{f}_-) - \psi'(\mathcal{L}_i \hat{f}_-)\mathcal{L}_i \hat{f}_- + \psi'(\mathcal{L}_i \hat{f}_-)\mathcal{L}_i \hat{f}, \tag{7.9}$$

the PLS (7.8) can be approximated by

$$\frac{1}{n}||\tilde{\boldsymbol{y}} - V(T\boldsymbol{d} + \Sigma \boldsymbol{c})||^2 + \lambda \boldsymbol{c}^T \Sigma \boldsymbol{c}, \tag{7.10}$$

where $\tilde{y}_i = y_i - \psi(\mathcal{L}_i \hat{f}_-) + \psi'(\mathcal{L}_i \hat{f}_-)\mathcal{L}_i \hat{f}_-$, $\tilde{\boldsymbol{y}} = (\tilde{y}_1, \ldots, \tilde{y}_n)^T$, $V = \text{diag}(\psi'(\mathcal{L}_1 \hat{f}_-), \ldots, \psi'(\mathcal{L}_n \hat{f}_-))$, and $T = \{\mathcal{L}_i \phi_\nu\}_{i=1}^{n}{}_{\nu=1}^{p}$. Assume that V is invertible, and let $\tilde{T} = VT$, $\tilde{\Sigma} = V\Sigma V$, $\tilde{\boldsymbol{c}} = V^{-1}\boldsymbol{c}$, and $\tilde{\boldsymbol{d}} = \boldsymbol{d}$. Then the approximated PLS (7.10) reduces to

$$\frac{1}{n}||\tilde{\boldsymbol{y}} - \tilde{T}\tilde{\boldsymbol{d}} - \tilde{\Sigma}\tilde{\boldsymbol{c}}||^2 + \lambda \tilde{\boldsymbol{c}}^T \tilde{\Sigma} \tilde{\boldsymbol{c}}, \tag{7.11}$$

which has the same form as (2.19). Thus the Gauss–Newton method updates $\boldsymbol{c}$ and $\boldsymbol{d}$ by solving

$$(\tilde{\Sigma} + n\lambda I)\tilde{\boldsymbol{c}} + \tilde{T}\tilde{\boldsymbol{d}} = \tilde{\boldsymbol{y}},$$
$$\tilde{T}^T \tilde{\boldsymbol{c}} = \boldsymbol{0}. \tag{7.12}$$

Equations in (7.12) have the same form as those in (2.21). Therefore, the same method in Section 2.4 can be used to compute $\tilde{\boldsymbol{c}}$ and $\tilde{\boldsymbol{d}}$. New estimates of $\boldsymbol{c}$ and $\boldsymbol{d}$ can be derived from $\boldsymbol{c} = V\tilde{\boldsymbol{c}}$ and $\boldsymbol{d} = \tilde{\boldsymbol{d}}$. It is easy to see that the equations in (7.12) are equivalent to

$$(\Sigma + n\lambda V^{-2})\boldsymbol{c} + T\boldsymbol{d} = V^{-1}\tilde{\boldsymbol{y}},$$
$$T^T\boldsymbol{c} = \boldsymbol{0}, \qquad (7.13)$$

which have the same form as those in (5.6).

We next describe the Newton–Raphson method. Let

$$I(\boldsymbol{c}, \boldsymbol{d}) = \frac{1}{2}\sum_{i=1}^{n}\Big\{y_i - \psi\Big(\sum_{\nu=1}^{p}d_\nu\mathcal{L}_i\phi_\nu + \sum_{j=1}^{n}c_j\mathcal{L}_i\xi_j\Big)\Big\}^2.$$

Let $\boldsymbol{\psi} = (\psi(\mathcal{L}_1\hat{f}_-), \ldots, \psi(\mathcal{L}_n\hat{f}_-))^T$, $\boldsymbol{u} = -V(\boldsymbol{y} - \boldsymbol{\psi})$, where V is defined above, $O = \text{diag}((y_1 - \psi(\mathcal{L}_1\hat{f}_-))\psi''(\mathcal{L}_1\hat{f}_-), \ldots, (y_n - \psi(\mathcal{L}_n\hat{f}_-))\psi''(\mathcal{L}_n\hat{f}_-))$, and $W = V^2 - O$. Then $\partial I/\partial\boldsymbol{c} = -\Sigma V(\boldsymbol{y} - \boldsymbol{\psi}) = \Sigma\boldsymbol{u}$, $\partial I/\partial\boldsymbol{d} = -T^TV(\boldsymbol{y}-\boldsymbol{\psi}) = T^T\boldsymbol{u}$, $\partial^2 I/\partial\boldsymbol{c}\partial\boldsymbol{c}^T = \Sigma V^2\Sigma - \Sigma O\Sigma = \Sigma W\Sigma$, $\partial^2 I/\partial\boldsymbol{c}\partial\boldsymbol{d}^T = \Sigma WT$, and $\partial^2 I/\partial\boldsymbol{d}\partial\boldsymbol{d}^T = T^TWT$. The Newton–Raphson iteration satisfies the following equations

$$\begin{pmatrix} \Sigma W\Sigma + n\lambda\Sigma & \Sigma WT \\ T^TW\Sigma & T^TWT \end{pmatrix}\begin{pmatrix} \boldsymbol{c} - \boldsymbol{c}_- \\ \boldsymbol{d} - \boldsymbol{d}_- \end{pmatrix} = \begin{pmatrix} -\Sigma\boldsymbol{u} - n\lambda\Sigma\boldsymbol{c}_- \\ -T^T\boldsymbol{u} \end{pmatrix}. \quad (7.14)$$

Assume that W is positive definite. It is easy to see that solutions to

$$(\Sigma + n\lambda W^{-1})\boldsymbol{c} + T\boldsymbol{d} = \hat{\boldsymbol{f}}_- - W^{-1}\boldsymbol{u},$$
$$T^T\boldsymbol{c} = \boldsymbol{0}, \qquad (7.15)$$

are also solutions to (7.14), where $\hat{\boldsymbol{f}}_- = (\mathcal{L}_1\hat{f}_-, \ldots, \mathcal{L}_n\hat{f}_-)^T$. Again, the equations in (7.15) have the same form as those in (5.6). Methods in Section 5.2.1 can be used to solve (7.15).

Note that, when O is ignored, $W = V^2$, and $V^{-1}\tilde{\boldsymbol{y}} = \hat{\boldsymbol{f}}_- - W^{-1}\boldsymbol{u}$. Then, equations in (7.13) are the same as those in (7.15), and the Newton–Raphson method is the same as the Gauss–Newton method.

7.3.2 Extended Gauss–Newton Method

We now consider the estimation of the general NNR model (7.4). In general, the solution to the PLS (7.6) $\hat{f}$ no longer falls in a finite dimensional space. Therefore, certain approximation is necessary.

Let f be a fixed element in $\mathcal{H}$. A nonlinear functional $\mathcal{N}$ on $\mathcal{H}$ is called *Fréchet differentiable* at f if there exists a bounded linear functional $D\mathcal{N}(f)$ such that (Debnath and Mikusiński 1999)

$$\lim_{||h||\to 0}\frac{|\mathcal{N}(f+h) - \mathcal{N}(f) - D\mathcal{N}(f)(h)|}{||h||} = 0.$$

Denote $\hat{f}_-$ as the current approximation of $\hat{f}$. Assume that $\mathcal{N}_i$ are Fréchet differentiable at $\hat{f}_-$ and denote $\mathcal{D}_i = D\mathcal{N}_i(\hat{f}_-)$. Note that $\mathcal{D}_i$ are known bounded linear functionals at the current approximation. The best linear approximation of $\mathcal{N}_i$ near $\hat{f}_-$ is (Debnath and Mikusiński 1999)

$$\mathcal{N}_i f \approx \mathcal{N}_i \hat{f}_- + \mathcal{D}_i(f - \hat{f}_-). \tag{7.16}$$

Based on the linear approximation (7.16), the NNR model can be approximated by

$$\tilde{y}_i = \mathcal{D}_i f + \epsilon_i, \quad i = 1, \ldots, n, \tag{7.17}$$

where $\tilde{y}_i = y_i - \mathcal{N}_i \hat{f}_- + \mathcal{D}_i \hat{f}_-$. We minimize

$$\frac{1}{n} \sum_{i=1}^{n} (\tilde{y}_i - \mathcal{D}_i f)^2 + \lambda \|P_1 f\|^2 \tag{7.18}$$

to get a new approximation of $\hat{f}$. Since $\mathcal{D}_i$ are bounded linear functionals, the solution to (7.18) has the form

$$\tilde{f} = \sum_{\nu=1}^{p} \tilde{d}_\nu \phi_i(x) + \sum_{i=1}^{n} \tilde{c}_i \tilde{\xi}_i(x), \tag{7.19}$$

where $\tilde{\xi}_i(x) = \mathcal{D}_{i(z)} R_1(x, z)$. Coefficients $\tilde{c}_i$ and $\tilde{d}_\nu$ can be calculated using the same method in Section 2.4. An iterative algorithm can then be formed with the convergent solution as the final approximation of $\hat{f}$. We refer to this algorithm as the *extended Gauss–Newton* (EGN) algorithm since it is an extension of the Gauss–Newton method to infinite dimensional spaces.

Note that the linear functionals $\mathcal{D}_i$ depend on the current approximation $\hat{f}_-$. Thus, representers $\tilde{\xi}_j$ change along iterations. This approach adaptively chooses representers to approximate the PLS estimate $\hat{f}$. As in nonlinear regression, the performance of this algorithm depends largely on the curvature of the nonlinear functional. Simulations indicate that the EGN algorithm works well and converges quickly for commonly used nonlinear functionals. For the special NNR model (7.5), it can be shown that the EGN algorithm is equivalent to the standard Gauss–Newton method presented in Section 7.3.1 (Ke and Wang 2004).

An alternative approach is to approximate $\hat{f}$ by a finite series and solve coefficients using the Gauss–Newton or the Newton–Raphson algorithm. When a finite series with good approximation property is available, this approach may be preferable since it is relatively easy to implement.

However, the choice of such a finite series may become difficult in certain situations. The EGN approach is fully automatic and adaptive. On the other hand, it is more difficult to implement and may become computationally intensive.

7.3.3 Smoothing Parameter Selection and Inference

The smoothing parameter λ is fixed in Sections 7.3.1 and 7.3.2. We now discuss methods for selecting λ. Similar to Section 3.4, we define the leaving-out-one *cross-validation* (CV) criterion as

$$\text{CV}(\lambda) = \frac{1}{n} \sum_{i=1}^{n} (\mathcal{N}_i \hat{f}^{[i]} - y_i)^2, \tag{7.20}$$

where $\hat{f}^{[i]}$ is the minimizer of the following PLS

$$\frac{1}{n} \sum_{j \neq i} (y_j - \mathcal{N}_j f)^2 + \lambda \|P_1 f\|^2. \tag{7.21}$$

Again, computation of $\hat{f}^{[i]}$ for each $i = 1, \ldots, n$ is costly. For fixed i and z, let $h(i, z)$ be the minimizer of

$$\frac{1}{n} (z - \mathcal{N}_i f)^2 + \sum_{j \neq i} (y_j - \mathcal{N}_j f)^2 + \lambda \|P_1 f\|^2.$$

That is, $h(i, z)$ is the solution to (7.6) when the ith observation is replaced by z. It is not difficult to check that arguments in Section 3.4 still hold for nonlinear functionals. Therefore, we have the following lemma.

Leaving-out-one Lemma *For any fixed i, $h(i, \mathcal{N}_i \hat{f}^{[i]}) = \hat{f}^{[i]}$.*

Note that $h(i, y_i) = \hat{f}$. Define

$$a_i \triangleq \frac{\partial \mathcal{N}_i h(i, y_i)}{\partial y_i} = D\mathcal{N}_i(h(i, y_i)) \frac{\partial h(i, y_i)}{\partial y_i}, \tag{7.22}$$

where $D\mathcal{N}_i(h(i, y_i))$ is the Fréchet differential of $\mathcal{N}_i$ at $h(i, y_i)$. Let $y_i^{[i]} = \mathcal{N}_i \hat{f}^{[i]}$. Applying the leaving-out-one lemma, we have

$$\Delta_i(\lambda) \triangleq \frac{\mathcal{N}_i \hat{f} - \mathcal{N}_i \hat{f}^{[i]}}{y_i - \mathcal{N}_i \hat{f}^{[i]}} = \frac{\mathcal{N}_i h(i, y_i) - \mathcal{N}_i h(i, y_i^{[i]})}{y_i - y_i^{[i]}} \approx \frac{\partial \mathcal{N}_i h(i, y_i)}{\partial y_i} = a_i.$$

Then

$$y_i - \mathcal{N}_i \hat{f}^{[i]} = \frac{y_i - \mathcal{N}_i \hat{f}}{1 - \Delta_i(\lambda)} \approx \frac{y_i - \mathcal{N}_i \hat{f}}{1 - a_i}.$$

Subsequently, the CV criterion (7.20) is approximated by

$$\mathrm{CV}(\lambda) \approx \frac{1}{n} \sum_{i=1}^{n} \frac{(y_i - \mathcal{N}_i \hat{f})^2}{(1 - a_i)^2}. \tag{7.23}$$

Replacing a_i by the average $\sum_{i=1}^{n} a_i/n$, we have the GCV criterion

$$\mathrm{GCV}(\lambda) = \frac{\frac{1}{n} \sum_{i=1}^{n} (y_i - \mathcal{N}_i \hat{f})^2}{(1 - \frac{1}{n} \sum_{i=1}^{n} a_i)^2}. \tag{7.24}$$

Note that a_i in (7.22) depends on $\hat{f}$, and $\hat{f}$ may depend on $\boldsymbol{y}$ nonlinearly. Therefore, unlike the linear case, in general, there is no explicit formula for a_i, and it is impossible to compute the CV or GCV estimate of λ by minimizing (7.24) directly. One approach is to replace $\hat{f}$ in (7.22) by its current approximation, $\hat{f}_-$. This suggests estimating λ at each iteration. Specifically, at each iteration, select λ for the approximating SSR model (7.17) by the standard GCV criterion (3.4). This iterative approach is easy to implement. However, it does not guarantee convergence. Simulations indicate that convergence is achieved in most cases (Ke and Wang 2004).

We use the following connections between an NNR model (7.5) and a *nonlinear mixed-effects* (NLME) model to extend the GML method. Consider the following NLME model

$$\begin{aligned}
\boldsymbol{y} &= \boldsymbol{\psi}(\boldsymbol{\gamma}) + \boldsymbol{\epsilon}, \quad \boldsymbol{\epsilon} \sim \mathrm{N}(0, \sigma^2 I), \\
\boldsymbol{\gamma} &= T\boldsymbol{d} + \Sigma\boldsymbol{c}, \quad \boldsymbol{c} \sim \mathrm{N}(0, \sigma^2 \Sigma^+/n\lambda),
\end{aligned} \tag{7.25}$$

where $\boldsymbol{\gamma} = (\gamma_1, \ldots, \gamma_n)^T$, $\boldsymbol{\psi}(\boldsymbol{\gamma}) = (\psi(\gamma_1), \ldots, \psi(\gamma_n))^T$, $\boldsymbol{d}$ are fixed effects, $\boldsymbol{c}$ are random effects, $\boldsymbol{\epsilon} = (\epsilon_1, \ldots, \epsilon_n)^T$ are random errors independent of $\boldsymbol{c}$, and Σ^+ is the Moore–Penrose inverse of Σ. It is common practice to estimate $\boldsymbol{c}$ and $\boldsymbol{d}$ as minimizers of the following joint negative log-likelihood (Lindstrom and Bates 1990)

$$\frac{1}{n} \|\boldsymbol{y} - \boldsymbol{\psi}(T\boldsymbol{d} + \Sigma\boldsymbol{c})\|^2 + \lambda \boldsymbol{c}^T \Sigma \boldsymbol{c}. \tag{7.26}$$

The joint negative log-likelihood (7.26) and the PLS (7.8) lead to the same estimates of $\boldsymbol{c}$ and $\boldsymbol{d}$. In their two-step procedure, at the LME step, Lindstrom and Bates (1990) approximated the NLME model (7.25) by the following linear mixed-effects model

$$\boldsymbol{w} = X\boldsymbol{d} + Z\boldsymbol{c} + \boldsymbol{\epsilon}, \tag{7.27}$$

where

$$X = \frac{\partial \psi(T\boldsymbol{d} + \Sigma \boldsymbol{c})}{\partial \boldsymbol{d}} \Big|_{\boldsymbol{c}=\boldsymbol{c}_-, \boldsymbol{d}=\boldsymbol{d}_-},$$

$$Z = \frac{\partial \psi(T\boldsymbol{d} + \Sigma \boldsymbol{c})}{\partial \boldsymbol{c}} \Big|_{\boldsymbol{c}=\boldsymbol{c}_-, \boldsymbol{d}=\boldsymbol{d}_-},$$

$$\boldsymbol{w} = \boldsymbol{y} - \psi(T\boldsymbol{d}_- + \Sigma \boldsymbol{c}_-) + X\boldsymbol{d}_- + Z\boldsymbol{c}_-.$$

The subscript minus indicates quantities evaluated at the current iteration. It is not difficult to see that $\boldsymbol{w} = \tilde{\boldsymbol{y}}$, $X = VT$, and $Z = V\Sigma$. Therefore, the REML estimate of λ based on the LME model (7.27) is the same as the GML estimate for the approximate SSR model corresponding to (7.12). This suggests an iterative approach by estimating λ for the approximate SSR model (7.17) at each iteration using the GML method.

The UBR method (3.3) may also be used to estimate λ at each iteration. The following algorithm summarizes the above procedures.

Linearization Algorithm

1. *Initialize*: $f = f_0$.

2. *Linearize*: Update f by finding the PLS estimate of an approximate SSR model with linear functionals. The smoothing parameter λ is estimated using the GCV, GML, or UBR method.

3. Repeat Step 2 until convergence.

For the special model (7.5), the Gauss–Newton method (i.e. solve (7.12)) or the Newton–Raphson method (i.e., solve (7.15)) may be used at Step 2. For the general NNR model (7.4), the EGN method (i.e., fit model (7.17)) may be used at Step 2.

We estimate σ^2 by

$$\hat{\sigma}^2 = \frac{\sum_{i=1}^n (y_i - \mathcal{N}_i f)^2}{n - \sum_{i=1}^n a_i},$$

where a_i are defined in (7.22) and computed at convergence.

We now discuss how to construct Bayesian confidence intervals. At convergence, the extended Gauss–Newton method approximates the original model by

$$\tilde{y}_i^* = \mathcal{D}_i^* f + \epsilon_i, \quad i = 1, \ldots, n, \tag{7.28}$$

where $\mathcal{D}_i^* = D\mathcal{N}_i(\hat{f})$ and $\tilde{y}_i^* = y_i - \mathcal{N}_i \hat{f} + \mathcal{D}_i^* \hat{f}$. Assume a prior distribution for f as

$$F(x) = \sum_{\nu=1}^p \zeta_\nu \phi_\nu(x) + \delta^{1/2} U(x),$$

where $\zeta_\nu \overset{iid}{\sim} N(0, \kappa)$ and $U(x)$ is a zero-mean Gaussian stochastic process with $\mathrm{Cov}(U(x), U(z)) = R_1(x, z)$. Assume that observations are generated from

$$\tilde{y}_i^* = \mathcal{D}_i^* F + \epsilon_i, \quad i = 1, \dots, n. \tag{7.29}$$

Since $\mathcal{D}_i^*$ are bounded linear functionals, the posterior mean of the Bayesian model (7.29) equals $\hat{f}$. Posterior variances and Bayesian confidence intervals for model (7.29) can be calculated as in Section 3.8. Based on the first-order approximation (7.28), the performances of these approximate Bayesian confidence intervals depend largely on the accuracy of the linear approximation at convergence. The bootstrap method may also be used to construct confidence intervals.

7.4 Estimation with Multiple Functions

We now consider the case when $r > 1$. The goal is to estimate nonparametric functions $\boldsymbol{f} = (f_1, \dots, f_r)$ in model (7.4). Note that $f_k \in \mathcal{H}_k$ for $k = 1, \dots, r$. Assume that $\mathcal{H}_k = \mathcal{H}_{k0} \oplus \mathcal{H}_{k1}$, where $\mathcal{H}_{k0} = \mathrm{span}\{\phi_{k1}, \dots, \phi_{kp_k}\}$ and $\mathcal{H}_{k1}$ is an RKHS with RK R_{k1}. We estimate $\boldsymbol{f}$ as minimizers of the following PLS

$$\frac{1}{n} \sum_{i=1}^{n} (y_i - \mathcal{N}_i(\boldsymbol{f}))^2 + \sum_{k=1}^{r} \lambda_k \|P_{1k} f_k\|^2, \tag{7.30}$$

where P_{k1} are projection operators from $\mathcal{H}_k$ onto $\mathcal{H}_{k1}$, and $\lambda_1, \dots, \lambda_r$ are smoothing parameters.

The linearization procedures in Section 7.3 may be applied to all functions simultaneously. However, it is usually computationally intensive when n and/or r are large. We use a Gauss–Seidel type algorithm to estimate functions iteratively one at a time.

Nonlinear Gauss–Seidel Algorithm

1. *Initialize:* $f_k = f_{k0}, \quad k = 1, \dots, r.$

2. *Cycle:* For $k = 1, \dots, r, 1, \dots, r, \dots$, conditional on the current approximations of $f_1, \dots, f_{k-1}, f_{k+1}, \dots, f_r$, update f_k using the linearization algorithm in Section 7.3.

3. Repeat Step 2 until convergence.

Step 2 involves an inner iteration of the linearization algorithm. The convergence of this inner iteration is usually unnecessary, and a small number of iterations is usually good enough. The nonlinear Gauss–Seidel procedure is an extension of the backfitting procedure.

Denote $\hat{f}$ as the estimate of f at convergence. Note here that $\hat{f}$ denotes the estimate of the vector of functions f rather than the fitted values of a single function. Assume that the Fréchet differentials of $\mathcal{N}_i$ with respect to f evaluated at $\hat{f}$, $\mathcal{D}_i^* = D\mathcal{N}_i(\hat{f})$, exist. Then $\mathcal{D}_i^* h = \sum_{k=1}^r \mathcal{D}_{ki}^* h_k$, where $\mathcal{D}_{ki}^*$ is the partial Fréchet differential of $\mathcal{N}_i$ with respect to f_k evaluated at $\hat{f}$, $h = (h_1, \ldots, h_r)$, and $h_k \in \mathcal{H}_k$ (Flett 1980). Using the linear approximation, the NNR model (7.4) may be approximated by

$$\tilde{y}_i^* = \sum_{k=1}^r \mathcal{D}_{ki}^* f_k + \epsilon_i, \quad i = 1, \ldots, n, \tag{7.31}$$

where $\tilde{y}_i^* = y_i - \mathcal{N}_i(\hat{f}) + \sum_{k=1}^r \mathcal{D}_{ki}^* \hat{f}_k$. The corresponding Bayes model for (7.31) is given in (8.12) and (8.13). Section 8.2.2 provides formulae for posterior means and covariances, and discussion about how to construct Bayesian confidence intervals for functions f_k in model (7.31). These Bayesian confidence intervals provide approximate confidence intervals for f_k in the NNR model. Again, the bootstrap method may also be used to construct confidence intervals.

7.5 The nnr Function

The function nnr in the assist package is designed to fit the special NNR model (7.5) and when $\mathcal{L}_i$ are evaluational functionals. Sections 7.6.2 and 7.6.3 provide two example implementations of the EGN method for two NNR models that cannot be fitted by the nnr function.

A typical call is

```
nnr(formula, func, start)
```

where formula is a two-sided formula specifying the response variable on the left side of a ~ operator and an expression for the function ψ on the right side with f_k treated as parameters. The argument func inputs a list of formulae, each specifying bases $\phi_{k1}, \ldots, \phi_{kp_k}$ for $\mathcal{H}_{k0}$, and RK R_{k1} for $\mathcal{H}_{k1}$. Each formula in this list has the form

$$f \sim \texttt{list}(\sim \phi_1 + \cdots + \phi_p, \texttt{rk}).$$

For example, suppose $\boldsymbol{f} = (f_1, f_2)$, where f_1 and f_2 are functions of an independent variable x. Suppose both f_1 and f_2 are modeled using cubic splines. Then the bases and RKs of f_1 and f_2 can be specified using

```
func=list(f1(x)~list(~x,cubic(x)),f2(x)~list(~x,cubic(x)))
```

or simply

```
func=list(f1(x)+f2(x)~list(~x, cubic(x)))
```

The argument `start` inputs a vector or an expression that specifies the initial values of the unknown functions.

The method of selecting smoothing parameters is specified by the argument `spar`. The options `spar=''v''`, `spar=''m''`, and `spar=''u''` correspond to the GCV, GML, and UBR methods, respectively, with GCV as the default. The option `method` in the argument `control` specifies the computational method with `method=''GN''` and `method=''NR''` corresponding to the Gauss–Newton and Newton–Raphson methods, respectively. The default is the Newton–Raphson method.

An object of nnr class is returned. The generic function `summary` can be applied to extract further information. Approximate Bayesian confidence intervals can be constructed based on the output of the `intervals` function.

7.6   Examples

7.6.1   Nonparametric Regression Subject to Positive Constraint

The exponential transformation in (7.2) may be used to relax the positive constraint for a univariate or multivariate regression function. In this section we use a simple simulation to illustrate how to fit model (7.2) and the advantage of the exponential transformation over simple cubic spline fit.

We generate $n = 50$ samples from model (7.1) with $g(x) = \exp(-6x)$, x_i equally spaced in $[0, 1]$, and $\epsilon_i \overset{iid}{\sim} N(0, 0.1^2)$. We fit g with a cubic spline and the exponential transformation model (7.2) with f modeled by a cubic spline:

```
> n <- 50
> x <- seq(0,1,len=n)
> y <- exp(-6*x)+.1*rnorm(n)
```

```
> ssrfit <- ssr(y~x, cubic(x))
> nnrfit <- nnr(y~exp(f(x)), func=f(u)~list(~u,cubic(u)),
    start=list(f=log(abs(y)+0.001)))
```

where, for simplicity, we used $\log(|y_i| + 0.001)$ as initial values.

We compute the mean squared error (MSE) as

$$\text{MSE} = \frac{1}{n} \sum_{i=1}^{n} (\hat{g}(x_i) - g(x_i))^2,$$

where $\hat{g}$ is either the cubic spline fit or the fit based on the exponential transformation model (7.2). The simulation was repeated 100 times. Ignoring the positive constraint, the cubic spline fits have larger MSEs than those based on the exponential transformation (Figure 7.1(a)). Figure 7.1(b) shows observations, the true function, and fits for a typical replication. The cubic spline fit has portions that are negative. The exponential transformation leads to a better fit.

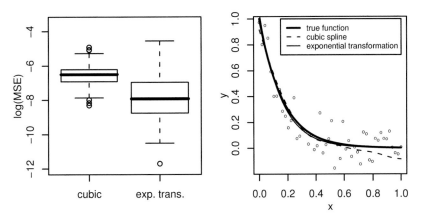

FIGURE 7.1 (a) Boxplots of MSEs on the logarithmic scale of the cubic spline fits and the fits based on the exponential transformation; and (b) observations (circles), the true function, and estimates for a typical replication.

7.6.2 Nonparametric Regression Subject to Monotone Constraint

Consider model (7.1) and suppose that g is known to be strictly increasing. We relax the constraint by considering the transformed model

(7.3). It is clear that model (7.3) cannot be written in the form of (7.5). Therefore, it cannot be fitted using the current version of the **nnr** function. We now illustrate how to apply the EGN method in Section 7.3.2 to estimate the function f in (7.3). A similar procedure may be derived for other situations. See Section 7.6.3 for another example.

For simplicity, we model f in (7.3) using the cubic spline model space $W_2^2[0,1]$. Let $\mathcal{N}_i f = \int_0^{x_i} \exp\{f(s)\}ds$. Then it can be shown that the Fréchet differential of $\mathcal{N}_i$ at $\hat{f}_-$ exists, and $\mathcal{D}_i f = D\mathcal{N}_i(\hat{f}_-)(f) = \int_0^{x_i} \exp\{\hat{f}_-(s)\}f(s)ds$ (Debnath and Mikusiński 1999). Then model (7.3) can be approximated by

$$\tilde{y}_i = \beta + \mathcal{D}_i f + \epsilon_i, \quad i = 1, \ldots, n, \qquad (7.32)$$

where $\tilde{y}_i = y_i - \mathcal{N}_i \hat{f}_- + \mathcal{D}_i \hat{f}_-$. Model (7.32) is a partial spline model. Suppose we use the construction of $W_2^2[0,1]$ in Section 2.6, where $\phi_1(x) = k_0(x) = 1$ and $\phi_2(x) = k_1(x) = x - 0.5$ are basis functions for $\mathcal{H}_0$, and $R_1(x,z) = k_2(x)k_2(z) - k_4(|x - z|)$ is the RK for $\mathcal{H}_1$. To fit model (7.32) we need to compute $\tilde{y}_i$, $T = \{\mathcal{D}_i\phi_\nu\}_{i=1}^n{}_{\nu=1}^2$ and $\Sigma = \{\mathcal{D}_{i(x)}\mathcal{D}_{j(z)}R_1(x,z)\}_{i,j=1}^n$. It can be shown that

$$\tilde{y}_i = y_i - \int_0^{x_i} \exp\{\hat{f}_-(s)\}\{1 - \hat{f}_-(s)\}ds,$$

$$\mathcal{D}_i\phi_1 = \int_0^{x_i} \exp\{\hat{f}_-(s)\}ds,$$

$$\mathcal{D}_i\phi_2 = \int_0^{x_i} \exp\{\hat{f}_-(s)\}(s - 0.5)ds,$$

$$\mathcal{D}_{i(x)}\mathcal{D}_{j(z)}R_1(x,z) = \int_0^{x_i}\int_0^{x_j} \exp\{\hat{f}_-(s) + \hat{f}_-(t)\}R_1(s,t)dsdt.$$

The above integrals do not have closed forms. We approximate them using the Gaussian quadrature method. For simplicity of notation, suppose that x values are distinct and ordered such that $x_1 < x_2 < \cdots < x_n$. Let $x_0 = 0$. Write

$$\int_0^{x_i} \exp\{\hat{f}_-(s)\}ds = \sum_{j=1}^i \int_{x_{j-1}}^{x_j} \exp\{\hat{f}_-(s)\}ds.$$

We then approximate each integral $\int_{x_{j-1}}^{x_j} \exp\{\hat{f}_-(s)\}ds$ using Gaussian quadrature with three points. The approximation is quite accurate when $x_j - x_{j-1}$ is small. More points may be added for wider intervals. We use the same method to approximate integrals $\int_0^{x_i} \exp\{\hat{f}_-(s)\}\{1 - \hat{f}_-(s)\}ds$

and $\int_0^{x_i} \exp\{\hat{f}_-(s)\}(s - 0.5)ds$. Write the double integral

$$\int_0^{x_i} \int_0^{x_j} \exp\{\hat{f}_-(s) + \hat{f}_-(t)\}R_1(s,t)dsdt$$

$$= \sum_{k=1}^{i} \sum_{l=1}^{j} \int_{x_{k-1}}^{x_k} \int_{x_{l-1}}^{x_l} \exp\{\hat{f}_-(s) + \hat{f}_-(t)\}R_1(s,t)dsdt.$$

We then approximate each integral

$$\int_{x_{k-1}}^{x_k} \int_{x_{l-1}}^{x_l} \exp\{\hat{f}_-(s) + \hat{f}_-(t)\}R_1(s,t)dsdt$$

using the simple product rule (Evans and Swartz 2000).

The R function `inc` in Appendix B implements the EGN procedure for model (7.3). The function `inc` is available in the `assist` package.

We now use a small-scale simulation to show the advantage of the transformed model (7.3) over simple cubic spline fit. We generate $n = 50$ samples from model (7.1) with $g(x) = 1 - \exp(-6x)$, x_i equally spaced in $[0, 1]$, and $\epsilon_i \overset{iid}{\sim} N(0, 0.1^2)$. We fit g with a cubic spline and the model (7.3) with f modeled by a cubic spline, both with the GML choice of the smoothing parameter:

```
> n <- 50; x <- seq(0,1,len=n)
> y <- 1-exp(-6*x)+.1*rnorm(n)
> ssrfit <- ssr(y~x, cubic(x), spar=''m'')
> incfit <- inc(y, x, spar=''m'')
```

MSEs are computed as in Section 7.6.1. The simulation was repeated 100 times. Ignoring the monotonicity constraint, cubic spline fits have larger MSEs than those based on model (7.3) (Figure 7.2(a)). Figure 7.2(b) shows observations, the true function, and fits for a typical replication. The cubic spline fit is not monotone. The fit based on model (7.3) is closer to the true function.

If the function g in model (7.1) is known to be both positive and strictly increasing, then we can consider model (7.3) with an additional constraint $\beta > 0$. Writing $\beta = \exp(\alpha)$, then model (7.3) becomes a semiparametric nonlinear regression model in Section 8.3. A similar EGN procedure can be derived to fit such a model.

Now consider the child growth data consisting of height measurements of a child over one school year. Observations are shown in Figure 7.3(a).

We first fit a cubic spline to model (7.1) with the GCV choice of the smoothing parameter. The cubic spline fit, shown in Figure 7.3(a) as the dashed line, is not monotone. It is reasonable to assume that the mean

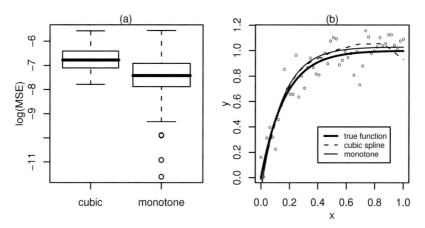

FIGURE 7.2 (a) Boxplots of MSEs on the logarithmic scale of the cubic spline fits and the fits based on model (7.3); and (b) observations (circles), the true function, and estimates for a typical replication.

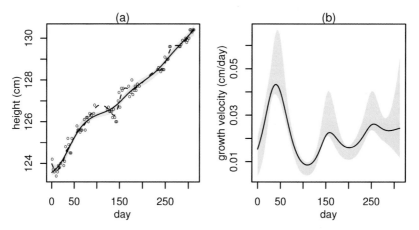

FIGURE 7.3 Child growth data, plots of (a) observations (circles), the cubic spline fit (dashed line), and the fit based on model (7.3) with 95% bootstrap confidence intervals (shaded region), and (b) estimate of the velocity of growth with 95% bootstrap confidence intervals (shaded region).

growth function g in (7.1) is a strictly increasing function. Therefore, we fit model (7.3) using the `inc` function and compute bootstrap confidence intervals with 10,000 bootstrap samples:

```
> library(fda); attach(onechild)
```

```
> x <- day/365; y <- height
> onechild.cub.fit <- ssr(y~x, cubic(x))
> onechild.inc.fit <- inc(y, x, spar=''m'')
> nboot <- 10000; set.seed <- 23057
> yhat <- onechild.inc.fit$y.fit; resi <- y-yhat
> fb <- gb <- NULL; totboot <- 0
> while (totboot<nboot) {
    yb <- yhat+sample(resi, replace=T)
    tmp <- try(inc(yb, x, spar=''m''))
    if(class(tmp)!=''try-error''&tmp$iter[2]<1.e-6) {
      fb <- cbind(fb,
        (tmp$pred$f-onechild.inc.fit$pred$f)/tmp$sigma)
      gb <- cbind(gb,
        (tmp$pred$y-onechild.inc.fit$pred$y)/tmp$sigma)
      totboot <- totboot+1
    }
  }
> shat <- onechild.inc.fit$sigma
> gl <- onechild.inc.fit$pred$y-
    shat*apply(gb,1,quantile,prob=.975)
> gu <- onechild.inc.fit$pred$y-
    shat*apply(gb,1,quantile,prob=.025)
> hl <- exp(onechild.inc.fit$pred$f-
    shat*apply(fb,1,quantile,prob=.975))/365
> hu <- exp(onechild.inc.fit$pred$f-
    shat*apply(fb,1,quantile,prob=.025))/365
```

where `gl` and `gu` are lower and upper bounds for the mean function g, and `hl` and `hu` are lower and upper bounds for the velocity defined as $h(x) = g'(x) = \exp\{f(x)\}$. The percentile-t bootstrap confidence intervals were constructed.

The constrained fit based on model (7.3) is shown in Figure 7.3(a) with 95% bootstrap confidence intervals. Figure 7.3(b) shows the estimate of $g'(x)$ with 95% bootstrap confidence intervals. As in Ramsay (1998), the velocity shows three short bursts. Note that Ramsay (1998) used a different transformation, $g(x) = \beta_1 + \beta_2 \int_0^x \exp\{\int_0^s w(u)du\}ds$, in model (7.1) to relax the monotone constraint. It is easy to see that $w(x) = f'(x)$, where $f(x) = \log g'(x)$. Therefore, with respect to the mean function g, the penalty in this section equals $\int_0^1 [\{\log g'(x)\}'']^2 dx$, and the penalty in Ramsay (1998) equals $\int_0^1 \{g''(x)/g'(x)\}^2 dx$.

7.6.3 Term Structure of Interest Rates

In this section we use the bond data to investigate the term structure of interest rate, a concept central to economic and financial theory.

Consider a set of n coupon bonds from which the interest rate term structure is to be inferred. Denote the current time as zero. Let y_i be the current price of bond i, r_{ij} be the payment paid at a future time x_{ij}, $0 < x_{i1} < \cdots < x_{im_i}$, $i = 1, \ldots, n$, $j = 1, \ldots, m_i$. The pricing model assumes that (Fisher, Nychka and Zervos 1995, Jarrow, Ruppert and Yu 2004)

$$y_i = \sum_{j=1}^{m_i} r_{ij} g(x_{ij}) + \epsilon_i, \quad i = 1, \ldots, n, \tag{7.33}$$

where g is a *discount function* and $g(x_{ij})$ represents the price of a dollar delivered at time x_{ij}, and ϵ_i are iid random errors with mean zero and variance σ^2. The goal is to estimate the discount function g from observations y_i. Assume that $g \in W_2^2[0, b]$ for a fixed time b and define $\mathcal{L}_i g = \sum_{j=1}^{m_i} r_{ij} g(x_{ij})$. Then, it is easy to see that $\mathcal{L}_i$ are bounded linear functionals on $W_2^2[0, b]$. Therefore, model (7.33) is a special case of the general SSR model (2.10). Consider the cubic spline construction in Section 2.2. Two basis functions of the null space are $\phi_1(x) = 1$ and $\phi_2(x) = x$, and denote the RK of $\mathcal{H}_1$ as R_1. To fit model (7.33) using the `ssr` function, we need to compute $T = \{\mathcal{L}_i \phi_\nu\}_{i=1}^{n}{}_{\nu=1}^{2}$ and $\Sigma = \{\mathcal{L}_i \mathcal{L}_j R_1\}_{i,j=1}^{n}$. Define $\boldsymbol{r}_i = (r_{i1}, \ldots, r_{im_i})$ and $X = \mathrm{diag}(\boldsymbol{r}_1, \ldots, \boldsymbol{r}_n)$. Then

$$T = \Big\{ \sum_{j=1}^{m_i} r_{ij} \phi_\nu(x_{ij}) \Big\}_{i=1}^{n}{}_{\nu=1}^{2} = XS,$$

where $S = (\boldsymbol{\phi}_1, \boldsymbol{\phi}_2)$ and $\boldsymbol{\phi}_\nu = (\phi_\nu(x_{11}), \ldots, \phi_\nu(x_{1m_1}), \ldots, \phi_\nu(x_{n1}), \ldots, \phi_\nu(x_{nm_n}))^T$ for $\nu = 1, 2$. Similarly, it can be shown that

$$\Sigma = \Big\{ \sum_{k=1}^{m_i} \sum_{l=1}^{m_j} r_{ik} r_{jl} R_1(x_{ik}, x_{jl}) \Big\}_{i,j=1}^{n} = X \Lambda X^T,$$

where $\Lambda = \{\Lambda_{ij}\}_{i,j=1}^{n}$ and $\Lambda_{ij} = \{R_1(x_{ik}, x_{jl})\}_{k=1}^{m_i}{}_{l=1}^{m_j}$. Note that R_1 can be calculated by the `cubic2` function.

The bond data set contains 78 Treasury bonds and 144 GE (General Electric Company) bonds. We first fit model (7.33) to the Treasury bond and compute estimates of g at grid points as follows:

```
> data(bond); attach(bond)
```

```
> group <- as.vector(table(name[type==''govt'']))
> y <- price[type==''govt''][cumsum(group)]
> x <- time[type==''govt'']
> r <- payment[type==''govt'']
> X <- assist:::diagComp(matrix(r,nrow=1),group)
> S <- cbind(1, x)
> T <- X%*%S
> Q <- X%*%cubic2(x)%*%t(X)
> bond.cub.fit.govt <- ssr(y~T-1, Q, spar=''m'',
                           limnla=c(6,10))
> grid1 <- seq(min(x),max(x),len=100)
> g.cub.p.govt <- cbind(1,grid1)%*%bond.cub.fit.govt$coef$d
  +cubic2(grid1,x)%*%t(X)%*%bond.cub.fit.govt$coef$c
```

where diagComp is a function in the assist package that constructs the matrix $X = \text{diag}(r_1, \ldots, r_n)$. We compute bootstrap confidence intervals for g at grid points as follows:

```
> boot.bond.one <- function(x, y, yhat, X, T, Q,
  spar=''m'', limnla=c(-3,6), grid, nboot, seed=0) {
  set.seed <- seed
  resi <- y-yhat
  gb <- NULL
  for(i in 1:nboot) {
   yb <- yhat + sample(resi, replace=TRUE)
   tmp <- try(ssr(yb~T-1, Q, spar=spar, limnla=c(-3,6)))
   if(class(tmp)!=''try-error'') gb <- cbind(gb,
    cbind(1,grid)%*%tmp$coef$d+
    cubic2(grid,x)%*%t(X)%*%tmp$coef$c)
  }
  return(gb)
  }
> g.cub.b.govt <- boot.bond.one(x, y,
  yhat=bond.cub.fit.govt$fit,
  X, T, Q, grid=grid1, nboot=5000, seed=3498)
> gl <- apply(g.cub.b.govt, 1, quantile, prob=.025)
> gu <- apply(g.cub.b.govt, 1, quantile, prob=.975)
```

where gl and gu are lower and upper bounds for the discount function g, and the 95% percentile bootstrap confidence intervals were computed based on 5,000 bootstrap samples. Model (7.33) for the GE bond can be fitted similarly. The estimates of discount functions and bootstrap confidence intervals for the Treasury and GE bonds are shown in Figure 7.4(a).

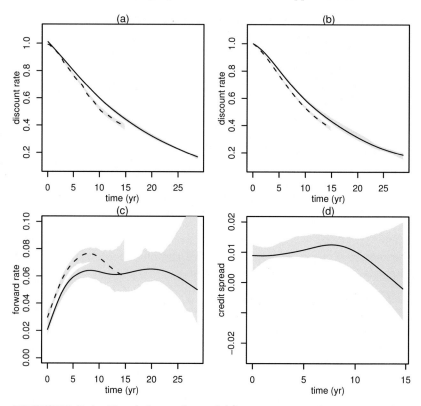

FIGURE 7.4 Bond data, plots of (a) unconstrained estimates of the discount function for Treasury bond (solid line) and GE bond (dashed line) based on model (7.33) with 95% bootstrap confidence intervals (shaded regions), (b) constrained estimates of the discount function for Treasury bond (solid line) and GE bond (dashed line) based on model (7.35) with 95% bootstrap confidence intervals (shaded regions), (c) estimates of the forward rate for Treasury bond (solid line) and GE bond (dashed line) based on model (7.35) with 95% bootstrap confidence intervals (shaded regions), and (d) estimates of the credit spread based on model (7.35) with 95% bootstrap confidence intervals (shaded region).

The discount function g is required to be positive, decreasing, and satisfy $g(0) = 1$. These constraints are ignored in the above direct estimation. One simple approach to dealing with these constraints is to represent g by

$$g(x) = \exp\left\{-\int_0^x f(s)ds\right\}, \tag{7.34}$$

where $f(s) \geq 0$ is the so-called *forward rate*. Reparametrization (7.34) takes care of all constraints on g. The goal now is to estimate the forward rate f. Assuming $f \in W_2^2[0, b]$ and replacing g in (7.33) by (7.34) leads to the following NNR model

$$y_i = \sum_{j=1}^{m_i} r_{ij} \exp\left\{-\int_0^{x_{ij}} f(s)ds\right\} + \epsilon_i, \quad i = 1, \ldots, n. \tag{7.35}$$

For simplicity, the nonnegative constraint on f is not enforced since its estimate is not close to zero. When necessary, the nonnegative constraint can be enforced with a further reparametrization of f such as $f = \exp(h)$. The R function one.bond in Appendix C implements the EGN algorithm to fit model (7.35) for a single bond. For example, we can fit model (7.35) to Treasury bond as follows:

```
bond.nnr.fit.govt <- one.bond(price= price[type==''govt''],
   payment=payment[type==''govt''],
   time=time[type==''govt''],
   name=name[type==''govt''])
```

In the following, we model Treasury and GE bonds jointly. Assume that

$$y_{ki} = \sum_{j=1}^{m_{ki}} r_{kij} \exp\left[-\int_0^{x_{kij}} \{f_1(s) + f_2(s)\delta_{k,2}\}ds\right] + \epsilon_{ki},$$

$$k = 1, 2; \ i = 1, \ldots, n_k, \tag{7.36}$$

where k represents bond type with $k = 1$ and $k = 2$ corresponding to Treasury and GE bonds, respectively, y_{ki} is the current price for bond i of type k, r_{kij} are future payments for bond i of type k, $\delta_{\nu,\mu}$ is the Kronecker delta, f_1 is the forward rate for Treasury bond, and f_2 represents the difference between GE and Treasury bonds (also called the *credit spread*). We assume that $f_1, f_2 \in W_2^2[0, b]$. Model (7.36) is a general NNR model with two functions f_1 and f_2. It cannot be fitted directly by the nnr function. We now describe how to implement the nonlinear Gauss–Seidel algorithm to fit model (7.36).

Denote the current estimates of f_1 and f_2 as $\hat{f}_{1-}$ and $\hat{f}_{2-}$, respectively. First consider updating f_1 for fixed f_2. The Fréchet differential of $\mathcal{N}_i$ with respect to f_1 at current estimates $\hat{f}_{1-}$ and $\hat{f}_{2-}$ is

$$\mathcal{D}_{ki}h = -\sum_{j=1}^{m_{ki}} r_{kij} \exp\left[-\int_0^{x_{kij}} \{\hat{f}_{1-}(s) + \hat{f}_{2-}(s)\delta_{k,2}\}ds\right] \int_0^{x_{kij}} h(s)ds$$

$$= -\sum_{j=1}^{m_{ki}} r_{kij} w_{kij} \int_0^{x_{kij}} h(s)ds,$$

where $w_{kij} = \exp\left[-\int_0^{x_{kij}}\{\hat{f}_{1-}(s) + \hat{f}_{2-}(s)\delta_{k,2}\}ds\right]$. Let

$$\tilde{y}_{ki,1} = \sum_{j=1}^{m_{ki}} r_{kij}w_{kij}(1 + f_{kij,1}) - y_{ki},$$

$$\mathcal{L}_{ki,1}f_1 = \sum_{j=1}^{m_{ki}} r_{kij}w_{kij}\int_0^{x_{kij}} f_1(s)ds,$$

where $f_{kij,1} = \int_0^{x_{kij}} \hat{f}_{1-}(s)ds$. We need to fit the following SSR model

$$\tilde{y}_{ki,1} = \mathcal{L}_{ki,1}f_1 + \epsilon_{ki}, \quad k = 1,2; \ i = 1,\ldots,n_k, \qquad (7.37)$$

to update f_1. Let $T_k = \{\mathcal{L}_{ki,1}\phi_\nu\}_{i=1\ \nu=1}^{n_k\ 2}$ for $k = 1,2$, $T = (T_1^T, T_2^T)^T$, $\Sigma_{uv} = \{\mathcal{L}_{ui,1}\mathcal{L}_{vj,1}R_1\}_{i=1\ j=1}^{n_u\ n_v}$ for $u,v = 1,2$, and $\Sigma = \{\Sigma_{uv}\}_{u,v=1}^2$. To fit model (7.37) using the ssr function, we need to compute matrices T and Σ. Define $\boldsymbol{b}_{ki} = (r_{ki1}w_{ki1},\ldots,r_{kim_i}w_{kim_i})$ and $X_k = \mathrm{diag}(\boldsymbol{b}_{k1},\ldots,\boldsymbol{b}_{kn_k})$ for $k = 1,2$. Define

$$\boldsymbol{\psi}_{k\nu} = \left(\int_0^{x_{k11}} \phi_\nu(s)ds,\ldots,\int_0^{x_{k1m_1}} \phi_\nu(s)ds,\ldots,\int_0^{x_{kn1}} \phi_\nu(s)ds,\ldots,\right.$$

$$\left.\int_0^{x_{knm_n}} \phi_\nu(s)ds\right)^T, \quad \nu = 1,2; \ k = 1,2,$$

$$S_k = (\boldsymbol{\psi}_{k1}, \boldsymbol{\psi}_{k2}), \quad k = 1,2,$$

$$\Lambda_{uv,ij} = \{\int_0^{x_{uik}}\int_0^{x_{vjl}} R_1(s,t)dsdt\}_{k=1\ l=1}^{m_{ui}\ m_{vj}}, \quad u,v = 1,2;$$

$$i = 1,\ldots,n_u; \ j = 1,\ldots,n_v,$$

$$\Lambda_{uv} = \{\Lambda_{uv,ij}\}_{i=1\ j=1}^{n_u\ n_v}, \quad u,v = 1,2.$$

Then it can be shown that

$$T_k = \{\sum_{j=1}^{m_{ki}} r_{kij}w_{kij}\int_0^{x_{kij}} \phi_\nu(s)ds\}_{i=1\ \nu=1}^{n_k\ 2} = X_k S_k, \quad k = 1,2,$$

$$\Sigma_{uv} = \{\sum_{k=1\ l=1}^{m_{ui}\ m_{vj}} r_{uik}w_{uik}r_{vjl}w_{vjl}\int_0^{x_{uik}}\int_0^{x_{vjl}} R_1(s,t)dsdt\}_{i=1\ j=1}^{n_u\ n_v}$$

$$= X_u \Lambda_{uv} X_v^T, \quad u,v = 1,2.$$

As in Section 7.6.2, integrals $\int_0^{x_{kij}} \phi_\nu(s)ds$ and $\int_0^{x_{uik}}\int_0^{x_{vjl}} R_1(s,t)dsdt$ are approximated using the Gaussian quadrature method. Note that these integrals do not change along iterations. Therefore, they only need to be computed once.

Next we consider updating f_2 with fixed f_1. The Fréchet differential of $\mathcal{N}_i$ at current estimates $\hat{f}_{1-}$ and $\hat{f}_{2-}$

$$\mathcal{D}_{ki}h = -\sum_{j=1}^{m_{ki}} r_{kij}\delta_{k,2}\exp\left[-\int_0^{x_{kij}}\{\hat{f}_{1-}(s) + \hat{f}_{2-}(s)\delta_{k,2}\}ds\right]$$
$$\times \int_0^{x_{kij}} h(s)ds.$$

Note that $\mathcal{D}_{ki}h = 0$ when $k = 1$. Therefore, we use observations with $k = 2$ only at this step. Let

$$\tilde{y}_{2i,2} = \sum_{j=1}^{m_{2i}} r_{2ij}w_{2ij}(1 + f_{2ij,2}) - y_{2i},$$

$$\mathcal{L}_{2i,2}f = \sum_{j=1}^{m_{2i}} r_{k2j}w_{2ij}\int_0^{x_{2ij}} f(s)ds,$$

where $f_{2ij,2} = \int_0^{x_{2ij}} \hat{f}_{2-}(s)ds$. We need to fit the following SSR model

$$\tilde{y}_{2i,2} = \mathcal{L}_{2i,2}f_2 + \epsilon_{2i}, \quad i = 1,\ldots,n_2, \tag{7.38}$$

to update f_2. It can be shown that

$$\{\mathcal{L}_{2i,2}\phi_\nu\}_{i=1 \ \nu=1}^{n_k \ \ 2} = T_2,$$
$$\{\mathcal{L}_{2i,2}\mathcal{L}_{2j,2}R_1\}_{i=1 \ j=1}^{n_u \ n_v} = \Sigma_{22}.$$

The R function `two.bond` in Appendix C implements the nonlinear Gauss–Seidel algorithm to fit model (7.36) for two bonds. For example, we can fit model (7.36) to the bond data and $5,000$ bootstrap samples as follows:

```
> bond.nnr.fit <- two.bond(price=price, payment=payment,
    time=time, name=name, type=type, spar=''m'')
> boot.bond.two <- function(y, yhat, price, payment, time,
    name, type, spar=''m'', limnla=c(-3,6), nboot, seed=0)
  {
  set.seed <- seed
  resi <- y-yhat
  group <- c(as.vector(table(name[type==''govt''])),
             as.vector(table(name[type==''ge''])))
  fb <- f2b <- gb <- NULL
  for(i in 1:nboot) {
  price.b <- rep(yhat+sample(resi,replace=TRUE), group)
```

```
    tmp <- try(two.bond(price=price.b, payment=payment,
                        time=time, name=name, type=type))
    if(class(tmp)!=''try-error''&
       tmp$iter[2]<1.e-4&tmp$iter[3]==0) {
      fb <- cbind(fb,c(tmp$f.val$f1,tmp$f.val$f2))
      f2b <- cbind(f2b,tmp$f2.val)
      gb <- cbind(gb,c(tmp$dc[[1]],tmp$dc[[2]]))
      }
    list(fb=fb, f2b=f2b, gb=gb)
    }
  }
> gf.nnr.b <- boot.bond.two(y=bond.nnr.fit$y$y,
    yhat=bond.nnr.fit$y$yhat, price=price, payment=payment,
    time=time, name=name, type=type, nboot=5000,
    seed=2394)
```

where the function `boot.bond.two` returns a list of estimates of the discount functions (`gb`), the forward rates (`fb`), and the credit spread (`f2b`). The 95% percentile bootstrap confidence intervals for the discount functions of the Treasury and GE bonds can be computed as follows:

```
> n1 <- nrow(bond[type=="govt",])
> gl1 <- apply(gf.nnr.b$gb[1:n1,],1,quantile,prob=.025)
> gu1 <- apply(gf.nnr.b$gb[1:n1,],1,quantile,prob=.975)
> gl2 <- apply(gf.nnr.b$gb[-(1:n1),],1,quantile,prob=.025)
> gu2 <- apply(gf.nnr.b$gb[-(1:n1),],1,quantile,prob=.975)
```

where `gl1` and `gu1` are lower and upper bounds for the discount function of the Treasury bond, and `gl2` and `gu2` are lower and upper bounds for the discount function of the GE bond. The 95% percentile bootstrap confidence intervals for the forward rates and credit spread can be computed similarly based on the objects `gf.nnr.b$fb` and `gf.nnr.b$fb2`.

Figures 7.4(b) and 7.4(c) show the estimates for the discount and forward rates, respectively. As expected, the GE discount rate is consistently smaller than that of Treasury bonds, representing a higher risk associated with corporate bonds. To assess the difference between the two forward rates, we plot the estimated credit spread and its 95% bootstrap confidence intervals in Figure 7.4(d). The credit spread is positive when time is smaller than 11 years.

7.6.4   A Multiplicative Model for Chickenpox Epidemic

Consider the chickenpox epidemic data consisting of monthly number of chickenpox in New York City during 1931–1972. Denote y as the square root of reported cases in month x_1 of year x_2. Both x_1 and x_2 are transformed into the interval $[0, 1]$. Figure 7.5 shows the time series plot of square root of the monthly case numbers.

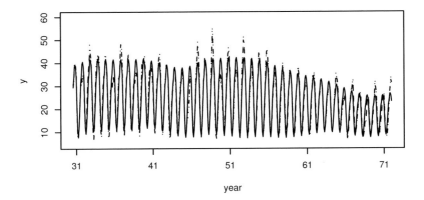

FIGURE 7.5 Chickenpox data, plots of the square root of the number of cases (dotted line), and the fits from the multiplicative model (7.39) (solid line) and SS ANOVA model (7.40) (dashed line).

There is a clear seasonal variation within a year that has long been recognized (Yorke and London 1973, Earn, Rohani, Bolker and Gernfell 2000). The seasonal variation was mainly caused by social behavior of children who made close contacts when school was in session, and temperature and humidity, that may affect the survival and transmission of dispersal stages. Thus the seasonal variations were similar over the years. The magnitude of this variation and the average number of cases may change over the years. Consequently, we consider the following *multiplicative* model

$$y(x_1, x_2) = g_1(x_2) + \exp\{g_2(x_2)\} \times g_3(x_1) + \epsilon(x_1, x_2), \qquad (7.39)$$

where $y(x_1, x_2)$ is the square root of reported cases in month x_1 of year x_2; and g_1, g_3 and $\exp(g_2)$ represent respectively the yearly mean, the seasonal trend in a year, and the magnitude of the seasonal variation for a particular year. The function $\exp(g_2)$ is referred to as the amplitude. A bigger amplitude corresponds to a bigger seasonal variation. For simplicity, we assume that random errors $\epsilon(x_1, x_2)$ are independent with a constant variance.

To make model (7.39) identifiable, we use the following two side conditions:

(a) $\int_0^1 g_2(x_2)dx_2 = 0$. The exponential transformation of g_2 and this condition make g_2 identifiable with g_3: the exponential transformation makes $\exp(g_2)$ free of a sign change, and $\int_0^1 g_2(x_2)dx_2 = 0$ makes $\exp(g_2)$ free of a positive multiplying constant. This condition can be fulfilled by removing the constant functions from the model space of g_2.

(b) $\int_0^1 g_3(x_1)dx_1 = 0$. This condition eliminates the additive constant making g_3 identifiable with g_1. This condition can be fulfilled by removing the constant functions from the model space of g_3.

We model g_1 and g_2 using cubic splines. Specifically, we assume that $g_1 \in W_2^2[0,1]$ and $g_2 \in W_2^2[0,1] \ominus \{1\}$, where constant functions are removed from the model space for g_2 to satisfy the side condition (a). Since the seasonal trend g_3 is close to a sinusoidal model, we model g_3 using a trigonometric spline with $\mathcal{L} = D^2 + (2\pi)^2$ ($m = 2$ in (2.70)). That is, $g_3 \in W_2^2(per) \ominus \{1\}$ where constant functions are removed from the model space for g_3 to satisfy the side condition (b). Obviously, the multiplicative model (7.39) is a special case of the NNR model with functions denoted as g_1, g_2, and g_3 instead of f_1, f_2, and f_3 (the notations of f_i are saved for an SS ANOVA model later). The multiplicative model (7.39) is fitted as follows:

```
> data(chickenpox)
> y <- sqrt(chickenpox$count)
> x1 <- (chickenpox$month-0.5)/12
> x2 <- ident(chickenpox$year)
> tmp <- ssr(y~1, rk=periodic(x1), spar=''m'')
> g3.ini <- predict(tmp, term=c(0,1))$fit
> chick.nnr <- nnr(y~g1(x2)+exp(g2(x2))*g3(x1),
    func=list(g1(x)~list(~I(x-.5),cubic(x)),
               g2(x)~list(~I(x-.5)-1,cubic(x)),
               g3(x)~list(~sin(2*pi*x)+cos(2*pi*x)-1,
                          lspline(x,type=''sine0''))),
    start=list(g1=mean(y),g2=0,g3=g3.ini),
    control=list(converg=''coef''),spar=''m'')
> grid <- data.frame(x1=seq(0,1,len=50),
                     x2=seq(0,1,len=50))
> chick.nnr.pred <- intervals(chick.nnr, newdata=grid,
    terms=list(g1=matrix(c(1,1,1,1,1,0,0,0,1),
```

```
                        nrow=3,byrow=T),
        g2=matrix(c(1,1,1,0,0,1),nrow=3,byrow=T),
        g3=matrix(c(1,1,1,1,1,0,0,0,1),nrow=3,byrow=T)))
```

We first fitted a periodic spline with variable x_1 only and used the fitted values to the smooth component as initial values for g_3. We used the average of y as the initial value for g_1 and constant zero as the initial value for g_2. The intervals function was used to compute approximate means and standard deviations for functions g_1, g_2, and g_3 and their projections. The estimates of g_1 and g_2 are shown in Figure 7.6. We also superimpose yearly averages in Figure 7.6(a) and the logarithm of scaled ranges in Figure 7.6(b). The scaled range of a specific year was calculated as the differences between the maximum and the minimum monthly number of cases divided by the range of the estimated seasonal trend g_3. It is clear that g_1 captures the long-term trend in the mean, and g_2 captures the long-term trend in the range of the seasonal variation. From the estimate of g_1 in Figure 7.6(a), we can see that yearly averages peaked in the 1930s and 1950s, and gradually decreased in the 1960s after the introduction of mass vaccination. The amplitude reflects the seasonal variation in the transmission rate (Yorke and London 1973). From the estimate of g_2 in Figure 7.6(b), we can see that the magnitude of the seasonal variation peaked in the 1950s and then declined in the 1960s, possibly as a result of changing public health conditions including mass vaccination. Figure 7.7 shows the estimate of the seasonal trend g_3, its projection onto the null space $\mathcal{H}_{30} = \text{span}\{\sin 2\pi x, \cos 2\pi x\}$ (the simple sinusoidal model), and its projection onto the orthogonal complement of the null space $\mathcal{H}_{31} = W_2^2(per) \ominus \text{span}\{1, \sin 2\pi x, \cos 2\pi x\}$. Since the projection onto the complement space is significantly different from zero, a simple sinusoidal model does not provide an accurate approximation.

To check the multiplicative model (7.39), we use the SS ANOVA decomposition in (4.21) and consider the following SS ANOVA model

$$y(x_1, x_2) = \mu + f_1(x_1) + f_2(x_2) + f_{12}(x_1, x_2) + \epsilon(x_1, x_2), \quad (7.40)$$

where μ is a constant, f_1 and f_2 are overall main effects of month and year, and f_{12} is the overall interaction between month and year. Note that side conditions $\int_0^1 f_1 dx_1 = \int_0^1 f_2 dx_2 = \int_0^1 f_{12} dx_1 = \int_0^1 f_{12} dx_2 = 0$ are satisfied by the SS ANOVA decomposition. It is not difficult to check

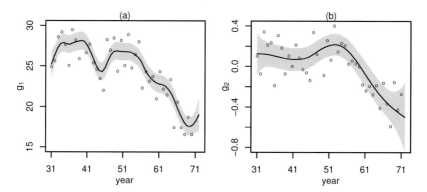

FIGURE 7.6   Chickenpox data, plots of (a) yearly averages (circles), estimate of g_1 (line), and 95% Bayesian confidence intervals (shaded region), and (b) yearly scaled ranges on logarithm scale (circles), estimate of g_2 (line), and 95% Bayesian confidence intervals (shaded region).

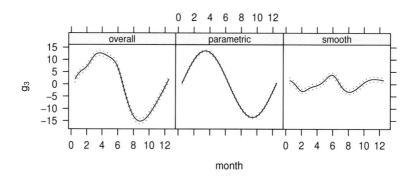

FIGURE 7.7   Chickenpox data, estimated g_3 (overall), and its projections to $\mathcal{H}_{30}$ (parametric) and $\mathcal{H}_{31}$ (smooth). Dotted lines are 95% Bayesian confidence intervals.

that model (7.40) reduces to model (7.39) iff

$$\mu = \int_0^1 g_1 dx_2,$$

$$f_1(x_1) = \left\{ \int_0^1 \exp(g_2) dx_2 \right\} g_3(x_1),$$

$$f_2(x_2) = g_1(x_2) - \int_0^1 g_1 dx_2,$$

$$f_{12}(x_1, x_2) = \left[\exp\{g_2(x_2)\} - \int_0^1 \exp(g_2) dx_2 \right] g_3(x_1).$$

Thus the multiplicative model assumes a multiplicative interaction.

We fit the SS ANOVA model (7.40) and compare it with the multiplicative model (7.39) using AIC, BIC, and GCV criteria:

```
> chick.ssanova <- ssr(y~x2,
    rk=list(periodic(x1),cubic(x2),
            rk.prod(periodic(x1),kron(x2-.5)),
            rk.prod(periodic(x1),cubic(x2))))
> n <- length(y)
> rss <- c(sum(chick.ssanova$resi**2),
            sum(chick.nnr$resi**2))/n
> df <- c(chick.ssanova$df, chick.nnr$df$f)
> gcv <- rss/(1-df/n)**2
> aic <- n*log(rss)+2*df
> bic <- n*log(rss)+log(n)*df
> print(round(rbind(gcv,aic,bic),2))
gcv     8.85   14.14
aic   826.75 1318.03
bic  1999.38 1422.83
```

The AIC and GCV criteria select the SS ANOVA model, while the BIC selects the multiplicative model. The SS ANOVA model captures local trend, particularly biennial pattern from 1945 to 1955, better than the multiplicative model (Figure 7.5).

7.6.5 A Multiplicative Model for Texas Weather

The domains for variables x_1 and x_2 in the multiplicative model (7.39) are not limited to continuous intervals. As the NNR model, these domains may be arbitrary sets. We now revisit the Texas weather data to which an SS ANOVA model has been fitted in Section 4.9.4. We have defined $x_1 = (\texttt{lat}, \texttt{long})$ as the geological `location` of a station, and x_2 as the `month` variable scaled into $[0, 1]$. For illustration purposes only, assuming that temperature profiles from all stations have the same shape except a vertical shift and scale transformation that may depend on geographical location, we consider the following multiplicative model

$$y(\boldsymbol{x}_1, x_2) = g_1(\boldsymbol{x}_1) + \exp\{g_2(\boldsymbol{x}_1)\} \times g_3(x_2) + \epsilon(\boldsymbol{x}_1, x_2), \quad (7.41)$$

where $y(\boldsymbol{x}_1, x_2)$ is the average temperature of month x_2 at location $\boldsymbol{x}_1$, g_1 and $\exp(g_2)$ represent, respectively, the mean and magnitude of the seasonal variation at location $\boldsymbol{x}_1$, and $g_3(x_2)$ represents the seasonal trend in a year. For simplicity, we assume that random errors $\epsilon(\boldsymbol{x}_1, x_2)$ are independent with a constant variance. We model functions g_1 and

g_2 using thin-plate splines, and g_3 using a periodic spline. Specifically, for identifiability, we assume that $g_1 \in W_2^2(\mathbb{R}^2)$, $g_2 \in W_2^2(\mathbb{R}^2) \ominus \{1\}$, and $g_3 \in W_2^2(per) \ominus \{1\}$. Model (7.41) is fitted as follows:

```
> g3.ini <- predict(ssr(y~1,rk=periodic(x2),data=tx.dat,
    spar=''m''),terms=c(0,1), pstd=F)$fit
> tx.nnr <- nnr(y~g1(x11,x12)+exp(g2(x11,x12))*g3(x2),
    func=list(g1(x,z)~list(~x+z,tp(list(x,z))),
              g2(x,z)~list(~x+z-1,tp(list(x,z))),
              g3(x)~list(periodic(x))),
    data=tx.dat, start=list(g1=mean(y),g2=0,g3=g3.ini))
```

The estimates of g_1 and g_2 are shown in Figure 7.8. From northwest to southeast, the temperature gets warmer and the variation during a year gets smaller.

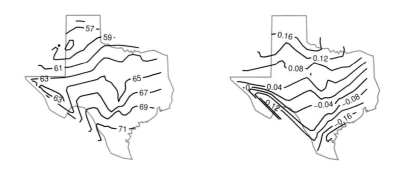

FIGURE 7.8 Texas weather data, plots of estimates of g_1 (left) and g_2 (right).

In Section 4.9.4 we have fitted the following SS ANOVA model

$$y(\boldsymbol{x}_1, x_2) = \mu + f_1(\boldsymbol{x}_1) + f_2(x_2) + f_{12}(\boldsymbol{x}_1, x_2) + \epsilon(\boldsymbol{x}_1, x_2), \quad (7.42)$$

where f_1 and f_2 are the overall main effects of location and month, and f_{12} is the overall interaction between location and month. The overall main effects and interaction are defined in Section 4.4.6. To compare the multiplicative model (7.41) with the SS ANOVA model fitted in Section 4.9.4, we compute GCV, AIC, and BIC criteria:

```
> n <- length(y)
> rss <- c(sum(tx.ssanova$resi**2),sum(tx.nnr$resi**2))
> df <- c(tx.ssanova$df,tx.nnr$df$f)
> gcv <- rss/(1-df/n)**2
> aic <- n*log(rss)+2*df
> bic <- n*log(rss)+log(n)*df
> print(rbind(gcv,aic,bic))
gcv  131.5558  434.5253
aic 2661.9293 3481.2111
bic 3730.4315 3894.1836
```

All criteria choose the SS ANOVA model. One possible reason is that the assumption of common temperature profile for all stations is too restrictive. To look at variation among temperature profiles, we fit each station separately using a periodic spline and compute normalized temperature profiles such that all profiles integrate to zero and have vertical range equal to one. These normalized temperature profiles are shown in Figure 7.9(a). Variation among these temperature profiles may be nonignorable. To further check if the variation among temperature profiles may be accounted for by a horizontal shift, we align all profiles such that all of them reach maximum at point 0.5 and plot the aligned profiles in Figure 7.9(b). Variation among these aligned temperature profiles is again nonignorable.

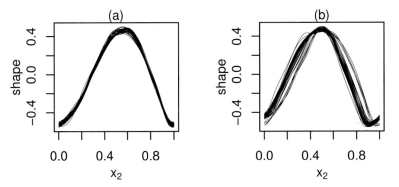

FIGURE 7.9 Texas weather data, plots of (a) normalized temperature profiles, and (b) aligned normalized temperature profiles.

It is easy to see that the SS ANOVA model (7.42) reduces to the multiplicative model (7.41) iff

$$f_{12}(\boldsymbol{x}_1, x_2) = \left[\frac{\exp\{g_2(\boldsymbol{x}_1)\}}{\sum_{j=1}^{J} w_j \exp\{g_2(\boldsymbol{u}_j)\}} - 1 \right] f_2(x_2), \qquad (7.43)$$

where $\boldsymbol{u}_j$ are fixed points in $\mathbb{R}^2$, and w_j are fixed positive weights such that $\sum_{j=1}^{J} w_j = 1$. Equation (7.43) suggests the following simple approach to check the multiplicative model (7.41): for a fixed station (thus a fixed $\boldsymbol{x}_1$), compute estimates of f_2 and f_{12} from the SS ANOVA fit and plot f_{12} against f_2 to see if the points fall on a straight line. Figure 7.10 shows plots of f_{12} against f_2 for two selected stations. The patterns are quite different from straight lines, especially for the Liberty station. Again, it indicates that the multiplicative model may not be appropriate for this case.

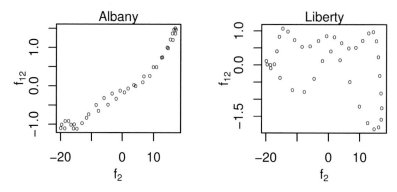

FIGURE 7.10 Texas weather data, plots of f_{12} against f_2 for two stations.

Chapter 8

Semiparametric Regression

8.1 Motivation

Postulating strict relationships between dependent and independent variables, parametric models are, in general, parsimonious. Parameters in these models often have meaningful interpretations. On the other hand, based on minimal assumptions about the relationship, nonparametric models are flexible. However, nonparametric models lose the advantage of having interpretable parameters and may suffer from the curse of dimensionality. Often, in practice, there is enough knowledge to model some components in the regression function parametrically. For other vague and/or nuisance components, one may want to leave them unspecified. Combining both parametric and nonparametric components, a semiparametric regression model can overcome limitations in parametric and nonparametric models while maintaining advantages of having interpretable parameters and flexibility.

Many specific semiparametric models have been proposed in the literature. The partial spline model (2.43) in Section 2.10 is perhaps the simplest semiparametric regression model. Other semiparametric models include the projection pursuit, single index, varying coefficients, functional linear, and shape invariant models. A *projection pursuit regression* model (Friedman and Stuetzle 1981) assumes that

$$y_i = \beta_0 + \sum_{k=1}^{r} f_k(\boldsymbol{\beta}_k^T \boldsymbol{x}_i) + \epsilon_i, \quad i = 1, \ldots, n, \tag{8.1}$$

where $\boldsymbol{x}$ are independent variables, β_0 and $\boldsymbol{\beta}_k$ are parameters, and f_k are nonparametric functions. A *partially linear single index* model (Carroll, Fan, Gijbels and Wand 1997, Yu and Ruppert 2002) assumes that

$$y_i = \boldsymbol{\beta}_1^T \boldsymbol{s}_i + f(\boldsymbol{\beta}_2^T \boldsymbol{t}_i) + \epsilon_i, \quad i = 1, \ldots, n, \tag{8.2}$$

where $\boldsymbol{s}$ and $\boldsymbol{t}$ are independent variables, $\boldsymbol{\beta}_1$ and $\boldsymbol{\beta}_2$ are parameters, and f is a nonparametric function. A varying-coefficient model (Hastie and

Tibshirani 1993) assumes that

$$y_i = \beta_1 + \sum_{k=1}^{r} u_{ik} f_k(x_{ik}) + \epsilon_i, \quad i = 1, \ldots, n, \tag{8.3}$$

where x_k and u_k are independent variables, β_1 is a parameter, and f_k are nonparametric functions.

The semiparametric linear and nonlinear regression models in this chapter include all the foregoing models as special cases. The general form of these models provides a framework for unified estimation, inference, and software development.

8.2   Semiparametric Linear Regression Models

8.2.1   The Model

Define a *semiparametric linear regression model* as

$$y_i = s_i^T \boldsymbol{\beta} + \sum_{k=1}^{r} \mathcal{L}_{ki} f_k + \epsilon_i, \quad i = 1, \ldots, n, \tag{8.4}$$

where s is a q-dimensional vector of independent variables, $\boldsymbol{\beta}$ is a vector of parameters, $\mathcal{L}_{ki}$ are bounded linear functionals, f_k are unknown functions, and $\boldsymbol{\epsilon} = (\epsilon_1, \ldots, \epsilon_n)^T \sim \mathrm{N}(\mathbf{0}, \sigma^2 W^{-1})$. We assume that W depends on an unknown vector of parameters $\boldsymbol{\tau}$. Let $\mathcal{X}_k$ be the domain of f_k and assume that $f_k \in \mathcal{H}_k$, where $\mathcal{H}_k$ is an RKHS on $\mathcal{X}_k$. Note that the domains $\mathcal{X}_k$ for different functions may be the same or different.

Model (8.4) is referred to as a semiparametric linear regression model since the mean response depends on $\boldsymbol{\beta}$ and f_k linearly. Extension to the nonlinear case will be introduced in Section 8.3. The semiparametric linear regression model (8.4) is an extension of the partial spline model (2.43) by allowing more than one nonparametric function. It is an extension of the additive models by including a parametric component and allowing general linear functionals. Note that some of the functions f_k may represent main effects and interactions in an SS ANOVA decomposition. Therefore, model (8.4) is also an extension of the SS ANOVA model (5.1) by allowing different linear functionals for different components. It is easy to see that the varying-coefficient model (8.3) is a special case of the semiparametric linear regression model with $q = 1$, $s_i = 1$, and $\mathcal{L}_{ki} f_k = u_{ik} f_k(x_{ik})$ for $k = 1, \ldots, r$. The functional linear models (FLM) are also a special case of the semiparametric linear regression model (see Section 8.4.1). In addition, random errors are allowed to be correlated and/or have unequal variances.

8.2.2 Estimation and Inference

Assume that $\mathcal{H}_k = \mathcal{H}_{k0} \oplus \mathcal{H}_{k1}$, where $\mathcal{H}_{k0} = \text{span}\{\phi_{k1}, \ldots, \phi_{kp_k}\}$ and $\mathcal{H}_{k1}$ is an RKHS with RK R_{k1}. Let $\boldsymbol{y} = (y_1, \ldots, y_n)^T$, $\boldsymbol{f} = (f_1, \ldots, f_r)$, and $\boldsymbol{\eta}(\boldsymbol{\beta}, \boldsymbol{f}) = (\boldsymbol{s}_1^T \boldsymbol{\beta} + \sum_{k=1}^r \mathcal{L}_{k1} f_k, \ldots, \boldsymbol{s}_n^T \boldsymbol{\beta} + \sum_{k=1}^r \mathcal{L}_{kn} f_k)^T$.

For a fixed W, we estimate $\boldsymbol{\beta}$ and $\boldsymbol{f}$ as minimizers to the following PWLS

$$\frac{1}{n}(\boldsymbol{y} - \boldsymbol{\eta}(\boldsymbol{\beta}, \boldsymbol{f}))^T W (\boldsymbol{y} - \boldsymbol{\eta}(\boldsymbol{\beta}, \boldsymbol{f})) + \lambda \sum_{k=1}^r \theta_k^{-1} \|P_{k1} f_k\|^2, \qquad (8.5)$$

where P_{k1} is the projection operator onto $\mathcal{H}_{k1}$ in $\mathcal{H}_k$. As in (4.32), different smoothing parameters $\lambda \theta_k^{-1}$ allow different penalties for each function.

For $k = 1, \ldots, r$, let $\xi_{ki}(x) = \mathcal{L}_{ki(z)} R_{k1}(x, z)$ and $\mathcal{S}_k = \mathcal{H}_{k0} \oplus \text{span}\{\xi_{k1}, \ldots, \xi_{kn}\}$. Let

$$\text{WLS}(\boldsymbol{\beta}, \mathcal{L}_1 \boldsymbol{f}, \ldots, \mathcal{L}_n \boldsymbol{f}) = \frac{1}{n}(\boldsymbol{y} - \boldsymbol{\eta}(\boldsymbol{\beta}, \boldsymbol{f}))^T W (\boldsymbol{y} - \boldsymbol{\eta}(\boldsymbol{\beta}, \boldsymbol{f}))$$

be the weighted LS where $\mathcal{L}_i \boldsymbol{f} = (\mathcal{L}_{1i} f_1, \ldots, \mathcal{L}_{ri} f_r)$. Then the PWLS (8.5) can be written as

$$\text{PWLS}(\boldsymbol{\beta}, \mathcal{L}_1 \boldsymbol{f}, \ldots, \mathcal{L}_n \boldsymbol{f}) = \text{WLS}(\boldsymbol{\beta}, \mathcal{L}_1 \boldsymbol{f}, \ldots, \mathcal{L}_n \boldsymbol{f}) + \lambda \sum_{k=1}^r \theta_k^{-1} \|P_{k1} f_k\|^2.$$

For any $f_k \in \mathcal{H}_k$, write $f_k = \varsigma_{k1} + \varsigma_{k2}$, where $\varsigma_{k1} \in \mathcal{S}_k$ and $\varsigma_{k2} \in \mathcal{S}_k^c$. As shown in Section 6.2.1, we have $\mathcal{L}_{ki} f_k = \mathcal{L}_{ki} \varsigma_{k1}$. Then for any $\boldsymbol{f}$,

$$\text{PWLS}(\boldsymbol{\beta}, \mathcal{L}_1 \boldsymbol{f}, \ldots, \mathcal{L}_n \boldsymbol{f})$$

$$= \text{WLS}(\boldsymbol{\beta}, \mathcal{L}_1 \varsigma_1, \ldots, \mathcal{L}_n \varsigma_1) + \lambda \sum_{k=1}^r \theta_k^{-1} (\|P_{k1} \varsigma_{k1}\|^2 + \|P_{k1} \varsigma_{k2}\|^2)$$

$$\geq \text{WLS}(\boldsymbol{\beta}, \mathcal{L}_1 \varsigma_1, \ldots, \mathcal{L}_n \varsigma_1) + \lambda \sum_{k=1}^r \theta_k^{-1} \|P_{k1} \varsigma_{k1}\|^2$$

$$= \text{PWLS}(\boldsymbol{\beta}, \mathcal{L}_1 \varsigma_1, \ldots, \mathcal{L}_n \varsigma_1),$$

where $\mathcal{L}_i \varsigma_1 = (\mathcal{L}_{1i} \varsigma_{11}, \ldots, \mathcal{L}_{ri} \varsigma_{r1})$ for $i = 1, \ldots, n$. Equality holds iff $\|P_{k1} \varsigma_{k2}\| = \|\varsigma_{k2}\| = 0$ for all $k = 1, \ldots, r$. Thus the minimizers $\hat{f}_k$ of the PWLS fall in the finite dimensional spaces $\mathcal{S}_k$, which can be represented as

$$\hat{f}_k = \sum_{\nu=1}^{p_k} d_{k\nu} \phi_{k\nu} + \theta_k \sum_{i=1}^{n} c_{ki} \xi_{ki}, \quad k = 1, \ldots, r, \tag{8.6}$$

where the multiplying constants θ_k make the solution and notations similar to those for the SS ANOVA models. Note that we only used the fact the WLS criterion depends on some bounded linear functionals $\mathcal{L}_{ki}$ in the foregoing arguments. Therefore, the Kimeldorf–Wahba representer theorem holds in general so long as the goodness-of-fit criterion depends on some bounded linear functionals.

Let $S = (s_1, \ldots, s_n)^T$, $T_k = \{\mathcal{L}_{ki} \phi_{k\nu}\}_{i=1}^{n} {}_{\nu=1}^{p_k}$ for $k = 1, \ldots, r$, $T = (T_1, \ldots, T_r)$, and $X = (S\ T)$. Let $\Sigma_k = \{\mathcal{L}_{ki} \mathcal{L}_{kj} R_{k1}\}_{i,j=1}^{n}$ for $k = 1, \ldots, r$, and $\Sigma_{\boldsymbol{\theta}} = \sum_{k=1}^{r} \theta_k \Sigma_k$, where $\boldsymbol{\theta} = (\theta_1, \ldots, \theta_r)$. Let $\hat{\boldsymbol{f}}_k = (\mathcal{L}_{k1} \hat{f}_k, \ldots, \mathcal{L}_{kn} \hat{f}_k)^T$, $\boldsymbol{c}_k = (c_{k1}, \ldots, c_{kn})^T$, and $\boldsymbol{d}_k = (d_{k1}, \ldots, d_{kp_k})^T$ for $k = 1, \ldots, r$. Based on (8.6), $\hat{\boldsymbol{f}}_k = T_k \boldsymbol{d}_k + \theta_k \Sigma_k \boldsymbol{c}_k$ for $k = 1, \ldots, r$. Let $\boldsymbol{d} = (\boldsymbol{d}_1^T, \ldots, \boldsymbol{d}_r^T)^T$. The overall fit

$$\hat{\boldsymbol{\eta}} = S\boldsymbol{\beta} + \sum_{k=1}^{r} \hat{\boldsymbol{f}}_k = S\boldsymbol{\beta} + T\boldsymbol{d} + \sum_{k=1}^{r} \theta_k \Sigma_k \boldsymbol{c}_k = X\boldsymbol{\alpha} + \sum_{k=1}^{r} \theta_k \Sigma_k \boldsymbol{c}_k,$$

where $\boldsymbol{\alpha} = (\boldsymbol{\beta}^T, \boldsymbol{d}^T)^T$. Plugging back into (8.5), we have

$$\frac{1}{n} \left(\boldsymbol{y} - X\boldsymbol{\alpha} - \sum_{k=1}^{r} \theta_k \Sigma_k \boldsymbol{c}_k \right)^T W \left(\boldsymbol{y} - X\boldsymbol{\alpha} - \sum_{k=1}^{r} \theta_k \Sigma_k \boldsymbol{c}_k \right) + \lambda \sum_{k=1}^{r} \theta_k \boldsymbol{c}_k^T \Sigma_k \boldsymbol{c}_k. \tag{8.7}$$

Taking the first derivatives with respect to $\boldsymbol{c}_k$ and $\boldsymbol{\alpha}$, we have

$$\Sigma_k W \left(\boldsymbol{y} - X\boldsymbol{\alpha} - \sum_{k=1}^{r} \theta_k \Sigma_k \boldsymbol{c}_k \right) - n\lambda \Sigma_k \boldsymbol{c}_k = \boldsymbol{0}, \quad k = 1, \ldots, r,$$

$$X^T W \left(\boldsymbol{y} - X\boldsymbol{\alpha} - \sum_{k=1}^{r} \theta_k \Sigma_k \boldsymbol{c}_k \right) = \boldsymbol{0}. \tag{8.8}$$

When all Σ_k are nonsingular, from the first equation in (8.8), we must have $\boldsymbol{c}_1 = \cdots = \boldsymbol{c}_r$. Setting $\boldsymbol{c}_1 = \cdots = \boldsymbol{c}_r = \boldsymbol{c}$, it is easy to see that solutions to

$$(\Sigma_{\boldsymbol{\theta}} + n\lambda W^{-1})\boldsymbol{c} + X\boldsymbol{\alpha} = \boldsymbol{y},$$

$$X^T \boldsymbol{c} = \boldsymbol{0}, \tag{8.9}$$

are also solutions to (8.8). Equations in (8.9) have the same form as those in (5.6). Therefore, a similar procedure as in Section 5.2 can be used to compute coefficients $\boldsymbol{c}$ and $\boldsymbol{d}$. Let the QR decomposition of X be

$$X = (Q_1 \ Q_2) \begin{pmatrix} R \\ 0 \end{pmatrix}$$

and $M = \Sigma_{\boldsymbol{\theta}} + n\lambda W^{-1}$. Then the solutions

$$\begin{aligned} \boldsymbol{c} &= Q_2(Q_2^T M Q_2)^{-1} Q_2^T \boldsymbol{y}, \\ \boldsymbol{d} &= R^{-1} Q_1^T (\boldsymbol{y} - M\boldsymbol{c}). \end{aligned} \quad (8.10)$$

Furthermore,

$$\hat{\boldsymbol{\eta}} = H(\lambda, \boldsymbol{\theta}, \boldsymbol{\tau})\boldsymbol{y},$$

where

$$H(\lambda, \boldsymbol{\theta}, \boldsymbol{\tau}) = I - n\lambda W^{-1} Q_2 (Q_2^T M Q_2)^{-1} Q_2^T \quad (8.11)$$

is the *hat* matrix.

When W is known, the UBR, GCV, and GML criteria presented in Section 5.2.3 can be used to estimate the smoothing parameters. We now extend the GML method to estimate the smoothing and covariance parameters simultaneously when W is unknown.

We first introduce a Bayes model for the semiparametric linear regression model (8.4). Since the parametric component $\boldsymbol{s}^T \boldsymbol{\beta}$ can be absorbed into any one of the null spaces $\mathcal{H}_{k0}$ in the setup of a Bayes model, for simplicity of notation, we drop this term in the following discussion. Assume priors for f_k as

$$F_k(x_k) = \sum_{\nu=1}^{p_k} \zeta_{k\nu} \phi_{k\nu}(x_k) + \sqrt{\delta\theta_k} U_k(x_k), \quad k = 1, \ldots, r, \quad (8.12)$$

where $\zeta_{k\nu} \overset{iid}{\sim} N(0, \kappa)$, $U_k(x_k)$ are independent zero-mean Gaussian stochastic processes with covariance function $R_{k1}(x_k, z_k)$, $\zeta_{k\nu}$, and U_k are mutually independent, and κ and δ are positive constants. Suppose that observations are generated by

$$y_i = \sum_{k=1}^r \mathcal{L}_{ki} F_k + \epsilon_i, \quad i = 1, \ldots, n. \quad (8.13)$$

Let $\mathcal{L}_{k0}$ be a bounded linear functional on $\mathcal{H}_k$. Let $\lambda = \sigma^2/n\delta$. The same arguments in Section 3.6 hold when $M = \Sigma + n\lambda I$ is replaced by $M = \Sigma_{\boldsymbol{\theta}} + n\lambda W^{-1}$ in this chapter. Therefore,

$$\lim_{\kappa \to \infty} \mathrm{E}(\mathcal{L}_{k0} F_k | \boldsymbol{y}) = \hat{f}_k, \quad k = 1, \ldots, r,$$

and an extension of the GML criterion is

$$\text{GML}(\lambda, \boldsymbol{\theta}, \boldsymbol{\tau}) = \frac{\boldsymbol{y}'W(I - H)\boldsymbol{y}}{[\det^+(W(I - H))]^{\frac{1}{n-p}}}, \tag{8.14}$$

where $\det^+$ is the product of the nonzero eigenvalues and $p = \sum_{k=1}^r p_k$. As in Section 5.2.4, we fit the corresponding LME model (5.27) with T being replaced by X in this section to compute the minimizers of the GML criterion (8.14).

For clustered data, the leaving-out-one-cluster approach presented in Section 5.2.2 may also be used to estimate the smoothing and covariance parameters.

We now discuss how to construct Bayesian confidence intervals. Any function $f_k \in \mathcal{H}_k$ can be represented as

$$f_k = \sum_{\nu=1}^{p_k} f_{0k\nu} + f_{1k}, \tag{8.15}$$

where $f_{0k\nu} \in \text{span}\{\phi_{k\nu}\}$ for $\nu = 1, \ldots, p_k$, and $f_{1k} \in \mathcal{H}_{k1}$. Our goal is to construct Bayesian confidence interval for

$$\mathcal{L}_{k0} f_{k, \boldsymbol{\gamma}_k} = \sum_{\nu=1}^{p_k} \gamma_{k,\nu} \mathcal{L}_{k0} f_{0k\nu} + \gamma_{k,p_k+1} \mathcal{L}_{k0} f_{1k} \tag{8.16}$$

for any bounded linear functional $\mathcal{L}_{k0}$ on $\mathcal{H}_k$ and any combination of $\boldsymbol{\gamma}_k = (\gamma_{k,1}, \ldots, \gamma_{k,p_k+1})^T$, where $\gamma_{k,j} = 1$ when the corresponding component in (8.15) is to be included and 0 otherwise.

Let $F_{0j\nu} = \zeta_{j\nu} \phi_{j\nu}$ for $\nu = 1, \ldots, p_j$, $F_{1j} = \sqrt{\delta \theta_j} U_j$, and $F_{1k} = \sqrt{\delta \theta_k} U_k$ for $j = 1, \ldots, r$ and $k = 1, \ldots, r$. Let $\mathcal{L}_{0j}$, $\mathcal{L}_{0j1}$ be bounded linear functionals on $\mathcal{H}_j$, and $\mathcal{L}_{0k2}$ be a bounded linear functional on $\mathcal{H}_k$.

Posterior means and covariances

For $j = 1, \ldots, r$, $\nu = 1, \ldots, p_j$, the posterior means are

$$\begin{aligned} E(\mathcal{L}_{0j} F_{0j\nu} | \boldsymbol{y}) &= (\mathcal{L}_{0j} \phi_{j\nu}) \boldsymbol{e}_{j,\nu}^T \boldsymbol{d}, \\ E(\mathcal{L}_{0j} F_{1j} | \boldsymbol{y}) &= \theta_j (\mathcal{L}_{0j} \boldsymbol{\xi}_j)^T \boldsymbol{c}. \end{aligned} \tag{8.17}$$

For $j = 1, \ldots, r$, $k = 1, \ldots, r$, $\nu = 1, \ldots, p_j$, $\mu = 1, \ldots, p_k$, the posterior covariances are

$$\begin{aligned} \delta^{-1} Cov(\mathcal{L}_{0j1} F_{0j\nu}, \mathcal{L}_{0k2} F_{0k\mu} | \boldsymbol{y}) &= (\mathcal{L}_{0j1} \phi_{j\nu})(\mathcal{L}_{0k2} \phi_{k\mu}) \boldsymbol{e}_{j,\nu}^T A \boldsymbol{e}_{k,\mu}, \\ \delta^{-1} Cov(\mathcal{L}_{0j1} F_{0j\nu}, \mathcal{L}_{0k2} F_{1k} | \boldsymbol{y}) &= -\theta_k (\mathcal{L}_{0j1} \phi_{j\nu}) \boldsymbol{e}_{j,\nu}^T B(\mathcal{L}_{0k2} \boldsymbol{\xi}_k), \\ \delta^{-1} Cov(\mathcal{L}_{0j1} F_{1j}, \mathcal{L}_{0k2} F_{1k} | \boldsymbol{y}) &= \delta_{j,k} \theta_k \mathcal{L}_{0j1} \mathcal{L}_{0k2} R_{k1} - \\ &\quad \theta_j \theta_k (\mathcal{L}_{0j1} \boldsymbol{\xi}_j)^T C(\mathcal{L}_{0k2} \boldsymbol{\xi}_k), \end{aligned} \tag{8.18}$$

where $e_{j,\mu}$ is a vector of dimension $\sum_{l=1}^{r} p_l$ with the $(\sum_{l=1}^{j-1} p_l + \mu)$th el-ement being one and all other elements being zero, c and d are given in the equation (8.10), $\mathcal{L}_{0j1}\xi_j = (\mathcal{L}_{0j1}\mathcal{L}_{j1}R_{j1}, \ldots, \mathcal{L}_{0j1}\mathcal{L}_{jn}R_{j1})^T$, $\mathcal{L}_{0k2}\xi_k = (\mathcal{L}_{0k2}\mathcal{L}_{k1}R_{k1}, \ldots, \mathcal{L}_{0k2}\mathcal{L}_{kn}R_{k1})^T$, $M = \Sigma_{\theta} + n\lambda W^{-1}$, $A = (T^T M^{-1} T)^{-1}$, $B = AT^T M^{-1}$, and $C = M^{-1}(I - B)$. For simplicity of notation, we define $\sum_{l=1}^{0} p_l = 0$.

Derivation of the above results can be found in Wang and Ke (2009). Posterior mean and variance of $\mathcal{L}_{j0}f_{j,\gamma_j}$ in (8.16) can be calculated using the above formulae. Bayesian confidence intervals for $\mathcal{L}_{j0}f_{j,\gamma}$ can then be constructed. Bootstrap confidence intervals can also be constructed as in previous chapters.

The semiparametric linear regression model(8.4) can be fitted by the `ssr` function. The independent variables s and null bases $\phi_{k1}, \ldots, \phi_{kp_k}$ for $k = 1, \ldots, r$ are specified on the right-hand side of the `formula` argument, and RKs R_{k1} for $k = 1, \ldots, r$ are specified in the `rk` argument as a list. For non-iid random errors, variance and/or correlation structures are specified using the arguments `weights` and `correlation`. The argument `spar` specifies a method for selecting the smoothing parameter(s). UBR, GCV, and GML methods are available for the case when W is known, and the GML method is available when W needs to be estimated. The `predict` function can be used to compute posterior means and standard deviations. See Section 8.4.1 for examples.

8.2.3   Vector Spline

Suppose we have observations on r dependent variables $z_1, \ldots, z_r$. Assume the following partial spline models:

$$z_{jk} = s_{jk}^T \beta_k + \mathcal{L}_{kj} f_k + \varepsilon_{jk}, \quad k = 1, \ldots, r; \quad j = 1, \ldots, n_k, \quad (8.19)$$

where z_{jk} is the jth observation on z_k, s_{jk} is the jth observation on a q_k-dimensional vector of independent variables s_k, $f_k \in \mathcal{H}_k$ is an unknown function, $\mathcal{H}_k$ is an RKHS on an arbitrary set $\mathcal{X}_k$, $\mathcal{L}_{kj}$ is a bounded linear functional, and ε_{jk} is a random error. Model (8.19) is a semiparametric extension of the linear *seemingly unrelated regression* model. For simplicity, it is assumed that the regression model for each dependent variable involves one nonparametric function only. The following discussions hold when the partial spline models in (8.19) are replaced by the semiparametric linear regression models (8.4).

There are two possible approaches to estimating the parameters $\beta = (\beta_1^T, \ldots, \beta_r^T)^T$ and the functions $f = (f_1, \ldots, f_r)$. The first approach is to fit the partial spline models in (8.19) separately, once for each dependent variable. The second approach is to fit all partial spline models in (8.19) simultaneously, which can be more efficient when the random errors are correlated (Wang, Guo and Brown 2000, Smith and Kohn 2000). We now discuss how to accomplish the second approach using the methods in this section.

Let $m_1 = 0$ and $m_k = \sum_{l=1}^{k-1} n_l$ for $k = 2, \ldots, r$. Let $i = m_k + j$ for $j = 1, \ldots, n_k$ and $k = 1, \ldots, r$. Then there is an one-to-one correspondence between i and (j, k). Define $y_i = z_{jk}$ for $i = 1, \ldots, n$, where $n = \sum_{l=1}^{r} n_l$. Then the partial spline models in (8.19) can be written jointly as

$$
\begin{aligned}
y_i &= \boldsymbol{s}_{jk}^T \boldsymbol{\beta}_k + \mathcal{L}_{kj} f_k + \varepsilon_{jk} \\
&= \sum_{l=1}^{r} \delta_{k,l} \boldsymbol{s}_{jl}^T \boldsymbol{\beta}_l + \sum_{l=1}^{r} \delta_{k,l} \mathcal{L}_{lj} f_l + \varepsilon_{jk} \\
&= \tilde{\boldsymbol{s}}_i^T \boldsymbol{\beta} + \sum_{l=1}^{r} \tilde{\mathcal{L}}_{li} f_l + \epsilon_i,
\end{aligned}
\tag{8.20}
$$

where $\delta_{k,l}$ is the Kronecker delta, $\tilde{\boldsymbol{s}}_i^T = (\delta_{k,1} \boldsymbol{s}_{j1}^T, \ldots, \delta_{k,r} \boldsymbol{s}_{jr}^T)$, $\tilde{\mathcal{L}}_{li} = \delta_{k,l} \mathcal{L}_{lj}$, and $\epsilon_i = \varepsilon_{jk}$. Assume that $\boldsymbol{\epsilon} = (\epsilon_1, \ldots, \epsilon_n)^T \sim N(\boldsymbol{0}, \sigma^2 W^{-1})$. It is easy to see that $\tilde{\mathcal{L}}_{li}$ are bounded linear functionals. Thus the model (8.20) for all dependent variables is a special case of the semiparametric linear regression model (8.4). Therefore, the estimation and inference methods described in Section 8.2.2 can be used. In particular, all parameters $\boldsymbol{\beta}$ and nonparametric functions $\boldsymbol{f}$ are estimated jointly based on the PWLS (8.5). In comparison, the first approach that fits model (8.19) separately for each dependent variable is equivalent to fitting model (8.20) based on the PLS.

As an interesting special case, consider the following SSR model for $r = 2$ dependent variables

$$
z_{jk} = f_k(x_{jk}) + \varepsilon_{jk}, \quad k = 1, 2; \ j = 1, \ldots, n_k,
\tag{8.21}
$$

where the model space of f_k is an RKHS $\mathcal{H}_k$ on $\mathcal{X}_k$. Assume that $\mathcal{H}_k = \mathcal{H}_{k0} \oplus \mathcal{H}_{k1}$, where $\mathcal{H}_{k0} = \text{span}\{\phi_{k1}, \ldots, \phi_{kp_k}\}$, and $\mathcal{H}_{k1}$ is an RKHS with RK R_{k1}. Then it is easy to check that

$$
\tilde{\mathcal{L}}_{ki} \phi_{k\nu} = \begin{cases} \mathcal{L}_{1i} \phi_{1\nu}, & 1 \leq i \leq n_1, \\ \mathcal{L}_{2i} \phi_{2\nu}, & n_1 < i \leq n_1 + n_2, \end{cases}
$$

and

$$
\tilde{\mathcal{L}}_{ki} \tilde{\mathcal{L}}_{kj} R_{k1} = \begin{cases} \mathcal{L}_{1i} \mathcal{L}_{1j} R_{11}, & 1 \leq i, j \leq n_1, \\ \mathcal{L}_{2i} \mathcal{L}_{2j} R_{21}, & n_1 < i, j \leq n_1 + n_2. \end{cases}
$$

For illustration, we generate a data set from model (8.21) with $\mathcal{X}_1 = \mathcal{X}_2 = [0, 1]$, $n_1 = n_2 = 100$, $x_{i1} = x_{i2} = i/n$, $f_1(x) = \sin(2\pi x)$, $f_2(x) =$

$\sin(2\pi x) + 2x$, and the paired random errors $(\varepsilon_{i1}, \varepsilon_{i2})$ are iid bivariate normal random variables with mean zero and $\text{Var}(\epsilon_{i1}) = 0.25$, $\text{Var}(\epsilon_{i2}) = 1$, and $\text{Cor}(\epsilon_{i1}, \epsilon_{i2}) = 0.8$.

```
> n <- 100; s1 <- .5; s2 <- 1; r <- .8
> A <- diag(c(s1,s2))%*%matrix(c(sqrt(1-r**2),0,r,1),2,2)
> e <- NULL; for (i in 1:n) e <- c(e,A%*%rnorm(2))
> x <- 1:n/n
> y1 <- sin(2*pi*x) + e[seq(1,2*n,by=2)]
> y2 <- sin(2*pi*x) + 2*x + e[seq(2,2*n,by=2)]
> bisp.dat <- data.frame(y=c(y1,y2),x=rep(x,2),
    id=as.factor(rep(c(0,1),rep(n,2))), pair=rep(1:n,2))
```

We model both f_1 and f_2 using the cubic spline space $W_2^2[0,1]$ under the construction in Section 2.6. We first fit each SSR model in (8.21) separately and compute posterior means and standard deviations:

```
> bisp.fit1 <- ssr(y~I(x-.5), rk=cubic(x), spar=''m'',
    data=bisp.dat[bisp.dat$id==0,])
> bisp.p1 <- predict(bisp.fit1)
> bisp.fit2 <- ssr(y~I(x-.5), rk=cubic(x), spar=''m'',
    data=bisp.dat[bisp.dat$id==1,])
> bisp.p2 <- predict(bisp.fit2)
```

The functions of f_1 and f_2 and their estimates and confidence intervals based on separate fits are shown in the top panel of Figure 8.1.

Next we fit the SSR models in (8.21) jointly, compute posterior means and standard deviations, and compare the posterior standard deviations with those based on separate fits:

```
> bisp.fit3 <- ssr(y~id/I(x-.5)-1,
    rk=list(rk.prod(cubic(x),kron(id==0)),
            rk.prod(cubic(x),kron(id==1))), spar=''m'',
    weights=varIdent(form=~1|id),
    correlation=corSymm(form=~1|pair), data=bisp.dat)
> summary(bisp.fit3)
...
Coefficients (d):
            id0               id1 id0:I(x - 0.5) id1:I(x - 0.5)
  -0.002441981      1.059626687   -0.440873366     2.086878037

GML estimate(s) of smoothing parameter(s) :
  8.358606e-06 5.381727e-06
Equivalent Degrees of Freedom (DF):   13.52413
Estimate of sigma:   0.483716
```

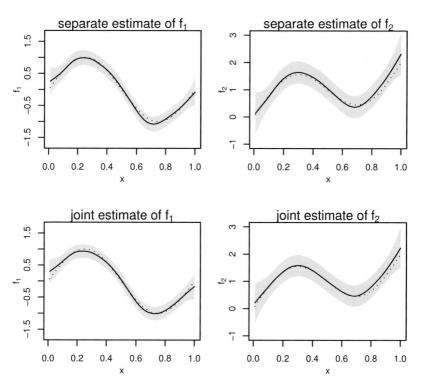

FIGURE 8.1 Plots of the true functions (dotted lines), cubic spline estimates (solid lines), and 95% Bayesian confidence intervals (shaded regions) of f_1 (left) and f_2 (right). Plots in the top panel are based on the separate fits and plots in the bottom panel are based on the joint fit.

```
Correlation structure of class corSymm representing
  Correlation:
    1
  2 0.777
Variance function structure of class varIdent representing
          0         1
  1.000000 1.981937

> bisp.p31 <- predict(bisp.fit3,
    newdata=bisp.dat[bisp.dat$id==0,],
    terms=c(1,0,1,0,1,0))
> bisp.p32 <- predict(bisp.fit3,
```

```
  newdata=bisp.dat[bisp.dat$id==1,],
  terms=c(0,1,0,1,0,1))
> mean((bisp.p1$pstd-bisp.p31$pstd)/bisp.p1$pstd)
  0.08417699
> mean((bisp.p2$pstd-bisp.p32$pstd)/bisp.p2$pstd)
  0.04500096
```

An arbitrary pairwise variance–covariance structure was assumed, and the variance–covariance structure was specified with the combination of the `weights` and `correlation` options. On average, the posterior standard deviations based on the joint fit are smaller than those based on separate fits. Estimates and confidence intervals based on the joint fit are shown in the bottom panel of Figure 8.1.

In many applications, the domains of f_1 and f_2 are the same. That is, $\mathcal{X}_1 = \mathcal{X}_2 = \mathcal{X}$. Then we can rewrite $f_j(x)$ as $f(j,x)$ and regard it as a bivariate function of j and x defined on the product domain $\{1,2\} \otimes \mathcal{X}$. The joint approach described above for fitting the SSR models in (8.21) is equivalent to representing the original functions as

$$f(j,x) = \delta_{j,1}f_1(x) + \delta_{j,2}f_2(x). \tag{8.22}$$

Sometimes the main interest is the difference between f_1 and f_2: $d(x) = f_2(x) - f_1(x)$. We may reparametrize $f(j,x)$ as

$$f(j,x) = f_1(x) + \delta_{j,2}d(x) \tag{8.23}$$

or

$$f(j,x) = \frac{1}{2}\{f_1(x) + f_2(x)\} + \frac{1}{2}(\delta_{j,2} - \delta_{j,1})d(x). \tag{8.24}$$

Models (8.23) and (8.24) correspond to the SS ANOVA decompositions of $f(j,t)$ with the set-to-zero and sum-to-zero side conditions, respectively. The following statements fit model (8.23) and compute posterior means and standard deviations of $d(x)$:

```
> bisp.fit4 <- update(bisp.fit3,
    y~I(x-.5)+I(id==1)+I((x-.5)*(id==1)),
    rk=list(cubic(x),rk.prod(cubic(x),kron(id==1))))
> bisp.p41 <- predict(bisp.fit4,
    newdata=bisp.dat[bisp.dat$id==1,], terms=c(0,0,1,1,0,1))
```

where $d(x)$ is modeled using the cubic spline space $W_2^2[0,1]$ under the construction in Section 2.6. The function of $d(x)$ and its estimate are shown in the left panel of Figure 8.2. Model (8.24) can be fitted similarly. Sometimes it is of interest to check if f_1 and f_2 are parallel rather than if $d(x) = 0$. Let $d_1(x)$ be the projection of $d(x)$ onto $W_2^2[0,1] \ominus \{1\}$. Then

f_1 and f_2 are parallel iff $d_1(x) = 0$. Similarly, the projection of $d(x)$ onto $W_2^2[0, 1] \ominus \{1\} \ominus \{x - .5\}$, $d_2(x)$, can be used to check if f_1 and f_2 differ by a linear function. We compute posterior means and standard deviations of $d_1(x)$ and $d_2(x)$ as follows:

```
> bisp.p42 <- predict(bisp.fit4,
    newdata=bisp.dat[bisp.dat$id==1,], terms=c(0,0,0,1,0,1))
> bisp.p43 <- predict(bisp.fit4,
    newdata=bisp.dat[bisp.dat$id==1,], terms=c(0,0,0,0,0,1))
```

The functions of $d_1(x)$ and $d_2(x)$, their estimates, and 95% Bayesian confidence intervals are shown in the middle and right panels of Figure 8.2. We can see that f_1 and f_2 are not parallel but differ by a linear function.

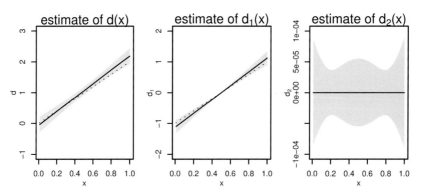

FIGURE 8.2   Plots of the functions $d(x)$ (left), $d_1(x)$ (middle), and $d_2(x)$ (right) as dotted lines. Estimates of these functions are plotted as solid lines, and 95% Bayesian confidence intervals are marked as the shaded regions.

We now introduce a more sophisticated SS ANOVA decomposition that can be used to investigate various relationships between f_1 and f_2. Define two averaging operators $\mathcal{A}_1^{(1)}$ and $\mathcal{A}_2^{(2)}$ such that

$$\mathcal{A}_1^{(1)} f = w_1 f(1, x) + w_2 f(2, x),$$
$$\mathcal{A}_1^{(2)} f = \int_{\mathcal{X}} f(j, x) dP(x),$$

where $w_1 + w_2 = 1$, and P is a probability measure on $\mathcal{X}$. Then we have the following SS ANOVA decomposition

$$f(j, x) = \mu + g_1(j) + g_2(x) + g_{12}(j, x), \tag{8.25}$$

where

$$\mu = \mathcal{A}_1^{(1)} \mathcal{A}_1^{(2)} f = \int_{\mathcal{X}} \{w_1 f_1(x) + w_2 f_2(x)\} dP(x),$$

$$g_1(j) = (I - \mathcal{A}_1^{(1)}) \mathcal{A}_1^{(2)} f = \int_{\mathcal{X}} f_j(x) dP(x) - \mu,$$

$$g_2(x) = \mathcal{A}_1^{(1)} (I - \mathcal{A}_1^{(2)}) f = w_1 f_1(x) + w_2 f_2(x) - \mu,$$

$$g_{12}(j, x) = (I - \mathcal{A}_1^{(1)})(I - \mathcal{A}_1^{(2)}) f = f_j(x) - \mu - g_1(j) - g_2(x).$$

The constant μ is the overall mean, the marginal functions g_1 and g_2 are the main effects, and the bivariate function g_{12} is the interaction.

The SS ANOVA decomposition (8.25) makes certain hypotheses more transparent. For example, it is easy to check that the following hypotheses are equivalent:

$$H_0 : f_1(x) = f_2(x) \iff H_0 : g_1(j) + g_{12}(j, x) = 0,$$

$$H_0 : f_1(x) - f_2(x) = \text{constant} \iff H_0 : g_{12}(j, x) = 0,$$

$$H_0 : \int_{\mathcal{X}} f_1(x) dP(x) = \int_{\mathcal{X}} f_2(x) dP(x) \iff H_0 : g_1(j) = 0,$$

$$H_0 : w_1 f_1(x) + w_2 f_2(x) = \text{constant} \iff H_0 : g_2(x) = 0.$$

Furthermore, if $g_1(j) \neq 0$ and $g_2(x) \neq 0$,

$$H_0 : a f_1(x) + b f_2(x) = c, \ |a| + |b| > 0 \iff H_0 : g_{12}(j, t) = \beta g_1(j) g_2(x).$$

Therefore, the hypothesis that f_1 and f_2 are equal is equivalent to $g_1(j) + g_{12}(j, x) = 0$. The hypothesis that f_1 and f_2 are parallel is equivalent to the hypothesis that the interaction $g_{12} = 0$. The hypothesis that the integrals of f_1 and f_2 with respect to the probability measure P are equal is equivalent to the hypothesis that the main effect $g_1(j) = 0$. The hypothesis that the weighted average of f_1 and f_2 is a constant is equivalent to the hypothesis that the main effect $g_2(x) = 0$. Note that the probability measure P and weights w_j are arbitrary, which can be selected for specific hypotheses. Under the specified conditions, the hypothesis that there exists a linear relationship between the functions f_1 and f_2 is equivalent to the hypothesis that the interaction is multiplicative. Thus, for these hypotheses, we can fit the SS ANOVA model (8.25) and perform tests on the corresponding components.

8.3 Semiparametric Nonlinear Regression Models

8.3.1 The Model

A *semiparametric nonlinear regression* (SNR) model assumes that

$$y_i = \mathcal{N}_i(\boldsymbol{\beta}, \boldsymbol{f}) + \epsilon_i, \quad i = 1, \ldots, n, \tag{8.26}$$

where $\boldsymbol{\beta} = (\beta_1, \ldots, \beta_q)^T \in \mathbb{R}^q$ is a vector of parameters, $\boldsymbol{f} = (f_1, \ldots, f_r)$ are unknown functions, f_k belongs to an RKHS $\mathcal{H}_k$ on an arbitrary domain $\mathcal{X}_k$ for $k = 1, \ldots, r$, $\mathcal{N}_i$ are known nonlinear functionals on $\mathbb{R}^q \times \mathcal{H}_1 \times \cdots \times \mathcal{H}_r$, and $\boldsymbol{\epsilon} = (\epsilon_1, \ldots, \epsilon_n)^T \sim \mathrm{N}(0, \sigma^2 W^{-1})$.

It is obvious that the SNR model (8.26) is an extension of the NNR model (7.4) to allow parameters in the model. The mean function may depend on both parameters $\boldsymbol{\beta}$ and nonparametric functions $\boldsymbol{f}$ nonlinearly. The nonparametric functions $\boldsymbol{f}$ are regarded as parameters just like $\boldsymbol{\beta}$. Certain constraints may be required to make an SNR model identifiable. Specific conditions depend on the form of a model and the purpose of an analysis. Often, identifiability can be achieved by adding constraints on parameters, absorbing some parameters into $\boldsymbol{f}$ and/or adding constraints on $\boldsymbol{f}$ by removing certain components from the model spaces. Illustrations of how to make an SNR model identifiable can be found in Section 8.4.

The semiparametric linear regression model (8.4) is a special case of SNR model when $\mathcal{N}_i$ are linear in $\boldsymbol{\beta}$ and $\boldsymbol{f}$. When $\mathcal{N}_i$ are linear in $\boldsymbol{f}$ for fixed $\boldsymbol{\beta}$, model (8.26) can be expressed as

$$y_i = \alpha(\boldsymbol{\beta}; \boldsymbol{x}_i) + \sum_{k=1}^{r} \mathcal{L}_{ki}(\boldsymbol{\beta}) f_k + \epsilon_i, \quad i = 1, \ldots, n, \tag{8.27}$$

where α is a known linear or nonlinear function of independent variables $\boldsymbol{x} = (x_1, \ldots, x_d)$, $\boldsymbol{x}_i = (x_{i1}, \ldots, x_{id})$, and $\mathcal{L}_{ki}(\boldsymbol{\beta})$ are bounded linear functionals that may depend on $\boldsymbol{\beta}$. Model (8.27) will be referred to as a *semiparametric conditional linear model*.

One special case of model (8.27) is

$$y_i = \alpha(\boldsymbol{\beta}; \boldsymbol{x}_i) + \sum_{k=1}^{r} \delta_k(\boldsymbol{\beta}; \boldsymbol{x}_i) f_k(\gamma_k(\boldsymbol{\beta}; \boldsymbol{x}_i)) + \epsilon_i, \quad i = 1, \ldots, n, \tag{8.28}$$

where α, δ_k, and γ_k are known functions. Containing many existing models as special cases, model (8.28) is interesting in its own right. It is obvious that both nonlinear regression and nonparametric regression models are special cases of model (8.28). The project pursuit regression model (8.1) is a special case with $\alpha(\boldsymbol{\beta}; \boldsymbol{x}) = \beta_0$, $\delta_k(\boldsymbol{\beta}; \boldsymbol{x}) \equiv 1$ and $\gamma_k(\boldsymbol{\beta}; \boldsymbol{x}) = \boldsymbol{\beta}_k^T \boldsymbol{x}$, where $\boldsymbol{\beta} = (\beta_0, \boldsymbol{\beta}_1^T, \ldots, \boldsymbol{\beta}_r^T)^T$. Partially linear single index model (8.2) is a special case with $r = 1$, $\boldsymbol{x} = (\boldsymbol{s}^T, \boldsymbol{t}^T)^T$, $\alpha(\boldsymbol{\beta}; \boldsymbol{x}) = \boldsymbol{\beta}_1^T \boldsymbol{s}$, $\delta_1(\boldsymbol{\beta}; \boldsymbol{x}) \equiv 1$, $\gamma_1(\boldsymbol{\beta}; \boldsymbol{x}) = \boldsymbol{\beta}_2^T \boldsymbol{t}$, and $\boldsymbol{\beta} = (\boldsymbol{\beta}_1^T, \boldsymbol{\beta}_2^T)^T$. Other special cases can be found in Section 8.4.

Sometimes one may want to investigate how parameters $\boldsymbol{\beta}$ depend on other covariates. One approach is to build a second-stage linear model, $\boldsymbol{\beta} = A\boldsymbol{\vartheta}$, where A is a known matrix. See Section 7.5 in Pinheiro and Bates (2000) for details. The general form of models (8.26), (8.27), and (8.28) remains the same when the second-stage model is plugged in. Therefore, the estimation procedures in Section 8.3.3 also apply to the SNR model combined with a second-stage model with $\boldsymbol{\vartheta}$ as parameters.

Let $\boldsymbol{y} = (y_1, \ldots, y_n)^T$, $\boldsymbol{\eta}(\boldsymbol{\beta}, \boldsymbol{f}) = (\mathcal{N}_1(\boldsymbol{\beta}, \boldsymbol{f}), \ldots, \mathcal{N}_n(\boldsymbol{\beta}, \boldsymbol{f}))^T$, and $\boldsymbol{\epsilon} = (\epsilon_1, \ldots, \epsilon_n)^T$. Then model (8.26) can be written in a vector form as

$$\boldsymbol{y} = \boldsymbol{\eta}(\boldsymbol{\beta}, \boldsymbol{f}) + \boldsymbol{\epsilon}. \tag{8.29}$$

8.3.2 SNR Models for Clustered Data

Clustered data such as repeated measures, longitudinal, and multilevel data are common in practice. The SNR model for clustered data assumes that

$$y_{ij} = \mathcal{N}_{ij}(\boldsymbol{\beta}_i, \boldsymbol{f}) + \epsilon_{ij}, \quad i = 1, \ldots, m; \quad j = 1, \ldots, n_i, \tag{8.30}$$

where y_{ij} is the jth observation in cluster i, $\boldsymbol{\beta}_i = (\beta_{i1}, \ldots, \beta_{iq})^T \in \mathbb{R}^q$ is a vector of parameters for cluster i, $\boldsymbol{f} = (f_1, \ldots, f_r)$ are unknown functions, f_k belongs to an RKHS $\mathcal{H}_k$ on an arbitrary domain $\mathcal{X}_k$ for $k = 1, \ldots, r$, $\mathcal{N}_{ij}$ are known nonlinear functionals on $\mathbb{R}^q \times \mathcal{H}_1 \times \cdots \times \mathcal{H}_r$, and ϵ_{ij} are random errors. Let $\boldsymbol{\epsilon}_i = (\epsilon_{i1}, \ldots, \epsilon_{in_i})^T$ and $\boldsymbol{\epsilon} = (\boldsymbol{\epsilon}_1^T, \ldots, \boldsymbol{\epsilon}_m^T)^T$. We assume that $\boldsymbol{\epsilon} \sim N(0, \sigma^2 W^{-1})$. Usually, observations are correlated within a cluster and independent between clusters. In this case, W^{-1} is block diagonal.

Again, when $\mathcal{N}_{ij}$ are linear in $\boldsymbol{f}$ for fixed $\boldsymbol{\beta}_i$, model (8.30) can be expressed as

$$y_{ij} = \alpha(\boldsymbol{\beta}_i; \boldsymbol{x}_{ij}) + \sum_{k=1}^r \mathcal{L}_{kij}(\boldsymbol{\beta}_i) f_k + \epsilon_{ij}, \quad i = 1, \ldots, m; \quad j = 1, \ldots, n_i, \tag{8.31}$$

where α is a known function of independent variables $\boldsymbol{x} = (x_1, \ldots, x_d)$, $\boldsymbol{x}_{ij} = (x_{ij1}, \ldots, x_{ijd})$, and $\mathcal{L}_{kij}(\boldsymbol{\beta})$ are bounded linear functionals that may depend on $\boldsymbol{\beta}$. Model (8.31) will be referred to as a *semiparametric conditional linear model* for clustered data.

Similar to model (8.28), one special case of model (8.31) is

$$y_{ij} = \alpha(\boldsymbol{\beta}_i; \boldsymbol{x}_{ij}) + \sum_{k=1}^{r} \delta_k(\boldsymbol{\beta}_i; \boldsymbol{x}_{ij}) f_k(\gamma_k(\boldsymbol{\beta}_i; \boldsymbol{x}_{ij})) + \epsilon_{ij},$$

$$i = 1, \ldots, m; \quad j = 1, \ldots, n_i, \tag{8.32}$$

where α, δ_k, and γ_k are known functions. Model (8.32) can be regarded as an extension of the *self-modeling nonlinear regression* (SEMOR) model proposed by Lawton, Sylvestre and Maggio (1972). In particular, the *shape invariant model* (SIM) (Lawton et al. 1972, Wang and Brown 1996),

$$y_{ij} = \beta_{i1} + \beta_{i2} f\left(\frac{x_{ij} - \beta_{i3}}{\beta_{i4}}\right) + \epsilon_{ij}, \quad i = 1, \ldots, m; \quad j = 1, \ldots, n_i, \tag{8.33}$$

is a special case of model (8.32) with $d = 1$, $r = 1$, $q = 4$, $\alpha(\boldsymbol{\beta}_i; x_{ij}) = \beta_{i1}$, $\delta_1(\boldsymbol{\beta}_i; x_{ij}) = \beta_{i2}$, and $\gamma_k(\boldsymbol{\beta}_i; x_{ij}) = (x_{ij} - \beta_{i3})/\beta_{i4}$. Again, a second-stage linear model may also be constructed for parameters $\boldsymbol{\beta}_i$, and the estimation procedures in Section 8.3.3 apply to the combined model.

Let $n = \sum_{i=1}^{m} n_i$, $\boldsymbol{y}_i = (y_{i1}, \ldots, y_{in_i})^T$, $\boldsymbol{y} = (\boldsymbol{y}_1^T, \ldots, \boldsymbol{y}_m^T)^T$, $\boldsymbol{\beta} = (\boldsymbol{\beta}_1^T, \ldots, \boldsymbol{\beta}_m^T)^T$, $\boldsymbol{\eta}_i(\boldsymbol{\beta}_i, \boldsymbol{f}) = (\mathcal{N}_{i1}(\boldsymbol{\beta}_i, \boldsymbol{f}), \ldots, \mathcal{N}_{in_i}(\boldsymbol{\beta}_i, \boldsymbol{f}))^T$, and $\boldsymbol{\eta}(\boldsymbol{\beta}, \boldsymbol{f}) = (\boldsymbol{\eta}_1^T(\boldsymbol{\beta}_1, \boldsymbol{f}), \ldots, \boldsymbol{\eta}_m^T(\boldsymbol{\beta}_m, \boldsymbol{f}))^T$. Then model (8.30) can be written in the vector form (8.29).

8.3.3   Estimation and Inference

For simplicity we present the estimation and inference procedures for SNR models in Section 8.3.1 only. The same methods apply to SNR models in Section 8.3.2 for clustered data with a slight modification of notation.

Consider the vector form (8.29) and assume that W depends on an unknown vector of parameters $\boldsymbol{\tau}$. Assume that $f_k \in \mathcal{H}_k$, and $\mathcal{H}_k = \mathcal{H}_{k0} \oplus \mathcal{H}_{k1}$, where $\mathcal{H}_{k0} = \text{span}\{\phi_{k1}, \ldots, \phi_{kp_k}\}$ and $\mathcal{H}_{k1}$ is an RKHS with RK R_{k1}. Our goal is to estimate parameters $\boldsymbol{\beta}, \boldsymbol{\tau}, \sigma^2$, and nonparametric functions $\boldsymbol{f}$.

Let

$$l(\boldsymbol{y}; \boldsymbol{\beta}, \boldsymbol{f}, \boldsymbol{\tau}, \sigma^2) = \log|\sigma^2 W^{-1}| + \frac{1}{\sigma^2}(\boldsymbol{y} - \boldsymbol{\eta})^T W(\boldsymbol{y} - \boldsymbol{\eta}) \tag{8.34}$$

be twice the negative log-likelihood where an additive constant is ignored for simplicity. We estimate $\boldsymbol{\beta}$, $\boldsymbol{\tau}$, and $\boldsymbol{f}$ as minimizers of the penalized likelihood (PL)

$$l(\boldsymbol{y}; \boldsymbol{\beta}, \boldsymbol{f}, \boldsymbol{\tau}, \sigma^2) + \frac{n\lambda}{\sigma^2} \sum_{k=1}^{r} \theta_k^{-1} \|P_{k1} f_k\|^2, \tag{8.35}$$

where P_{k1} is the projection operator onto $\mathcal{H}_{k1}$ in $\mathcal{H}_k$, and $\lambda\theta_k^{-1}$ are smoothing parameters. The multiplying constant n/σ^2 is introduced in the penalty term such that, ignoring an additive constant, the PL (8.35) has the same form as the PWLS (8.5) when $\mathcal{N}$ is linear in both $\boldsymbol{\beta}$ and $\boldsymbol{f}$.

We will first develop a backfitting procedure for the semiparametric conditional linear model (8.27) and then develop an algorithm for the general SNR model (8.26).

Consider the semiparametric conditional linear model (8.27). We first consider the estimation of $\boldsymbol{f}$ with fixed $\boldsymbol{\beta}$ and $\boldsymbol{\tau}$. When $\boldsymbol{\beta}$ and $\boldsymbol{\tau}$ are fixed, the PL (8.35) is equivalent to the PWLS (8.5). Therefore, the solutions of $\boldsymbol{f}$ to (8.35) can be represented as those in (8.6). We will use the same notations as in Section 8.2.2. Note that both T and $\Sigma_{\boldsymbol{\theta}}$ may depend on $\boldsymbol{\beta}$ even though the dependence is not expressed explicitly for simplicity. Let $\boldsymbol{\alpha} = (\alpha(\boldsymbol{\beta}; \boldsymbol{x}_1), \ldots, \alpha(\boldsymbol{\beta}; \boldsymbol{x}_n))^T$. We need to solve equations

$$
\begin{aligned}
(\Sigma_{\boldsymbol{\theta}} + n\lambda W^{-1})\boldsymbol{c} + T\boldsymbol{d} &= \boldsymbol{y} - \boldsymbol{\alpha}, \\
T^T\boldsymbol{c} &= \boldsymbol{0},
\end{aligned}
\tag{8.36}
$$

for coefficients $\boldsymbol{c}$ and $\boldsymbol{d}$. Note that $\boldsymbol{\alpha}$ and W are fixed since $\boldsymbol{\beta}$ and $\boldsymbol{\tau}$ are fixed, and the equations in (8.36) have the same form as those in (5.6). Therefore, methods in Section 5.2.1 can be used to solve (8.36), and the UBR, GCV, and GML methods in Section 5.2.3 can be used to estimate smoothing parameters λ and $\boldsymbol{\theta}$.

Next we consider the estimation of $\boldsymbol{\beta}$ and $\boldsymbol{\tau}$ with fixed $\boldsymbol{f}$. When $\boldsymbol{f}$ is fixed, the PL (8.35) is equivalent to

$$
\log|\sigma^2 W^{-1}| + \frac{1}{\sigma^2}\{\boldsymbol{y} - \boldsymbol{\eta}(\boldsymbol{\beta}, \boldsymbol{f})\}^T W\{\boldsymbol{y} - \boldsymbol{\eta}(\boldsymbol{\beta}, \boldsymbol{f})\}.
\tag{8.37}
$$

We use the backfitting and Gauss–Newton algorithms in Pinheiro and Bates (2000) to find minimizers of (8.37) by updating $\boldsymbol{\beta}$ and $\boldsymbol{\tau}$ iteratively. Details about the backfitting and Gauss–Newton algorithms can be found in Section 7.5 of Pinheiro and Bates (2000).

Putting pieces together, we have the following algorithm.

Algorithm for semiparametric conditional linear models

1. *Initialize*: Set initial values for $\boldsymbol{\beta}$ and $\boldsymbol{\tau}$.

2. *Cycle*: Alternate between (a) and (b) until convergence:

   (a) Conditional on current estimates of $\boldsymbol{\beta}$ and $\boldsymbol{\tau}$, update $\boldsymbol{f}$ using methods in Section 5.2.1 with smoothing parameters selected by the UBR, GCV, or GML method in Section 5.2.3.

(b) Conditional on current estimates of f, update β and τ by solving (8.37) alternatively using the backfitting and Gauss–Newton algorithms.

Note that smoothing parameters are estimated iteratively with fixed τ at step 2(a). The parameters τ are estimated at step 2(b), which makes the algorithm relatively easy to implement. An alternative computationally more expensive approach is to estimate smoothing parameters and τ jointly at step (a).

Finally, we consider the estimation for the general SNR model (8.26). When η is nonlinear in f, the solutions of f to (8.35) usually do not fall in finite dimensional spaces. Therefore, certain approximations are necessary. Again, first consider estimating f with fixed β and τ. We now extend the EGN procedure in Section 7.3 to multiple functions. Let f_- be the current estimate of f. For any fixed β, $\mathcal{N}_i$ is a functional on $\mathcal{H}_1 \times \cdots \times \mathcal{H}_r$. We assume that $\mathcal{N}_i$ is Fréchet differentiable at f_- and write $\mathcal{D}_i = D\mathcal{N}_i(f_-)$. Then $\mathcal{D}_i h = \sum_{k=1}^r \mathcal{D}_{ki} h_k$, where $\mathcal{D}_{ki}$ is the partial Fréchet differential of $\mathcal{N}_i$ with respect to f_k evaluated at f_-, $h = (h_1, \ldots, h_r)$ and $h_k \in \mathcal{H}_k$ (Flett 1980). For $k = 1, \ldots, r$, $\mathcal{L}_{ki}$ is a bounded linear functional on $\mathcal{H}_k$. Approximating $\mathcal{N}_i(\beta, f)$ by its linear approximation

$$\mathcal{N}_i(\beta, f) \approx \mathcal{N}_i(\beta, f_-) + \sum_{k=1}^r \mathcal{D}_{ki}(f_k - f_{k-}), \qquad (8.38)$$

we have an approximate semiparametric conditional linear model

$$\tilde{y}_i = \sum_{k=1}^r \mathcal{D}_{ki} f_k + \epsilon_i, \quad i = 1, \ldots, n, \qquad (8.39)$$

where $\tilde{y}_i = y_i - \mathcal{N}_i(\beta, f_-) + \sum_{k=1}^r \mathcal{D}_{ki} f_{k-}$. Functions f in model (8.39) can be estimated using the method in Section 8.2.2. Consequently, we have the following algorithm for the general SNR model.

Algorithm for general SNR model

1. *Initialize*: Set initial values for β, τ, and f.

2. *Cycle*: Alternate between (a) and (b) until convergence:

    (a) Conditional on current estimates of β, τ, and f, compute $\mathcal{D}_{ki}$ and $\tilde{y}_i$, and update f by applying step 2(a) in the Algorithm for semiparametric conditional linear models to the approximate model (8.39). Repeat this step until convergence.

(b) Conditional on current estimates of $\boldsymbol{f}$, update $\boldsymbol{\beta}$ and $\boldsymbol{\tau}$ by solving (8.37) alternatively using the backfitting and Gauss–Newton algorithms.

Denote the final estimates of $\boldsymbol{\beta}$, $\boldsymbol{\tau}$, and $\boldsymbol{f}$ as $\hat{\boldsymbol{\beta}}$, $\hat{\boldsymbol{\tau}}$, and $\hat{\boldsymbol{f}}$. We estimate σ^2 by

$$\hat{\sigma}^2 = \frac{(\boldsymbol{y} - \hat{\boldsymbol{\eta}})^T \hat{W} (\boldsymbol{y} - \hat{\boldsymbol{\eta}})}{n - d - \operatorname{tr}(\tilde{H}^*)}, \tag{8.40}$$

where $\hat{\boldsymbol{\eta}} = \boldsymbol{\eta}(\hat{\boldsymbol{\beta}}, \hat{\boldsymbol{f}})$, $\hat{W}$ is the estimate of W with $\boldsymbol{\tau}$ being replaced by $\hat{\boldsymbol{\tau}}$, d is the degrees of freedom for parameters, which is usually taken as the total number of parameters, and $\tilde{H}^*$ is the hat matrix for model (8.39) computed at convergence.

Conditional on $\boldsymbol{f}$, inference for $\boldsymbol{\beta}$ and $\boldsymbol{\tau}$ can be made based on the approximate distributions of the maximum likelihood estimates. Conditional on $\boldsymbol{\beta}$ and $\boldsymbol{\tau}$, model (8.27) is a special case of the semiparametric linear regression model (8.4). Therefore, Bayesian confidence intervals can be constructed as in Section 8.2 for semiparametric conditional linear models. For the general SNR model, approximate Bayesian confidence intervals can be constructed based on the linear approximation (8.39) at convergence. The bootstrap approach may also be used to construct confidence intervals.

8.3.4 The `snr` Function

The function `snr` in the `assist` package is designed to fit the following special SNR models

$$y_i = \psi(\boldsymbol{\beta}, \mathcal{L}_{1i}(\boldsymbol{\beta})f_1, \ldots, \mathcal{L}_{ri}(\boldsymbol{\beta})f_r) + \epsilon_i, \quad i = 1, \ldots, n, \tag{8.41}$$

for cross-sectional data and

$$y_{ij} = \psi(\boldsymbol{\beta}_i, \mathcal{L}_{1ij}(\boldsymbol{\beta}_i)f_1, \ldots, \mathcal{L}_{rij}(\boldsymbol{\beta}_i)f_r) + \epsilon_{ij},$$
$$i = 1, \ldots, m; \quad j = 1, \ldots, n_i, \tag{8.42}$$

for clustered data, where ψ is a known nonlinear function, and $\mathcal{L}_{ki}(\boldsymbol{\beta})$ and $\mathcal{L}_{kij}(\boldsymbol{\beta}_i)$ are evaluational functionals on $\mathcal{H}_k$. Obviously, (8.41) includes the model (8.28) as a special case, and (8.42) includes the model (8.32) as a special case.

A modified procedure is implemented in the `snr` function. Note that models (8.41) and (8.42) reduce to the NNR model (7.5) when $\boldsymbol{\beta}$ is fixed and random errors are iid. For non-iid random errors, when both $\boldsymbol{\beta}$ and $\boldsymbol{\tau}$ are fixed, similar transformations as in Section 5.2.1 may be used. The nonlinear Gauss–Seidel algorithm in Section 7.4 can then be used to update nonparametric functions $\boldsymbol{f}$. Therefore, we implement the following procedure in the $\texttt{snr}$ function.

Algorithm in the $\texttt{snr}$ function

1. *Initialize*: Set initial values for $\boldsymbol{\beta}$, $\boldsymbol{\tau}$, and $\boldsymbol{f}$.

2. *Cycle*: Alternate between (a) and (b) until convergence:

    (a) Conditional on current estimates of $\boldsymbol{\beta}$, $\boldsymbol{\tau}$ and $\boldsymbol{f}$, apply transformations as in Section 5.2.1 when random errors are non-iid, and use the nonlinear Gauss–Seidel algorithm to update $\boldsymbol{f}$.

    (b) Conditional on current estimates of $\boldsymbol{f}$, update $\boldsymbol{\beta}$ and $\boldsymbol{\tau}$ by solving (8.37) alternatively using the backfitting and Gauss–Newton algorithms.

No initial values for f_k are necessary if ψ depends on f_k linearly in (8.41) or (8.42). The initial $\boldsymbol{\tau}$ is set such that W equals the identity matrix. Step (b) is implemented using the $\texttt{gnls}$ function in the $\texttt{nlme}$ package.

A typical call is

```
snr(formula, func, params, start)
```

where $\texttt{formula}$ is a two-sided formula specifying the response variable on the left side of a ˜ operator and an expression for the function ψ in the model (8.41) or (8.42) on the right side with $\boldsymbol{\beta}$ and f_k treated as parameters. The argument $\texttt{func}$ inputs a list of formulae, each specifying bases $\phi_{k1}, \ldots, \phi_{kp_k}$ for $\mathcal{H}_{k0}$ and RK R_{k1} for $\mathcal{H}_{k1}$ in the same way as the $\texttt{func}$ argument in the $\texttt{nnr}$ function. The argument $\texttt{params}$ inputs a list of two-sided linear formulae specifying second-stage models for $\boldsymbol{\beta}$. When there is no second-stage model for a parameter, it is specified as ˜1. The argument $\texttt{start}$ inputs initial values for all parameters. When ψ depends on functions f_k nonlinearly, initial values for those functions should also be provided in the $\texttt{start}$ argument.

An object of snr class is returned. The generic function $\texttt{summary}$ can be applied to extract further information. Predictions at covariate values can be computed using the $\texttt{predict}$ function. Posterior means and standard deviations for $\boldsymbol{f}$ can be computed using the $\texttt{intervals}$ function. See Sections 8.4.2–8.4.6 for examples.

8.4 Examples

8.4.1 Canadian Weather — Revisit

Consider the Canadian weather data again with annual temperature and precipitation profiles from all 35 stations as functional data. We now investigate how the monthly logarithm of rainfall depends on climate regions and temperature. Let y be the logarithm of `prec`, x_1 be the `region` indicator, and x_2 be the `month` variable scaled into $[0, 1]$. Consider the following FLM

$$y_{k,x_1}(x_2) = f_1(x_1, x_2) + w_{k,x_1}(x_2)f_2(x_2) + \epsilon_{k,x_1}(x_2), \qquad (8.43)$$

where $y_{k,x_1}(x_2)$, $w_{k,x_1}(x_2)$, and $\epsilon_{k,x_1}(x_2)$ are profiles of log precipitation, residual temperature after removing the region effect, and random error of station k in climate region x_1, respectively. Both annual log precipitation and temperature profiles can be regarded as functional data. Therefore, model (8.43) is an example of situation (iii) in Section 2.10 where both the independent and dependent variables involve functional data. We model the bivariate function $f_1(x_1, x_2)$ using the tensor product space $\mathbb{R}^4 \otimes W_2^2(per)$. Then, as in (4.22), f_1 admits the SS ANOVA decomposition

$$f_1(x_1, x_2) = \mu + f_{1,1}(x_1) + f_{1,2}(x_2) + f_{1,12}(x_1, x_2).$$

Model (8.43) is the same as model (14.1) in Ramsay and Silverman (2005) with $\mu(x_2) = \mu + f_{1,2}(x_2)$, $\alpha_{x_1}(x_2) = f_{1,1}(x_1) + f_{1,12}(x_1, x_2)$, and $\beta(x_2) = f_2(x_2)$. The function f_2 is the varying coefficient function for the temperature effect. We model f_2 using the periodic spline space $W_2^2(per)$. There are 12 monthly observations for each station. Collect all observations on y, x_1, x_2, and w for all 35 stations and denote them as $\{(y_i, x_{i1}, x_{i2}, w_i), \ i = 1, \dots, n\}$ where $n = 420$. Denote the collection of random errors as $\epsilon_1, \dots, \epsilon_n$. Then model (8.43) can be rewritten as

$$y_i = \mu + f_{1,1}(x_{i1}) + f_{1,2}(x_{i2}) + f_{1,12}(x_{i1}, x_{i2}) + \mathcal{L}_{2i}f_2 + \epsilon_i, \qquad (8.44)$$

where $\mathcal{L}_{2i}f_2 = w_i f_2(x_{i2})$. Model (8.44) is a special case of the semiparametric linear regression model (8.4). Model spaces for $f_{1,1}$, $f_{1,2}$, and $f_{1,12}$ are $\mathcal{H}_1 = \mathcal{H}^1$, $\mathcal{H}_2 = \mathcal{H}^2$, and $\mathcal{H}_3 = \mathcal{H}^3$, where $\mathcal{H}^1$, $\mathcal{H}^2$, and $\mathcal{H}^3$ are defined in Section 4.4.4. The RKs R_1, R_2, and R_3 of $\mathcal{H}_1$, $\mathcal{H}_2$, and $\mathcal{H}_3$ can be calculated as products of the RKs of the involved marginal spaces. Define $\Sigma_1 = \{R_1(x_{i1}, x_{j1})\}_{i,j=1}^n$, $\Sigma_2 = \{R_2(x_{i2}, x_{j2})\}_{i,j=1}^n$, and $\Sigma_3 = \{R_3((x_{i1}, x_{i2}), (x_{j1}, x_{j2}))\}_{i,j=1}^n$. For f_2, the model space $\mathcal{H}_4 =$

$W_2^2(per)$ where the construction of $W_2^2(per)$ is given in Section 2.7. Specifically, write $W_2^2(per) = \mathcal{H}_{40} \oplus \mathcal{H}_{41}$, where $\mathcal{H}_{40} = \{1\}$ and $\mathcal{H}_{41} = W_2^2(per) \ominus \{1\}$. Denote $\phi_4(x_2) = 1$ as the basis of $\mathcal{H}_{40}$ and R_4 as the RK for $\mathcal{H}_{41}$. Then, $T_4 = (\mathcal{L}_{21}\phi_4, \dots, \mathcal{L}_{2n}\phi_4)^T = (w_1, \dots, w_n)^T \triangleq \boldsymbol{w}$, and $\Sigma_4 = \{\mathcal{L}_{2i}\mathcal{L}_{j2}R_4\}_{i,j=1}^n = \{w_i w_j R_4(x_{i2}, x_{j2})\}_{i,j=1}^n = \boldsymbol{w}\boldsymbol{w}^T \circ \Lambda$ where $\Lambda = \{R_4(x_{i2}, x_{j2})\}_{i,j=1}^n$ and $\circ$ represents elementwise multiplication of two matrices. Therefore, model (8.44) can be fitted as follows:

```
> x1 <- rep(as.factor(region),rep(12,35))
> x2 <- (rep(1:12,35)-.5)/12
> y <- log(as.vector(monthlyPrecip))
> w <- canada.fit2$resi
> canada.fit3 <- ssr(y~w,
     rk=list(shrink1(x1),
             periodic(x2),
             rk.prod(shrink1(x1),periodic(x2)),
             rk.prod(kron(w),periodic(x2))))
```

Estimates of region effects $\alpha_{x_1}(x_2)$ and coefficient function f_2 evaluated at grid points are computed as follows:

```
> xgrid <- seq(0,1,len=50)
> zone <- c(''Atlantic'',''Pacific'',
            ''Continental'',''Arctic'')
> grid <- data.frame(x1=rep(zone,rep(50,4)),
                x2=rep(xgrid,4), w=rep(0,200))
> alpha <- predict(canada.fit3, newdata=grid,
                terms=c(0,0,1,0,1,0))
> grid <- data.frame(x1=rep(zone[1],50), x2=xgrid,
                w=rep(1,50))
> f2 <- predict(canada.fit3, newdata=grid,
                terms=c(0,0,0,0,0,1))
```

Those estimates of $\alpha_{x_1}(x_2)$ and f_2 are shown in Figures 8.3 and 8.4, respectively.

Ramsay and Silverman (2005) also considered the following FLM (model (16.1) in Ramsay and Silverman (2005)):

$$y_k(x_2) = f_1(x_2) + \int_0^1 w_k(u) f_2(u, x_2) du + \epsilon_k(x_2), \tag{8.45}$$
$$k = 1, \dots, 35,$$

where $y_k(x_2)$ is the logarithm of annual precipitation profile at station k, $f_1(x_2)$ plays the part of an intercept as in the standard regression, $w_k(u)$ is the temperature profile at station k, $f_2(u, x_2)$ is an unknown

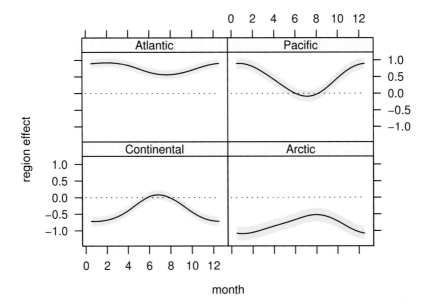

FIGURE 8.3 Canadian weather data, plots of estimated region effects to precipitation, and 95% Bayesian confidence intervals.

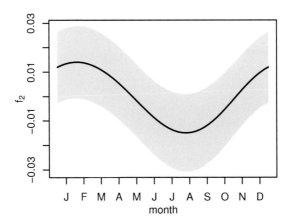

FIGURE 8.4 Canadian weather data, estimate of the coefficient function for temperature effect f_2, and 95% Bayesian confidence intervals.

weight function at month x_2, and $\epsilon_k(x_2)$ are random error processes. Comparing to model (8.44), the whole temperature profile is used to predict the current precipitation in (8.45). Note that $w_k(x_2)$ in model (8.45) is the actual temperature, and $w_k(x_2)$ in (8.44) is the residual temperature after the removing the region effect.

The goal is to model and estimate functions f_1 and f_2. It is reasonable to assume that f_1 and f_2 are smooth periodic functions. Specifically, we assume that $f_1 \in W_2^2(per)$, and $f_2 \in W_2^2(per) \otimes W_2^2(per)$. Let $\mathcal{H}_0 = \{1\}$ and $\mathcal{H}_1 = W_2^2(per) \ominus \{1\}$. Then, we have an one-way SS ANOVA decomposition for $W_2^2(per)$,

$$W_2^2(per) = \mathcal{H}_0 \oplus \mathcal{H}_1,$$

and a two-way SS ANOVA decomposition for $W_2^2(per) \otimes W_2^2(per)$,

$$
\begin{aligned}
&W_2^2(per) \otimes W_2^2(per) \\
&= \left\{ \mathcal{H}_0^{(1)} \oplus \mathcal{H}_1^{(1)} \right\} \otimes \left\{ \mathcal{H}_0^{(2)} \oplus \mathcal{H}_1^{(2)} \right\} \\
&= \left\{ \mathcal{H}_0^{(1)} \otimes \mathcal{H}_0^{(2)} \right\} \oplus \left\{ \mathcal{H}_1^{(1)} \otimes \mathcal{H}_0^{(2)} \right\} \oplus \left\{ \mathcal{H}_0^{(1)} \otimes \mathcal{H}_1^{(2)} \right\} \oplus \left\{ \mathcal{H}_1^{(1)} \otimes \mathcal{H}_1^{(2)} \right\} \\
&\triangleq \mathcal{H}^0 \oplus \mathcal{H}_2 \oplus \mathcal{H}_3 \oplus \mathcal{H}_4,
\end{aligned}
\tag{8.46}
$$

where $\mathcal{H}_0^{(1)} = \mathcal{H}_0^{(2)} = \{1\}$, and $\mathcal{H}_1^{(1)} = \mathcal{H}_1^{(2)} = W_2^2(per) \ominus \{1\}$. Equivalently, we have the following SS ANOVA decomposition for f_1 and f_2:

$$
\begin{aligned}
f_1(x_2) &= \mu_1 + f_{1,1}(x_2), \\
f_2(u, x_2) &= \mu_2 + f_{2,1}(u) + f_{2,2}(x_2) + f_{2,12}(u, x_2),
\end{aligned}
$$

where $f_{1,1} \in \mathcal{H}_1$, $f_{2,1} \in \mathcal{H}_2$, $f_{2,2} \in \mathcal{H}_3$, and $f_{2,12} \in \mathcal{H}_4$. Then model (8.45) can be rewritten as

$$
\begin{aligned}
y_k(x_2) &= \mu_1 + \mu_2 z_k + f_{1,1}(x_2) + \int_0^1 w_k(u) f_{2,1}(u) du + z_k f_{2,2}(x_2) \\
&\quad + \int_0^1 w_k(u) f_{2,12}(u, x_2) du + \epsilon_k(x_2),
\end{aligned}
\tag{8.47}
$$

where $z_k = \int_0^1 w_k(u) du$. Let $\boldsymbol{z} = (z_1, \ldots, z_{35})^T$ and $\boldsymbol{s} = (s_1, \ldots, s_n) \triangleq \boldsymbol{z} \otimes \boldsymbol{1}_{12}$, where $\boldsymbol{1}_k$ is a k-vector of all ones and $\otimes$ represents the Kronecker product. Denote $\{(y_i, x_{2i}), \ i = 1, \ldots, n\}$ as the collection of all observations on y and x_2, and $\epsilon_1, \ldots, \epsilon_n$ as the collection of random errors. Then model (8.47) can be rewritten as

$$
\begin{aligned}
y_i &= \mu_1 + \mu_2 s_i + f_{1,1}(x_{i2}) + \int_0^1 w_{[i]}(u) f_{2,1}(u) du + s_i f_{2,2}(x_{i2}) \\
&\quad + \int_0^1 w_{[i]}(u) f_{2,12}(u, x_{i2}) du + \epsilon_i, \quad i = 1, \ldots, n,
\end{aligned}
\tag{8.48}
$$

where $[i]$ represents the integer part of $(11 + i)/12$. Define a linear operator $\mathcal{L}_{1i}$ as the evaluational functional on $\mathcal{H}_1 = W_2^2(per) \ominus \{1\}$ such that $\mathcal{L}_{1i}f_{1,1} = f_{1,1}(x_{i2})$. Define linear operators $\mathcal{L}_{2i}$, $\mathcal{L}_{3i}$, and $\mathcal{L}_{4i}$ on subspaces $\mathcal{H}_2$, $\mathcal{H}_3$, and $\mathcal{H}_4$ in (8.46) such that

$$\mathcal{L}_{2i}f_{2,1} = \int_0^1 w_{[i]}(u)f_{2,1}(u)du,$$

$$\mathcal{L}_{3i}f_{2,2} = s_i f_{2,2}(x_{i2}),$$

$$\mathcal{L}_{4i}f_{2,12} = \int_0^1 w_{[i]}(u)f_{2,12}(u, x_{i2})du.$$

Assume that functions $w_k(u)$ for $k = 1, \ldots, 35$ are square integrable. Then $\mathcal{L}_{2i}$, $\mathcal{L}_{3i}$, and $\mathcal{L}_{4i}$ are bounded linear functionals, and model (8.48) is a special case of the semiparametric linear regression model (8.4). Let $t_i = (i - 0.5)/12$ for $i = 1, \ldots, 12$ be the middle point of month i, $\boldsymbol{x}_2 = (t_1, \ldots, t_{12})^T$, $\boldsymbol{w}_k = (w_k(x_{12}), \ldots, w_k(x_{m2}))^T$, where $m = 12$, and $W = (\boldsymbol{w}_1, \ldots, \boldsymbol{w}_{35})$. Let R_1 be the RK of $\mathcal{H}_1 = W_2^2(per) \ominus \{1\}$. Introduce the notation $R_1(\boldsymbol{u}, \boldsymbol{v}) = \{R_1(u_k, v_l)\}_{k=1}^{K}{}_{l=1}^{L}$ for any vectors $\boldsymbol{u} = (u_1, \ldots, u_K)^T$ and $\boldsymbol{v} = (v_1, \ldots, v_L)^T$. It is easy to check that $S = (\boldsymbol{1}_n, \boldsymbol{s})$, $\Sigma_1 = \boldsymbol{1}_{35} \otimes \boldsymbol{1}_{35}^T \otimes R_1(\boldsymbol{x}_2, \boldsymbol{x}_2)$, and $\Sigma_3 = \boldsymbol{z} \otimes \boldsymbol{z}^T \otimes R_1(\boldsymbol{x}_2, \boldsymbol{x}_2)$. A similar approximation as in Section 2.10 leads to $\Sigma_2 \approx \{W^T R_1(\boldsymbol{x}_2, \boldsymbol{x}_2)W\} \otimes \boldsymbol{1}_{12} \otimes \boldsymbol{1}_{12}^T/144$. Note that the RK of $\mathcal{H}_4$ in (8.46) equals $R_1(s,t)R_1(x,z)$. Then the (i,j)th element of the matrix Σ_4

$$\Sigma_4(i,j) = \mathcal{L}_{4i}\mathcal{L}_{4j}R_1(u,v)R_1(x,z)$$

$$= R_1(x_{i2}, x_{j2}) \int_0^1 \int_0^1 w_{[i]}(u)w_{[j]}(v)R_1(u,v)dudv$$

$$\approx \frac{1}{144}R_1(x_{i2}, x_{j2})\boldsymbol{w}_i^T R_1(\boldsymbol{x}_2, \boldsymbol{x}_2)\boldsymbol{w}_j.$$

Thus, $\Sigma_4 \approx \Sigma_1 \circ \Sigma_2$. We fit model (8.48) as follows:

```
> W <- monthlyTemp; z <- apply(W,2,mean)
> s <- rep(z,rep(12,35)); x <- seq(0.5,11.5,1)/12
> y <- log(as.vector(monthlyPrecip))
> Q1 <- kronecker(matrix(1,35,35),periodic(x))
> Q2 <- kronecker(t(W)%*%periodic(x)%*%W,
                  matrix(1,12,12))/144
> Q3 <- kronecker(z%*%t(z),periodic(x))
> Q4 <- rk.prod(Q1,Q2)
> canada.fit4 <- ssr(y~s, rk=list(Q1,Q2,Q3,Q4))
```

We now show how to compute the estimated functions evaluated at a set of points. From (8.6), the estimated functions are represented by

$$\hat{f}_1(x_2) = d_1 + \theta_1 \sum_{i=1}^{n} c_i R_1(x_2, x_{i2}),$$

$$\hat{f}_{2,1}(u) = \theta_2 \sum_{i=1}^{n} c_i \int_0^1 w_{[i]}(v) R_1(u, v) dv,$$

$$\hat{f}_{2,2}(x_2) = \theta_3 \sum_{i=1}^{n} c_i \left\{ \int_0^1 w_{[i]}(u) du \right\} R_1(x_2, x_{i2}),$$

$$\hat{f}_{2,12}(u, x_2) = \theta_4 \sum_{i=1}^{n} c_i \left\{ \int_0^1 w_{[i]}(v) R_1(u, v) dv \right\} R_1(x_2, x_{i2}).$$

Let $\boldsymbol{u}_0$ and $\boldsymbol{x}_0$ be a set of points in $[0, 1]$ for the variables u and x_2, respectively. For simplicity, assume that both $\boldsymbol{u}_0$ and $\boldsymbol{x}_0$ have length n_0. The following calculations can be extended to the case when $\boldsymbol{u}_0$ and $\boldsymbol{x}_0$ have different lengths. It is not difficult to check that

$$\hat{f}_1(\boldsymbol{x}_0) = d_1 \mathbf{1}_{n_0} + \theta_1 \sum_{i=1}^{n} c_i R_1(\boldsymbol{x}_0, x_{i2}) = d_1 \mathbf{1}_{n_0} + \theta_1 S_1 \boldsymbol{c},$$

$$\hat{f}_{2,1}(\boldsymbol{u}_0) = \theta_2 \sum_{i=1}^{n} c_i \int_0^1 w_{[i]}(u) R_1(\boldsymbol{u}_0, u) du$$

$$\approx \frac{1}{12} \theta_2 \sum_{i=1}^{n} c_i \sum_{j=1}^{12} R_1(\boldsymbol{u}_0, x_{j2}) w_{[i]}(x_{j2})$$

$$= \frac{1}{12} \theta_2 \sum_{i=1}^{n} c_i R_1(\boldsymbol{u}_0, \boldsymbol{x}_2) w_{[i]} = \theta_2 S_2 \boldsymbol{c},$$

$$\hat{f}_{2,2}(\boldsymbol{x}_0) = \theta_3 \sum_{i=1}^{n} c_i \left\{ \int_0^1 w_{[i]}(u) du \right\} R_1(\boldsymbol{x}_0, x_{i2}) = \theta_3 S_3 \boldsymbol{c},$$

where $S_1 = \mathbf{1}_{35}^T \otimes R_1(\boldsymbol{x}_0, \boldsymbol{x}_2)$, $S_2 = \{R_1(\boldsymbol{u}_0, \boldsymbol{x}_2) W\} \otimes \mathbf{1}_{12}^T / 12$, and $S_3 = \boldsymbol{z}^T \otimes R_1(\boldsymbol{x}_0, \boldsymbol{x}_2)$. The interaction $f_{2,12}$ is a bivariate function. Thus, we evaluate it at a bivariate grid $\{(u_{0k}, x_{0l}) : k, l = 1, \ldots, n_0\}$:

$$\hat{f}_{2,12}(u_{0k}, x_{0l}) = \theta_4 \sum_{i=1}^{n} c_i \left\{ \int_0^1 w_{[i]}(v) R_1(u_{0k}, v) dv \right\} R_1(x_{0l}, x_{i2})$$

$$\approx \theta_4 \sum_{i=1}^{n} c_i S_2[k, i] S_1[l, i].$$

Then $(\hat{f}_{2,12}(u_{01}, x_{01}), \ldots, \hat{f}_{2,12}(u_{01}, x_{0n_0}), \ldots, \hat{f}_{2,12}(u_{0n_0}, x_{01}), \ldots,$
$\hat{f}_{2,12}(u_{0n_0}, x_{0n_0})) = \theta_4 S_4 c$, where S_4 is an $n_0^2 \times n$ matrix with elements
$S_4[(k-1)n_0 + l, i] = S_2[k, i]S_1[l, i]$ for $k, l = 1, \ldots, n_0$ and $i = 1, \ldots, n$.

```
> ngrid <- 40; xgrid <- seq(0,1,len=ngrid)
> S1 <- kronecker(t(rep(1,35)), periodic(xgrid,x))
> S2 <- kronecker(periodic(xgrid,x)%*%W, t(rep(1,12)))/12
> S3 <- kronecker(t(z), periodic(xgrid,x))
> S4 <- NULL
> for (k in 1:ngrid) {
    for (l in 1:ngrid) S4 <- rbind(S4, S1[l,]*S2[k,])}
> the <- 10^canada.fit4$rkpk.obj$theta
> f1 <- canada.fit4$coef$d[1]
    +the[1]*S1%*%canada.fit4$coef$c
> mu2 <- canada.fit4$coef$d[2]
> f21 <- the[2]*S2%*%canada.fit4$coef$c
> f22 <- the[3]*S3%*%canada.fit4$coef$c
> f212 <- the[4]*S4%*%canada.fit4$coef$c
> f2 <- mu2+rep(f21,rep(ngrid,ngrid))+rep(f22,ngrid)+f212
```

Figures 8.5 displays the estimates of f_1 and f_2.

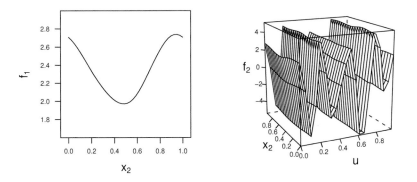

FIGURE 8.5 Canadian weather data, estimates of f_1 (left) and f_2 (right).

8.4.2 Superconductivity Magnetization Modeling

In this section we use the superconductivity data to illustrate how to check a nonlinear regression model. Figure 8.6 displays magnetization versus time on logarithmic scale.

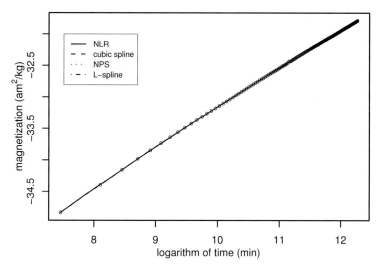

FIGURE 8.6 Superconductivity data, observations (circles), and the fits by nonlinear regression (NLR), cubic spline, nonlinear partial spline (NPS), and L-spline.

It seems that a straight line (Anderson and Kim model) can fit data well (Yeshurun, Malozemoff and Shaulov 1996). Let y be magnetism values, and x be logarithm of time scaled into the interval $[0, 1]$. To check the Anderson and Kim model, we fit a cubic spline with the GML choice of the smoothing parameter:

```
> library(NISTnls)
> a <- Bennett5; x <- ident(a$x); y <- a$y
> super.cub <- ssr(y~x, cubic(x), spar=''m'')
> anova(super.cub, simu.size=1000)
Testing H_0: f in the NULL space
      test.value  simu.size simu.p-value approximate.p-value
LMP 0.1512787         1000            0
GML 0.01622123        1000            0                    0
```

Let f_1 be the projection onto the space $W_2^2[0, 1] \ominus \{1, x - 0.5\}$ which represents the systematic departure from the straight line model. Both the LMP and GML tests for the hypothesis $H_0 : f_1(x) = 0$ conclude that the departure from the straight line model is statistically significant. Figure 8.7(a) shows the estimate of f_1 with 95% Bayesian confidence intervals. Those confidence intervals also indicate that, though small, the departure from a straight line is statistically significant. The deviation from the Anderson and Kim model has been noticed for high-temperature superconductors, and the following "interpolation formula" was proposed (Bennett, Swartzendruber, Blendell, Habib and Seyoum 1994, Yeshurun et al. 1996):

$$y = \beta_1(\beta_2 + x)^{-\frac{1}{\beta_3}} + \epsilon. \tag{8.49}$$

The nonlinear regression model (8.49) is fitted as follows:

```
> b10 <- -1500*(max(a$x)-min(a$x))**(-1/.85)
> b20 <- (45+min(a$x))/(max(a$x)-min(a$x))
> super.nls <- nls(y~b1*(b2+x)**(-1/b3),
    start=list(b1=b10,b2=b20,b3=.85))
```

The initial values were computed based on one set of initial values provided in the help file of **Bennett5**. The fit of the above NLR model is shown in Figure 8.6. To check the "interpolation formula", we can fit a nonlinear partial spline model

$$y = \beta_1(\beta_2 + x)^{-\frac{1}{\beta_3}} + f_2(x) + \epsilon, \tag{8.50}$$

with $f_2 \in W_2^2[0,1]$. Model (8.50) is a special case of the SNR model, which can be fitted as follows:

```
> bh <- coef(super.nls)
> super.snr <- snr(y~b1*(b2+x)**(-1/b3)+f(x),
    params=list(b1+b2+b3~1),
    func=f(u)~list(~I(u-.5), cubic(u)),
    start=list(params=bh), spar=''m'')
> summary(super.snr)
...
Coefficients:
        Value Std.Error   t-value p-value
b1 -466.4197 32.05006 -14.55285       0
b2   11.2296  0.21877  51.33033       0
b3    0.9322  0.01719  54.22932       0
...
GML estimate(s) of smoothing spline parameter(s):
  0.0001381760
Equivalent Degrees of Freedom (DF) for spline function:
```

```
10.96524
Residual standard error: 0.001703358
```

We used the estimates of parameters in (8.49), bh, as initial values for β_1, β_2, and β_3. The fit of model (8.50) is shown in Figure 8.6. To check if the departure from model (8.49), f_2, is significant, we compute posterior means and standard deviations for f_2:

```
> super.snr.pred <- intervals(super.snr)
```

The estimate of f_2 and its 95% Bayesian confidence intervals are shown in Figure 8.7(b). The magnitude of f_2 is very small. The zero line is outside the confidence intervals in some regions.

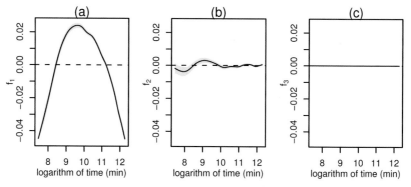

FIGURE 8.7 Superconductivity data, (a) estimate of the departure from the straight line model, (b) estimate of the departure from the "interpolation formula" based on the nonlinear partial spline, and (c) estimate of the departure from the "interpolation formula" based on the L-spline. Shaded regions are 95% Bayesian confidence intervals.

An alternative approach to checking the NLR model (8.49) is to use the L-spline introduced in Section 2.11. Consider the regression model (2.1) where $f \in W_2^2[0, 1]$. For fixed β_2 and β_3, it is clear that the space corresponding to the "interpolation formula"

$$\mathcal{H}_0 = \text{span}\{(\beta_2 + x)^{-\frac{1}{\beta_3}}\}$$

is the kernel of the differential operator

$$L = D + \frac{1}{\beta_3(\beta_2 + x)}.$$

Let $\mathcal{H}_1 = W_2^2[0,1] \ominus \mathcal{H}_0$. The Green function

$$G(x,s) = \begin{cases} \left(\frac{\beta_2+s}{\beta_2+x}\right)^{\frac{1}{\beta_3}}, & s \leq x, \\ 0, & s > x. \end{cases}$$

Therefore, the RK of $\mathcal{H}_1$

$$\begin{aligned} R_1(x,z) &= \int_0^1 \left(\frac{\beta_2+s}{\beta_2+x}\right)^{\frac{1}{\beta_3}} \left(\frac{\beta_2+s}{\beta_2+z}\right)^{\frac{1}{\beta_3}} ds \\ &= C(\beta_2+x)^{-\frac{1}{\beta_3}}(\beta_2+z)^{-\frac{1}{\beta_3}}, \end{aligned} \tag{8.51}$$

where $C = \int_0^1 (\beta_2+s)^{\frac{2}{\beta_3}} ds$. We fit the L-spline model as follows:

```
> ip.rk <- function(x,b2,b3) ((b2+x)%o%(b2+x))**(-1/b3)
> super.l <- ssr(y~I((bh[2]+x)**(-1/bh[3]))-1,
    rk=ip.rk(x,bh[2],bh[3]), spar=''m'')
```

The function `ip.rk` computes the RK R_1 in (8.51) with the constant C being ignored since it can be absorbed by the smoothing parameter. The values of β_2 and β_3 are fixed as the estimates from the nonlinear regression model (8.49). Let f_3 be the projection onto the space $\mathcal{H}_1$ which represents the systematic departure from the "interpolation formula". Figure 8.7(c) shows the estimate of f_3 with 95% Bayesian confidence intervals. The systematic departure from the "interpolation formula" is essentially zero.

8.4.3 Oil-Bearing Rocks

The rock data set contains measurements on four cross sections of each of 12 oil-bearing rocks. The aim is to predict permeability (`perm`) from three other measurements: the total area (`area`), total perimeter (`peri`) and a measure of "roundness" of the pores in the rock cross section (`shape`). Let $y = \log(\text{perm})$, $x_1 = \text{area}/10000$, $x_2 = \text{peri}/10000$ and $x_3 = \text{shape}$. A full multivariate nonparametric model such as an SS ANOVA model is not desirable in this case since there are only 48 observations. Consider the projection pursuit regression model (8.1) with $r = 2$. Let $\boldsymbol{\beta}_k = (\beta_{k1}, \beta_{k2}, \beta_{k3})^T$ for $k = 1,2$. For identifiability, we use spherical coordinates $\beta_{k1} = \sin(\alpha_{k1})\cos(\alpha_{k2})$, $\beta_{k2} = \sin(\alpha_{k1})\sin(\alpha_{k2})$, and $\beta_{k3} = \cos(\alpha_{k1})$, which satisfy the side condition $\beta_{k1}^2 + \beta_{k2}^2 + \beta_{k3}^2 = 1$. Then we have the following SNR model:

$$\begin{aligned} y = \beta_0 &+ f_1\big(\sin(\alpha_{11})\cos(\alpha_{12})x_1 + \sin(\alpha_{11})\sin(\alpha_{12})x_2 + \cos(\alpha_{11})x_3\big) \\ &+ f_2(\sin(\alpha_{21})\cos(\alpha_{22})x_1 + \sin(\alpha_{21})\sin(\alpha_{22})x_2 + \cos(\alpha_{21})x_3) + \epsilon. \end{aligned} \tag{8.52}$$

Note that the domains of f_1 and f_2 are not fixed intervals since they depend on unknown parameters. We model f_1 and f_2 using the one-dimensional thin-plate spline space $W_2^2(\mathbb{R})$. To make f_1 and f_2 identifiable with the constant β_0, we remove the constant functions from the model space. That is, we assume that $f_1, f_2 \in W_2^2(\mathbb{R}) \ominus \{1\}$. Random errors are assumed to be iid. Bounds on parameters α_{kj} are ignored for simplicity.

```
> attach(rock)
> y <- log(perm)
> x1 <- area/10000; x2 <- peri/10000; x3 <- shape
> rock.ppr <- ppr(y~x1+x2+x3, nterms=2, max.terms=5)
> b.ppr <- rock.ppr$alpha
> a11ini <- acos(b.ppr[3,1]/sqrt(sum(b.ppr[,1]**2)))
> a12ini <- atan(b.ppr[2,1]/b.ppr[1,1])
> a21ini <- acos(b.ppr[3,2]/sqrt(sum(b.ppr[,2]**2)))
> a22ini <- atan(b.ppr[2,2]/b.ppr[1,2])
> rock.snr <- snr(y~b0
    +f1(sin(a11)*cos(a12)*x1+sin(a11)*sin(a12)*x2
       +cos(a11)*x3)
    +f2(sin(a21)*cos(a22)*x1+sin(a21)*sin(a22)*x2
       +cos(a21)*x3),
    func=list(f1(u)~list(~u-1,tp(u)),
              f2(v)~list(~v-1,tp(v))),
    params=list(b0+a11+a12+a21+a22~1), spar=''m'',
    start=list(params=c(mean(y),a11ini,a12ini,
                        a21ini,a22ini)),
    control=list(prec.out=1.e-3,maxit.out=50))
> summary(rock.snr)
...
Coefficients:
        Value  Std.Error   t-value p-value
b0    5.341747 0.31181572  17.13110  0.0000
a11   1.574432 0.04597406  34.24610  0.0000
a12  -1.221826 0.01709925 -71.45495  0.0000
a21   0.836215 0.26074376   3.20704  0.0025
a22  -1.010785 0.02919986 -34.61607  0.0000
...
GML estimate(s) of smoothing spline parameter(s):
  1.173571e-05 1.297070e-05
Equivalent Degrees of Freedom (DF) for spline function:
  8.559332
Residual standard error: 0.7337079
```

```
> a <- rock.snr$coef[-1]
> u <- sin(a[1])*cos(a[2])*x1+sin(a[1])*sin(a[2])*x2
        +cos(a[1])*x3
> v <- sin(a[3])*cos(a[4])*x1+sin(a[3])*sin(a[4])*x2
        +cos(a[3])*x3
> ugrid <- seq(min(u),max(u),len=50)
> vgrid <- seq(min(v),max(v),len=50)
> rock.snr.ci <- intervals(rock.snr,
                  newdata=data.frame(u=ugrid,v=vgrid))
```

We fitted the projection pursuit regression model using the `ppr` function in R first (Venables and Ripley 2002) and used the estimates to compute initial values for spherical coordinates. The estimates and posterior standard deviations of f_1 and f_2 were computed at grid points using the `intervals` function. Figure 8.8 shows the estimates of f_1 and f_2 with 95% Bayesian confidence intervals. It is interesting to note that the overall shapes of the estimated functions from `ppr` (Figure 8.9 in Venables and Ripley (2002)) and `snr` are comparable even though the estimation methods are quite different.

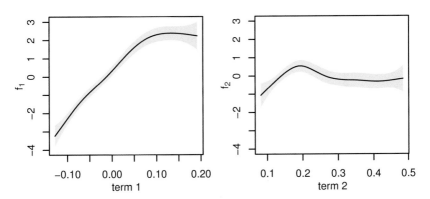

FIGURE 8.8 Rock data, estimates of f_1 (left) and f_2 (right). Shaded regions are 95% Bayesian confidence intervals.

8.4.4 Air Quality

The air quality data set contains daily measurements of the ozone concentration (`Ozone`) in parts per million and three meteorological variables: wind speed (`Wind`) in miles per hour, temperature (`Temp`) in degrees Fahrenheit, and solar radiation (`Solar.R`) in Langleys. The goal is to investigate how the air pollutant ozone concentration depends on three meteorological variables. Let $y = \texttt{Ozone}^{1/3}$, $x_1 = \texttt{Wind}$, $x_2 = \texttt{Temp}$, and $x_3 = \texttt{Solar.R}$. Yu and Ruppert (2002) considered the following partially linear single index model

$$y = f_1\big(\beta_1 x_1 + \beta_2 x_2 + \sqrt{1 - \beta_1^2 - \beta_2^2}\, x_3\big) + \epsilon. \qquad (8.53)$$

Note that, for identifiability, $\sqrt{1 - \beta_1^2 - \beta_2^2}$ is used to represent the coefficient for x_3 such that it is positive and the summation of all squared coefficients equals 1. Random errors are assumed to be iid. We model f_1 using the one-dimensional thin-plate spline space $W_2^2(\mathbb{R})$. The single index model (8.53) is an SNR model that can be fitted as follows:

```
> air <- na.omit(airquality)
> attach(air)
> y <- Ozone^(1/3)
> x1 <- Wind; x2 <- Temp; x3 <- ident(Solar.R)
> air.snr.1 <- snr(y~f1(b1*x1+b2*x2+sqrt(1-b1^2-b2^2)*x3),
             func=f1(u)~list(~u, rk=tp(u)),
             params=list(b1+b2~1), spar=''m'',
             start=list(params=c(-0.8,.5)),
             control=list(maxit.out=50))
```

The condition $\beta_1^2 + \beta_2^2 < 1$ is ignored for simplicity. The estimate of f_1 and its 95% Bayesian confidence intervals are shown in Figure 8.9(a). The estimate of f_1 is similar to that in Yu and Ruppert (2002).

Yu and Ruppert (2002) also considered the following partially linear single index model

$$y = f_2\big(\beta_1 x_1 + \sqrt{1 - \beta_1^2}\, x_2\big) + \beta_2 x_3 + \epsilon. \qquad (8.54)$$

The following statement fits model (8.54) with $f_2 \in W_2^2(\mathbb{R})$:

```
> air.snr.2 <- snr(y~f2(b1*x1+sqrt(1-b1^2)*x2)+b2*x3,
             func=f2(u)~list(~u, rk=tp(u)),
             params=list(b1+b2~1), spar=''m'',
             start=list(params=c(-0.8,1.3)))
```

The estimates of f_2 and $\beta_2 x_3$ are shown in Figure 8.9(b)(c). The effect of radiation may be nonlinear. So, we further fit the following SNR model

$$y = f_3\big(\beta_1 x_1 + \sqrt{1 - \beta_1^2}\, x_2\big) + f_4(x_3) + \epsilon. \qquad (8.55)$$

Again, we model f_3 using the space $W_2^2(\mathbb{R})$. Since the domain of x_3 is $[0, 1]$, we model f_4 using the cubic spline space $W_2^2[0, 1] \ominus \{1\}$ where constant functions are removed for identifiability.

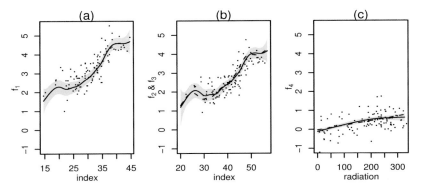

FIGURE 8.9 Air quality data, (a) observations (dots), and the estimate of f_1 (solid line) with 95% Bayesian confidence intervals; (b) partial residuals after removing the radiation effect (dots), the estimates of f_2 (dashed line) and f_3 (solid line), and 95% Bayesian confidence intervals (shaded region) for f_3; and (c) partial residuals after removing the index effects based on wind speed and temperature (dots), the estimates of $\beta_2 x_3$ in model (8.54) (dashed line) and f_4 (solid line), and 95% Bayesian confidence intervals (shaded region) for f_4.

```
> air.snr.3 <- snr(y~f3(a1*x1+sqrt(1-a1^2)*x2)+f4(x3),
        func=list(f3(u)~list(~u, rk=tp(u)),
                  f4(v)~list(~v-1, rk=cubic(v))),
        params=list(a1~1), spar=''m'',
        start=list(params=c(-0.8)))
```

Estimates of f_3 and f_4 are shown in Figure 8.9(b)(c). The estimate of f_4 increases with increasing radiation until a value of about 250 Langleys, after which it is flat. The difference between $\hat{f}_4$ and the linear estimate is not significant based on confidence intervals, perhaps due to the small sample size. To compare three models (8.53), (8.54), and (8.55), we compute AIC, BIC, and GCV criteria:

```
> n <- 111
> rss <- c(sum((y-air.snr.1$fitted)**2),
           sum((y-air.snr.2$fitted)**2),
           sum((y-air.snr.3$fitted)**2))/n
> df <- c(air.snr.1$df$f+air.snr.1$df$para,
          air.snr.2$df$f+air.snr.2$df$para,
          air.snr.3$df$f+air.snr.3$df$para)
> gcv <- rss/(1-df/n)**2
> aic <- n*log(rss)+2*df
> bic <- n*log(rss)+log(n)*df
```

```
> print(round(rbind(aic, bic, gcv),4))
aic -156.9928 -176.5754 -177.3056
bic -139.6195 -157.8198 -155.7608
gcv    0.2439    0.2046    0.2035
```

AIC and GCV select model (8.55), while BIC selects model (8.54).

8.4.5   The Evolution of the Mira Variable R Hydrae

The star data set contains magnitude (brightness) of the Mira variable R Hydrae during 1900–1950. Figure 8.10 displays observations over time. The Mira variable R Hydrae is well known for its declining period and amplitude. We will consider three SNR models in this section to investigate the pattern of the decline.

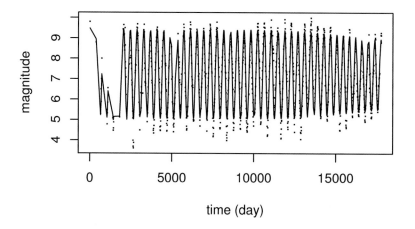

FIGURE 8.10 Star data, plot of observations (points), and the fit based on model (8.58) (solid line).

Let y = magnitude and x = time. We first consider the following SNR model (Genton and Hall 2007)

$$y = a(x)f_1(t(x)) + \epsilon, \tag{8.56}$$

where y is the magnitude on day x, $a(x) = 1 + \beta_1 x$ is the amplitude function, f_1 is the common periodic shape function with unit period, $t(x) = \log(1 + \beta_3 x/\beta_2)/\beta_3$ is a time transformation function, and ϵ is a random error. Random errors are assumed to be iid. The function $1/t'(x)$ can be regarded as the period function. Therefore, model (8.56) assumes that the amplitude and period evolve linearly. Since f_1 is close to a sinusoidal function, we model f_1 using the trigonometric spline with $\mathcal{L} = D\{D^2 + (2\pi)^2\}$ ($m = 2$ in (2.67)) and $f_1 \in W_2^3(per)$. Model (8.56) can be fitted as follows:

```
> data(star); attach(star)
> star.fit.1 <- snr(y~(1+b1*x)*f1(log(1+b3*x/b2)/b3),
    func=list(f1(u)~list(~sin(2*pi*u)+cos(2*pi*u),
                 lspline(u,type=''sine1''))),
    params=list(b1+b2+b3~1), spar=''m'',
    start=list(params=c(0.0000003694342,419.2645,
                   -0.00144125)))
> summary(star.fit.1)
...
Coefficients:
      Value  Std.Error   t-value p-value
b1   0.0000 0.00000035    1.0163  0.3097
b2 419.2485 0.22553623 1858.8966  0.0000
b3  -0.0014 0.00003203  -44.8946  0.0000
...

GML estimate(s) of smoothing spline parameter(s):
 0.9999997
Equivalent Degrees of Freedom (DF) for spline function:
 3.000037
Residual standard error: 0.925996
> grid <- seq(0,1,len=100)
> star.p.1 <- intervals(star.fit.1,
    newdata=data.frame(u=grid),
    terms=list(f1=matrix(c(1,1,1,1,0,0,0,1),
                   ncol=4,byrow=T)))
> co <- star.fit.1$coef
> tx <- log(1+co[3]*x/co[2])/co[3]
> xfold <- tx-floor(tx)
> yfold <- y/(1+co[1]*x)
```

We computed posterior means and standard deviations of f_1 and its projection onto the subspace $W_2^3(per) \ominus \{1, \sin 2\pi x, \cos 2\pi x\}$, $P_1 f_1$, using the `intervals` function. Estimate of f_1 and its 95% Bayesian confidence intervals are shown in Figure 8.11(a). We computed folded observations at day x as $\tilde{y} = y/(1 + \hat{\beta}_1 x)$ and $\tilde{x} = \hat{t}(x) - \lfloor \hat{t}(x) \rfloor$, where $z - \lfloor z \rfloor$ represents the fractional part of z. The folded observations are shown in Figure 8.11(a). The projection $P_1 f_1$ represents the departure from the sinusoidal model space $\mathrm{span}\{1, \sin 2\pi x, \cos 2\pi x\}$. Figure 8.11(b) indicates that the function f_1 is not significantly different from a sinusoidal model since $P_1 f_1$ is not significantly different from zero.

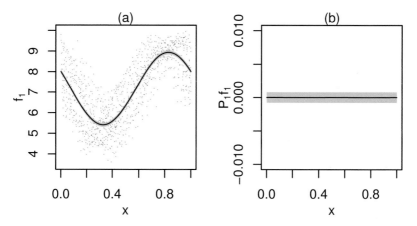

FIGURE 8.11 Star data, (a) folded observations (points), estimate of f_1 (solid line), and 95% Bayesian confidence intervals (shaded region); and (b) estimate of $P_1 f_1$ (solid line) and 95% Bayesian confidence intervals (shaded region).

Assuming that the periodic shape function can be well modeled by a sinusoidal function, we can investigate the evolving amplitude and period nonparametrically. We first consider the following SNR model

$$y = \beta_1 + \exp\{f_2(x)\} \sin[2\pi\{\beta_2 + \log(1 + \beta_4 x/\beta_3)/\beta_4\}] + \epsilon, \quad (8.57)$$

where β_1 is a parameter of the mean, $\exp\{f_2(x)\}$ is the amplitude function, and β_2 is a phase parameter. Note that the exponential transformation is used to enforce the positive constraint on the amplitude. We model f_2 using the cubic spline model space $W_2^2[0, b]$, where b equals the maximum of x. Model (8.57) can be fitted as follows:

```
> star.fit.2 <- snr(y~b1+
    exp(f2(x))*sin(2*pi*(b2+log(1+b4*x/b3)/b4)),
    func=list(f2(u)~list(~u,cubic2(u))),
    params=list(b1+b2+b3+b4~1), spar=''m'',
```

```
      start=list(params=c(7.1726,.4215,419.2645,-0.00144125),
       f=list(f2=log(diff(range(star.p.1$f1$fit[,1]))))),
      control=list(prec.out=.001,
                   rkpk.control=list(limnla=c(12,14))))
> summary(star.fit.2)
...
Coefficients:
      Value Std.Error   t-value p-value
b1   7.2004 0.0273582 263.1886       0
b2   0.3769 0.0143172  26.3278       0
b3 416.8002 0.5796450 719.0612       0
b4  -0.0011 0.0000617 -18.4201       0

GML estimate(s) of smoothing spline parameter(s):
 1.000000e+12
Equivalent Degrees of Freedom (DF) for spline function:
 9.829385
Residual standard error: 0.9093264
```

We used the logarithm of the range of $\hat{f}_1$ in model (8.56) as initial values for f_2. We also limited the search range for $\log_{10}(n\lambda)$, where n is the number of observations. To look at the general pattern of the amplitudes, we compute the amplitude for each period by fitting a sinusoidal model for each period of the folded data containing more than five observations. The logarithm of these estimated amplitudes and estimates of the amplitude functions based on models (8.56) and (8.57) are shown in Figure 8.12(a). Apparently, the amplitudes are underestimated based on these models.

Finally, we consider the following SNR model

$$y = \beta + \exp\{f_3(x)\} \sin\{2\pi f_4(x)\} + \epsilon, \qquad (8.58)$$

where both the amplitude and period functions are modeled nonparametrically. We model f_3 and f_4 using the cubic spline model space $W_2^2[0, b]$. Model (8.58) can be fitted as follows:

```
> co <- star.fit.2$coef
> f4ini <- co[2]+log(1+co[4]*x/co[3])/co[4]
> star.fit.3 <- snr(y~b+exp(f3(x))*sin(2*pi*f4(x)),
    func=list(f3(u)+f4(u)~list(~u,cubic2(u))),
    params=list(b~1), spar=''m'',
    start=list(params=c(7.1726),
               f=list(f3=star.fit.2$funcFitted,f4=f4ini)),
    control=list(prec.out=.001,
```

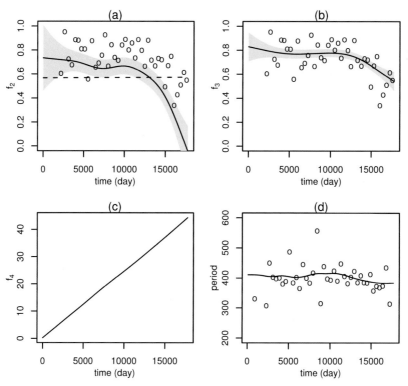

FIGURE 8.12 Star data, (a) estimated amplitudes based on folded data (circles), estimate of the amplitude function $a(x)$ in model (8.56) rescaled to match the amplitude function in model (8.57) (dashed line), estimate of the amplitude function $\exp(f_2(x))$ in model (8.57) (solid line), and 95% Bayesian confidence intervals (shaded region), all on logarithmic scale; (b) estimated amplitudes based on folded data (circles), estimate of the amplitude function $\exp(f_3(x))$ in model (8.58) (solid line), and 95% Bayesian confidence intervals (shaded region), all on logarithmic scale; (c) estimate of f_4 based on model (8.58) (solid line) and 95% Bayesian confidence intervals (shaded region); and (d) estimated periods based on the CLUSTER method (circles) and estimate of the period function $1/f_4'(x)$.

```
                    rkpk.control=list(limnla=c(12,14))))
> n <- length(y)
> rss <- c(sum((y-star.fit.1$fitted)**2),
          sum((y-star.fit.2$fitted)**2),
          sum((y-star.fit.3$fitted)**2))/n
```

```
> df <- c(star.fit.1$df$f+star.fit.1$df$para,
          star.fit.2$df$f+star.fit.2$df$para,
          star.fit.3$df$f+star.fit.3$df$para)
> aic <- n*log(rss)+2*df
> bic <- n*log(rss)+log(n)*df
> gcv <- rss/(1-df/n)**2
> print(round(rbind(aic, bic, gcv),4))
aic -164.0240 -202.6215 -1520.0960
bic -134.0823 -133.6093 -1197.6866
gcv    0.8598    0.8299     0.2476
```

We used the fitted values of the corresponding components from models (8.57) and (8.56) as initial values for f_3 and f_4. We also computed AIC, BIC, and GCV criteria for models (8.56), (8.57), and (8.58). Model (8.58) fits data much better, and the overall fit is shown in Figure 8.10. AIC, BIC, and GCV all select model (8.58). Estimates of functions f_3 and f_4 are shown in Figure 8.12(b)(c). The confidence intervals for f_4 are so narrow that they are indistinguishable from the estimate of f_4. The amplitude function f_3 fits data much better. To look at the general pattern of the periods, we first identify peaks using the CLUSTER method (Yang, Liu and Wang 2005). Observed periods are estimated as the lengths between peaks, and they are shown in Figure 8.12(d). The estimate of period function $1/f_4'$ in Figure 8.12(d) indicates that the evolution of the period may be nonlinear. By allowing a nonlinear period function, the model (8.58) leads to a much improved overall fit with less biased estimate of the amplitude function.

8.4.6 Circadian Rhythm

Many biochemical, physiological, and behavioral processes of living matters follow a roughly 24-hour cycle known as the *circadian rhythm*. We use the hormone data to illustrate how to fit an SIM to investigate circadian rhythms. The hormone data set contains cortisol concentration measured every 2 hours for a period of 24 hours from multiple subjects. In this section we use observations from normal subjects only. Cortisol concentrations on the $\log_{10}$ scale from nine normal subjects are shown in Figure 8.13.

It is usually assumed that there is a common shape function for all individuals. The time axis may be shifted and the magnitude of variation may differ between subjects; that is, there may be phase and amplitude differences between subjects. Therefore, we consider the following SIM

$$\text{conc}_{ij} = \beta_{i1} + \exp(\beta_{i2})f(\text{time}_{ij} - \text{alogit}(\beta_{i3})) + \epsilon_{ij},$$
$$i = 1,\ldots,m, \quad j = 1,\ldots,n_i, \tag{8.59}$$

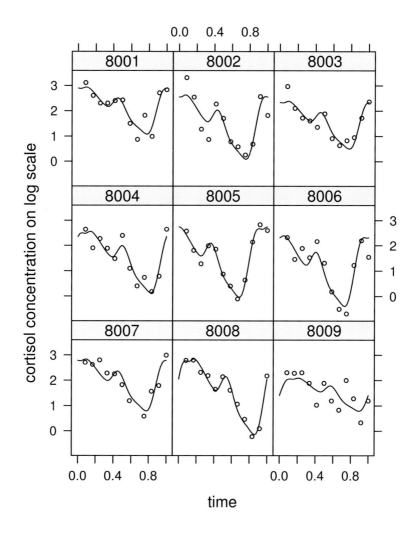

FIGURE 8.13 Hormone data, plots of cortisol concentrations (circles), and the fitted curves (solid lines) based on model (8.59). Subjects' ID are shown in the strip.

where m is the total number of subjects, n_i is the number of observations from subject i, $conc_{ij}$ is the cortisol concentration (on $\log_{10}$ scale) of the ith subject at the jth time point $time_{ij}$, β_{i1} is the 24-hour mean of the ith subject, $\exp(\beta_{i2})$ is the amplitude of the ith subject, $alogit(\beta_{i3}) = \exp(\beta_{3i})/\{1 + \exp(\beta_{i3})\}$ is the phase of the ith subject, and ϵ_{ij} are random errors. Note that the variable $time$ is transformed into the interval $[0, 1]$, the exponential transformation is used to enforce positive constraint on the amplitude, and the inverse logistic transformation *alogit* is used such that the phase is inside the interval $[0, 1]$. Comparing with the SIM model (8.33), there is no scale parameter β_4 in model (8.59) since the period is fixed to be 1. The function f is the common shape function. Since it is a periodic function with period 1, we model f using the trigonometric spline with $\mathcal{L} = D^2 + (2\pi)^2$ ($m = 2$ in (2.70)) and $f \in W_2^2(per) \ominus \{1\}$ where constant functions are removed from the model space to make f identifiable with β_{i1}. In order to make β_{i2} and β_{i3} identifiable with f, we add constraints: $\beta_{21} = \beta_{31} = 0$. Model (8.59) is an SNR model for clustered data. Assuming random errors are iid, model (8.59) can be fitted as follows:

```
> data(horm.cort)
> nor <- horm.cort[horm.cort$type==''normal'',]
> M <- model.matrix(~as.factor(ID), data=nor)
> nor.snr.fit1 <- snr(conc~b1+exp(b2)*f(time-alogit(b3)),
    func=f(u)~list(~sin(2*pi*u)+cos(2*pi*u)-1,
                lspline(u,type=''sine0'')),
    params=list(b1~M-1, b2+b3~M[,-1]-1),
    start=list(params=c(mean(nor$conc),rep(0,24))),
    data=nor, spar=''m'',
    control=list(prec.out=0.001,converg=''PRSS''))
```

Note that the second-stage models for parameters were specified by the `params` argument. We removed the first column in the design matrix M to satisfy the side condition $\beta_{21} = \beta_{31} = 0$. We used the option `converg=''PRSS''` instead of the default `converg=''COEF''` because this option usually requires fewer number of iterations. We compute fitted curves for all subjects evaluated at grid points:

```
> nor.grid <- data.frame(ID=rep(unique(nor$ID),rep(50,9)),
                    time=rep(seq(0,1,len=50),9))
> M <- model.matrix(~as.factor(ID), data=nor.grid)
> nor.snr.p <- predict(nor.snr.fit1,newdata=nor.grid)
```

Note that the matrix M needs to be generated again for the grid points. The fitted curves for all subjects are shown in Figure 8.13. We further compute the posterior means and standard deviations of f and its projection onto $W_2^2(per) \ominus \{1, \sin 2\pi x, \cos 2\pi x\}$, $P_1 f$, as follows:

```
> grid <- seq(0,1,len=100)
> nor.snr.p.f <- intervals(nor.snr.fit1,
    newdata=data.frame(u=grid),
    terms=list(f=matrix(c(1,1,1,0,0,1),nrow=2,byrow=T)))
```

The estimates of f and its projection $P_1 f$ are shown in Figure 8.14. It is obvious that $P_1 f$ is significantly different from zero. Thus a simple sinusoidal model may not be appropriate for this data.

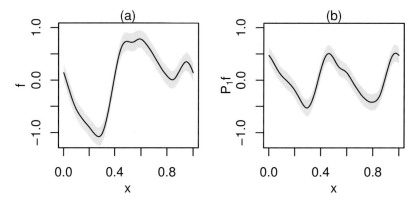

FIGURE 8.14 Hormone data, (a) estimate of f (solid line), and 95% Bayesian confidence intervals (shaded region); and (b) estimate of $P_1 f$ (solid line) and 95% Bayesian confidence intervals (shaded region).

Random errors in model (8.59) may be correlated. In the following we fit with an AR(1) within-subject correlation structure:

```
> M <- model.matrix(~as.factor(subject))
> nor.snr.fit2 <- update(nor.snr.fit1,
    cor=corAR1(form=~1|subject))
> summary(nor.snr.fit2)
...
Correlation Structure: AR(1)
 Formula: ~1 | subject
 Parameter estimate(s):
        Phi
-0.1557776
...
```

The lag 1 autocorrelation coefficient is small. We will further discuss how to deal with possible correlation within each subject in Section 9.4.5.

Chapter 9

Semiparametric Mixed-Effects Models

9.1 Linear Mixed-Effects Models

Mixed-effects models include both *fixed effects* and *random effects*, where random effects are usually introduced to model correlation within a cluster and/or spatial correlations. They provide flexible tools to model both the mean and the covariance structures simultaneously.

The simplest mixed-effects model is perhaps the classical two-way mixed model. Suppose A is a fixed factor with a levels, B is a random factor with b levels, and the design is balanced. The two-way mixed model assumes that

$$y_{ijk} = \mu + \alpha_i + \beta_j + (\alpha\beta)_{ij} + \epsilon_{ijk},$$
$$i = 1, \ldots, a; \quad j = 1, \ldots, b; \quad k = 1, \ldots, m, \tag{9.1}$$

where y_{ijk} is the kth observation at level i of factor A and level j of factor B, μ is the overall mean, α_i and β_j are main effects, $(\alpha\beta)_{ij}$ is the interaction, and ϵ_{ijk} are random errors. Since factor B is random, β_j and $(\alpha\beta)_{ij}$ are random effects. It is usually assumed that $\beta_j \stackrel{iid}{\sim} N(0, \sigma_b^2)$, $(\alpha\beta)_{ij} \stackrel{iid}{\sim} N(0, \sigma_{ab}^2)$, $\epsilon_{ijk} \stackrel{iid}{\sim} N(0, \sigma^2)$, and they are mutually independent.

Another simple mixed-effects model is the linear growth curve model. Suppose a response variable is measured repeatedly over a period of time or a sequence of doses from multiple individuals. Assume that responses across time or doses for each individual can be described by a simple straight line, while the intercepts and slopes for different individuals may differ. Then, one may consider the following linear growth curve model

$$y_{ij} = \beta_1 + \beta_2 x_{ij} + b_{1i} + b_{2i} x_{ij} + \epsilon_{ij},$$
$$i = 1, \ldots, m; \quad j = 1, \ldots, n_i, \tag{9.2}$$

where y_{ij} is the observation from individual i at time (or does) x_{ij}, β_1 and β_2 are population intercept and slope, b_{1i} and b_{2i} are random effects representing individual i's departures in intercept and slope from the population parameters, and ϵ_{ij} are random errors. Let $\boldsymbol{b}_i = (b_{1i}, b_{2i})^T$. It is usually assumed that $\boldsymbol{b}_i \overset{iid}{\sim} \mathrm{N}(\boldsymbol{0}, \sigma^2 D)$ for certain covariance matrix D, and random effects and random errors are mutually independent.

A general *linear mixed-effects* (LME) model assumes that

$$\boldsymbol{y} = S\boldsymbol{\beta} + Z\boldsymbol{b} + \boldsymbol{\epsilon}, \tag{9.3}$$

where $\boldsymbol{y}$ is an n-vector of observations on the response variable, S and Z are design matrices for fixed and random effects, respectively, $\boldsymbol{\beta}$ is a q_1-vector of unknown fixed effects, $\boldsymbol{b}$ is a q_2-vector of unobservable random effects, and $\boldsymbol{\epsilon}$ is an n-vector of random errors. We assume that $\boldsymbol{b} \sim \mathrm{N}(\boldsymbol{0}, \sigma^2 D)$, $\boldsymbol{\epsilon} \sim \mathrm{N}(\boldsymbol{0}, \sigma^2 \Lambda)$, and $\boldsymbol{b}$ and $\boldsymbol{\epsilon}$ are independent. Note that $\boldsymbol{y} \sim \mathrm{N}(S\boldsymbol{\beta}, \sigma^2(ZDZ^T + \Lambda))$. Therefore, the mean structure is modeled by the fixed effects, and the covariance structure is modeled by the random effects and random errors.

As discussed in Sections 3.5, 4.7, 5.2.2, and 7.3.3, smoothing spline estimates can be regarded as the BLUP estimates of the corresponding LME and NLME (nonlinear mixed-effects) models. Furthermore, the GML (generalized maximum likelihood) estimates of smoothing parameters are the REML (restricted maximum likelihood) estimates of variance components in the corresponding LME and NLME models. These connections between smoothing spline models and mixed-effects models will be utilized again in this chapter. In particular, details for the connection between semiparametric linear mixed-effects models and LME models will be given in Section 9.2.2.

9.2 Semiparametric Linear Mixed-Effects Models

9.2.1 The Model

A *semiparametric linear mixed-effects* (SLM) model assumes that

$$y_i = \boldsymbol{s}_i^T \boldsymbol{\beta} + \sum_{k=1}^{r} \mathcal{L}_{ki} f_k + \boldsymbol{z}_i^T \boldsymbol{b} + \epsilon_i, \quad i = 1, \ldots, n, \tag{9.4}$$

where $\boldsymbol{s}$ and $\boldsymbol{z}$ are independent variables for fixed and random effects respectively, $\boldsymbol{\beta}$ is a q_1-vector of parameters, $\boldsymbol{b}$ is a q_2-vector of random effects, $\mathcal{L}_{ki}$ are bounded linear functionals, f_k are unknown functions, and ϵ_i are random errors. For $k = 1, \ldots, r$, denote $\mathcal{X}_k$ as the domain of f_k and assume that $f_k \in \mathcal{H}_k$, where $\mathcal{H}_k$ is an RKHS on $\mathcal{X}_k$.

Let $\boldsymbol{y} = (y_1, \ldots, y_n)^T$, $S = (\boldsymbol{s}_1, \ldots, \boldsymbol{s}_n)^T$, $Z = (\boldsymbol{z}_1, \ldots, \boldsymbol{z}_n)^T$, $\boldsymbol{f} = (f_1, \ldots, f_r)$, $\boldsymbol{\gamma}(\boldsymbol{f}) = (\sum_{k=1}^{r} \mathcal{L}_{k1} f_k, \ldots, \sum_{k=1}^{r} \mathcal{L}_{kn} f_k)^T$, and $\boldsymbol{\epsilon} = (\epsilon_1, \ldots, \epsilon_n)^T$. Then model (9.4) can be written in the vector form

$$\boldsymbol{y} = S\boldsymbol{\beta} + \boldsymbol{\gamma}(\boldsymbol{f}) + Z\boldsymbol{b} + \boldsymbol{\epsilon}. \tag{9.5}$$

We assume that $\boldsymbol{b} \sim \mathrm{N}(\boldsymbol{0}, \sigma^2 D)$, $\boldsymbol{\epsilon} \sim \mathrm{N}(\boldsymbol{0}, \sigma^2 \Lambda)$, and they are mutually independent. It is clear that model (9.5) is an extension of the LME model with an additional term for nonparametric fixed effects. We note that the random effects are general. Stochastic processes including those based on smoothing splines can be used to construct models for random effects. Therefore, in a sense, the random effects may also be modeled nonparametrically. See Section 9.2.4 for an example.

For clustered data, an SLM model assumes that

$$y_{ij} = \boldsymbol{s}_{ij}^T \boldsymbol{\beta} + \sum_{k=1}^{r} \mathcal{L}_{kij} f_k + \boldsymbol{z}_{ij}^T \boldsymbol{b}_i + \epsilon_{ij}, \quad i = 1, \ldots, m; \ j = 1, \ldots, n_i, \tag{9.6}$$

where y_{ij} is the jth observation in cluster i, and $\boldsymbol{b}_i$ are random effects for cluster i. Let $n = \sum_{i=1}^{m} n_i$, $\boldsymbol{y}_i = (y_{i1}, \ldots, y_{in_i})^T$, $\boldsymbol{y} = (\boldsymbol{y}_1^T, \ldots, \boldsymbol{y}_m^T)^T$, $S_i = (\boldsymbol{s}_{i1}, \ldots, \boldsymbol{s}_{in_i})^T$, $S = (S_1^T, \ldots, S_m^T)^T$, $Z_i = (\boldsymbol{z}_{i1}, \ldots, \boldsymbol{z}_{in_i})^T$, $Z = \mathrm{diag}(Z_1, \ldots, Z_m)$, $\boldsymbol{b} = (\boldsymbol{b}_1^T, \ldots, \boldsymbol{b}_m^T)^T$, $\boldsymbol{\gamma}_i(\boldsymbol{f}) = (\sum_{k=1}^{r} \mathcal{L}_{ki1} f_k, \ldots, \sum_{k=1}^{r} \mathcal{L}_{kin_i} f_k)^T$, $\boldsymbol{\gamma}(\boldsymbol{f}) = (\boldsymbol{\gamma}_1^T(\boldsymbol{f}), \ldots, \boldsymbol{\gamma}_m^T(\boldsymbol{f}))^T$, $\boldsymbol{\epsilon}_i = (\epsilon_{i1}, \ldots, \epsilon_{in_i})^T$, and $\boldsymbol{\epsilon} = (\boldsymbol{\epsilon}_1^T, \ldots, \boldsymbol{\epsilon}_m^T)^T$. Then model (9.6) can be written in the same vector form as (9.5).

Similar SLM models may be constructed for multiple levels of grouping and other situations.

9.2.2 Estimation and Inference

For simplicity we present the estimation and inference procedures for the SLM model (9.4). The same methods apply to the SLM model (9.6) with a slight modification of notation.

We assume that the covariance matrices D and Λ depend on an unknown vector of covariance parameters $\boldsymbol{\tau}$. Our goal is to estimate $\boldsymbol{\beta}$, $\boldsymbol{f}$, $\boldsymbol{b}$, $\boldsymbol{\tau}$, and σ^2. Assume that $\mathcal{H}_k = \mathcal{H}_{k0} \oplus \mathcal{H}_{k1}$, where $\mathcal{H}_{k0} = \mathrm{span}\{\phi_{k1}, \ldots, \phi_{kp_k}\}$ and $\mathcal{H}_{k1}$ is an RKHS with RK R_{k1}. Let $\boldsymbol{\eta}(\boldsymbol{\beta}, \boldsymbol{f}) = S\boldsymbol{\beta} + \boldsymbol{\gamma}(\boldsymbol{f})$ and $W^{-1} = ZDZ^T + \Lambda$. The marginal distribution of $\boldsymbol{y}$ is $\boldsymbol{y} \sim \mathrm{N}(\boldsymbol{\eta}(\boldsymbol{\beta}, \boldsymbol{f}), \sigma^2 W^{-1})$. For fixed $\boldsymbol{\tau}$, we estimate $\boldsymbol{\beta}$ and $\boldsymbol{f}$ as minimizers of the PWLS (penalized weighted least squares)

$$\frac{1}{n}(\boldsymbol{y} - \boldsymbol{\eta}(\boldsymbol{\beta}, \boldsymbol{f}))^T W (\boldsymbol{y} - \boldsymbol{\eta}(\boldsymbol{\beta}, \boldsymbol{f})) + \lambda \sum_{k=1}^{r} \theta_k^{-1} \|P_{k1} f_k\|^2, \tag{9.7}$$

where P_{k1} is the projection operator onto $\mathcal{H}_{k1}$ in $\mathcal{H}_k$, and $\lambda\theta_k^{-1}$ are smoothing parameters. Note that the PWLS (9.7) has the same form as (8.5). Therefore, the results in Section 8.2.2 hold for the SLM models. In particular,

$$\hat{f}_k = \sum_{\nu=1}^{p_k} d_{k\nu}\phi_{k\nu} + \theta_k \sum_{i=1}^{n} c_i\xi_{ki}, \quad k = 1,\ldots,r, \tag{9.8}$$

where $\xi_{ki}(x) = \mathcal{L}_{ki(z)}R_{k1}(x,z)$.

For $k = 1,\ldots,r$, let $\boldsymbol{d}_k = (d_{k1},\ldots,d_{kp_k})^T$, $T_k = \{\mathcal{L}_{ki}\phi_{k\nu}\}_{i=1\,\nu=1}^{n\,\,p_k}$, and $\Sigma_k = \{\mathcal{L}_{ki}\mathcal{L}_{kj}R_{k1}\}_{i,j=1}^{n}$. Let $\boldsymbol{c} = (c_1,\ldots,c_n)^T$, $\boldsymbol{d} = (\boldsymbol{d}_1^T,\ldots,\boldsymbol{d}_r^T)^T$, $\boldsymbol{\alpha} = (\boldsymbol{\beta}^T,\boldsymbol{d}^T)^T$, $T = (T_1,\ldots,T_r)$, $X = (S\ T)$, $\boldsymbol{\theta} = (\theta_1,\ldots,\theta_r)$, and $\Sigma_{\boldsymbol{\theta}} = \sum_{k=1}^{r}\theta_k\Sigma_k$. We have $\mathcal{L}_{ki}\hat{f}_k = \sum_{\nu=1}^{p_k}d_{k\nu}\mathcal{L}_{ki}\phi_{k\nu}+\theta_k\sum_{j=1}^{n}c_j\mathcal{L}_{ki}\xi_{kj}$ and $\hat{\boldsymbol{f}}_k = (\mathcal{L}_{k1}\hat{f}_k,\ldots,\mathcal{L}_{kn}\hat{f}_k)^T = T_k\boldsymbol{d}_k + \theta_k\Sigma_k\boldsymbol{c}$. Consequently,

$$\boldsymbol{\gamma}(\boldsymbol{f}) = \sum_{k=1}^{r}(T_k\boldsymbol{d}_k + \theta_k\Sigma_k\boldsymbol{c}) = T\boldsymbol{d} + \Sigma_{\boldsymbol{\theta}}\boldsymbol{c}$$

and

$$\boldsymbol{\eta}(\boldsymbol{\beta},\boldsymbol{f}) = S\boldsymbol{\beta} + T\boldsymbol{d} + \Sigma_{\boldsymbol{\theta}}\boldsymbol{c} = X\boldsymbol{\alpha} + \Sigma_{\boldsymbol{\theta}}\boldsymbol{c}.$$

Furthermore,

$$\sum_{k=1}^{r}\theta_k^{-1}\|P_{k1}\hat{f}_k\|^2 = \sum_{k=1}^{r}\theta_k\boldsymbol{c}^T\Sigma_k\boldsymbol{c} = \boldsymbol{c}^T\Sigma_{\boldsymbol{\theta}}\boldsymbol{c}.$$

Then the PWLS (9.7) reduces to

$$(\boldsymbol{y} - X\boldsymbol{\alpha} - \Sigma_{\boldsymbol{\theta}}\boldsymbol{c})^T W(\boldsymbol{y} - X\boldsymbol{\alpha} - \Sigma_{\boldsymbol{\theta}}\boldsymbol{c}) + n\lambda\boldsymbol{c}^T\Sigma_{\boldsymbol{\theta}}\boldsymbol{c}. \tag{9.9}$$

Differentiating (9.9) with respect to $\boldsymbol{\alpha}$ and $\boldsymbol{c}$, we have

$$\begin{aligned} X^T WX\boldsymbol{\alpha} + X^T W\Sigma_{\boldsymbol{\theta}}\boldsymbol{c} &= X^T W\boldsymbol{y}, \\ \Sigma_{\boldsymbol{\theta}}WX\boldsymbol{\alpha} + (\Sigma_{\boldsymbol{\theta}}W\Sigma_{\boldsymbol{\theta}} + n\lambda\Sigma_{\boldsymbol{\theta}})\boldsymbol{c} &= \Sigma_{\boldsymbol{\theta}}W\boldsymbol{y}. \end{aligned} \tag{9.10}$$

We estimate random effects by (Wang 1998a)

$$\hat{\boldsymbol{b}} = DZ^T W(\boldsymbol{y} - X\boldsymbol{\alpha} - \Sigma_{\boldsymbol{\theta}}\boldsymbol{c}). \tag{9.11}$$

Henderson (hierarchical) likelihood is often used to justify the estimation of fixed and random effects in an LME model (Robinson 1991, Lee, Nelder and Pawitan 2006). We now show that the above PWLS estimates of $\boldsymbol{\beta}$, $\boldsymbol{f}$, and $\boldsymbol{b}$ can also be derived from the following penalized

Henderson (hierarchical) likelihood of $\boldsymbol{y}$ and $\boldsymbol{b}$

$$(\boldsymbol{y}-\boldsymbol{\eta}(\boldsymbol{\beta},\boldsymbol{f})-Z\boldsymbol{b})^T\Lambda^{-1}(\boldsymbol{y}-\boldsymbol{\eta}(\boldsymbol{\beta},\boldsymbol{f})-Z\boldsymbol{b})+\boldsymbol{b}^T D^{-1}\boldsymbol{b}+n\lambda\sum_{k=1}^{r}\theta_k^{-1}\|P_{k1}f_k\|^2,$$
(9.12)

where, ignoring some constant terms, the first two components in (9.12) equal twice the negative logarithm of the joint density function of $\boldsymbol{y}$ and $\boldsymbol{b}$. Again, the solution of $\boldsymbol{f}$ in (9.12) can be represented by (9.8). Then the penalized Henderson (hierarchical) likelihood (9.12) reduces to

$$(\boldsymbol{y}-X\boldsymbol{\alpha}-\Sigma_{\boldsymbol{\theta}}\boldsymbol{c}-Z\boldsymbol{b})^T\Lambda^{-1}(\boldsymbol{y}-X\boldsymbol{\alpha}-\Sigma_{\boldsymbol{\theta}}\boldsymbol{c}-Z\boldsymbol{b})+\boldsymbol{b}^T D^{-1}\boldsymbol{b}+n\lambda\boldsymbol{c}^T\Sigma_{\boldsymbol{\theta}}\boldsymbol{c}.$$
(9.13)

Differentiating (9.13) with respect to $\boldsymbol{\alpha}$, $\boldsymbol{c}$, and $\boldsymbol{b}$, we have

$$X^T\Lambda^{-1}X\boldsymbol{\alpha} + X^T\Lambda^{-1}\Sigma_{\boldsymbol{\theta}}\boldsymbol{c} + X^T\Lambda^{-1}Z\boldsymbol{b} = X^T\Lambda^{-1}\boldsymbol{y},$$

$$\Sigma_{\boldsymbol{\theta}}\Lambda^{-1}X\boldsymbol{\alpha} + (\Sigma_{\boldsymbol{\theta}}\Lambda^{-1}\Sigma_{\boldsymbol{\theta}} + n\lambda\Sigma_{\boldsymbol{\theta}})\boldsymbol{c} + \Sigma_{\boldsymbol{\theta}}\Lambda^{-1}Z\boldsymbol{b} = \Sigma_{\boldsymbol{\theta}}\Lambda^{-1}\boldsymbol{y}, \quad (9.14)$$

$$Z^T\Lambda^{-1}X\boldsymbol{\alpha} + Z^T\Lambda^{-1}\Sigma_{\boldsymbol{\theta}}\boldsymbol{c} + (D^{-1} + Z^T\Lambda^{-1}Z)\boldsymbol{b} = Z^T\Lambda^{-1}\boldsymbol{y}.$$

It is easy to check that

$$W = \Lambda^{-1} - \Lambda^{-1}Z(Z^T\Lambda^{-1}Z + D^{-1})^{-1}Z^T\Lambda^{-1}.$$
(9.15)

Since $I = (ZDZ^T + \Lambda)W = ZDZ^TW + \Lambda W$, then

$$ZDZ^TW = I - \Lambda\{\Lambda^{-1} - \Lambda^{-1}Z(D^{-1} + Z^T\Lambda^{-1}Z)^{-1}Z^T\Lambda^{-1}\}$$
$$= Z(D^{-1} + Z^T\Lambda^{-1}Z)^{-1}Z^T\Lambda^{-1}.$$
(9.16)

From (9.11) and (9.16), we have

$$Z\boldsymbol{b} = Z(D^{-1} + Z^T\Lambda^{-1}Z)^{-1}Z^T\Lambda^{-1}(\boldsymbol{y} - X\boldsymbol{\alpha} - \Sigma_{\boldsymbol{\theta}}\boldsymbol{c}).$$
(9.17)

From (9.15) and (9.17), the first equation in (9.10) is equivalent to

$$0 = X^TWX\boldsymbol{\alpha} + X^TW\Sigma_{\boldsymbol{\theta}}\boldsymbol{c} - X^TW\boldsymbol{y}$$
$$= X^T\Lambda^{-1}X\boldsymbol{\alpha} + X^T\Lambda^{-1}\Sigma_{\boldsymbol{\theta}}\boldsymbol{c} - X^T\Lambda^{-1}\boldsymbol{y}$$
$$\quad + X^T\Lambda^{-1}Z(D^{-1} + Z^T\Lambda^{-1}Z)^{-1}Z^T\Lambda^{-1}(\boldsymbol{y} - X\boldsymbol{\alpha} - \Sigma_{\boldsymbol{\theta}}\boldsymbol{c})$$
$$= X^T\Lambda^{-1}X\boldsymbol{\alpha} + X^T\Lambda^{-1}\Sigma_{\boldsymbol{\theta}}\boldsymbol{c} - X^T\Lambda^{-1}\boldsymbol{y} + X^T\Lambda^{-1}Z\boldsymbol{b},$$

which is the same as the first equation in (9.14). Similarly, the second equation in (9.10) is equivalent to the second equation in (9.14). Equation (9.11) is equivalent to

$$D^{-1}\boldsymbol{b} = Z^T\Lambda^{-1}(\boldsymbol{y} - X\boldsymbol{\alpha} - \Sigma_{\boldsymbol{\theta}}\boldsymbol{c})$$
$$\quad - Z^T\Lambda^{-1}Z(D^{-1} + Z^T\Lambda^{-1}Z)^{-1}Z^T\Lambda^{-1}(\boldsymbol{y} - X\boldsymbol{\alpha} - \Sigma_{\boldsymbol{\theta}}\boldsymbol{c})$$
$$= Z^T\Lambda^{-1}(\boldsymbol{y} - X\boldsymbol{\alpha} - \Sigma_{\boldsymbol{\theta}}\boldsymbol{c}) - Z^T\Lambda^{-1}Z\boldsymbol{b},$$

which is the same as the third equation in (9.14). Therefore, the PWLS estimates of $\boldsymbol{\beta}$, $\boldsymbol{f}$, and $\boldsymbol{b}$ based on (9.7) and (9.11) can be regarded as the penalized Henderson (hierarchical) likelihood estimates.

Let $Z_s = (I_n, \ldots, I_n)$, where I_n denotes an $n \times n$ identity matrix. Consider the following LME model

$$\boldsymbol{y} = S\boldsymbol{\beta} + T\boldsymbol{d} + \sum_{k=1}^{r} \boldsymbol{u}_k + Z\boldsymbol{b} + \boldsymbol{\epsilon} = X\boldsymbol{\alpha} + Z_s\boldsymbol{u} + Z\boldsymbol{b} + \boldsymbol{\epsilon}, \quad (9.18)$$

where $\boldsymbol{\beta}$ and $\boldsymbol{d}$ are fixed effects, $\boldsymbol{\alpha} = (\boldsymbol{\beta}^T, \boldsymbol{d}^T)^T$, $\boldsymbol{u}_k$ are random effects with $\boldsymbol{u}_k \sim \mathrm{N}(\boldsymbol{0}, \sigma^2 \theta_k \Sigma_k / n\lambda)$, $\boldsymbol{u} = (\boldsymbol{u}_1^T, \ldots, \boldsymbol{u}_r^T)^T$, $\boldsymbol{b}$ are random effects with $\boldsymbol{b} \sim \mathrm{N}(\boldsymbol{0}, \sigma^2 D)$, $\boldsymbol{\epsilon}$ are random errors with $\boldsymbol{\epsilon} \sim \mathrm{N}(\boldsymbol{0}, \sigma^2 \Lambda)$, and random effects and random errors are mutually independent. Let $D_s = \mathrm{diag}(\theta_1 \Sigma_1, \ldots, \theta_r \Sigma_r)$ and $\tilde{\boldsymbol{b}} = (\boldsymbol{u}^T, \boldsymbol{b}^T)^T$. Write

$$\sigma^{-2}\mathrm{Cov}(\tilde{\boldsymbol{b}}) = \mathrm{diag}\left((n\lambda)^{-1}D_s, D\right)$$
$$= \left\{I_{nr+q_2}\right\}\left\{\mathrm{diag}(D_s, D)\right\}\left\{\mathrm{diag}\left((n\lambda)^{-1}I_{nr}, I_{q_2}\right)\right\}.$$

Then equation (3.3) in Harville (1976) can be written as

$$X^T\Lambda^{-1}X\boldsymbol{\alpha} + X^T\Lambda^{-1}Z_sD_s\boldsymbol{\phi}_1 + X^T\Lambda^{-1}ZD\boldsymbol{\phi}$$
$$= X^T\Lambda^{-1}\boldsymbol{y},$$
$$D_sZ_s^T\Lambda^{-1}X\boldsymbol{\alpha} + (n\lambda D_s + D_sZ_s^T\Lambda^{-1}Z_sD_s)\boldsymbol{\phi}_1 + D_sZ_s^T\Lambda^{-1}ZD\boldsymbol{\phi}$$
$$= D_sZ_s^T\Lambda^{-1}\boldsymbol{y},$$
$$DZ^T\Lambda^{-1}X\boldsymbol{\alpha} + DZ^T\Lambda^{-1}Z_sD_s\boldsymbol{\phi}_1 + (D + DZ^T\Lambda^{-1}ZD)\boldsymbol{\phi}$$
$$= DZ^T\Lambda^{-1}\boldsymbol{y}.$$
$$(9.19)$$

Suppose $\boldsymbol{\alpha}$, $\boldsymbol{c}$, and $\boldsymbol{b}$ are solutions to (9.14). Note that $Z_sD_sZ_s^T = \Sigma_{\boldsymbol{\theta}}$. When $\Sigma_{\boldsymbol{\theta}}$ is invertible, multiplying $D_sZ_s^T\Sigma_{\boldsymbol{\theta}}^{-1}$ to both side of the second equation in (9.14), it is not difficult to see that $\boldsymbol{\alpha}$, $\boldsymbol{\phi}_1 = Z_s^T\boldsymbol{c}$, and $\boldsymbol{\phi} = D^{-1}\boldsymbol{b}$ are solutions to (9.19). From Theorem 2 in Harville (1976), the linear system (9.19) is consistent, and the BLUP estimate of $\boldsymbol{u}$ is $\hat{\boldsymbol{u}} = D_s\boldsymbol{\phi}_1 = D_sZ_s^T\boldsymbol{c} = (\theta_1\boldsymbol{c}^T\Sigma_1, \ldots, \theta_r\boldsymbol{c}^T\Sigma_r)^T$. Therefore, the PWLS estimate of each component, $\hat{\boldsymbol{u}}_k = \theta_k\Sigma_k\boldsymbol{c}$, is a BLUP. When $\Sigma_{\boldsymbol{\theta}}$ is not invertible, consider $Z_s\boldsymbol{u} = \sum_{k=1}^{r}\boldsymbol{u}_k$ instead of each individual $\boldsymbol{u}_k$. Letting $\tilde{\boldsymbol{b}} = (\boldsymbol{u}^TZ_s^T, \boldsymbol{b}^T)^T$ and following the same arguments, it can be shown that the overall fit $\Sigma_{\boldsymbol{\theta}}\boldsymbol{c}$ is a BLUP. Similarly, the estimate of $\boldsymbol{b}$ is also a BLUP.

We now discuss how to estimate smoothing parameters λ and $\boldsymbol{\theta}$ and variance–covariance parameters σ^2 and $\boldsymbol{\tau}$. Let the QR decomposition of

X be

$$X = (Q_1 \ Q_2) \begin{pmatrix} R \\ 0 \end{pmatrix}.$$

Consider the LME model (9.18) and orthogonal contrast $\boldsymbol{w}_1 = Q_2^T \boldsymbol{y}$. Then $\boldsymbol{w}_1 \sim N(\boldsymbol{0}, \delta Q_2^T M Q_2)$, where $\delta = \sigma^2/n\lambda$ and $M = \Sigma_{\boldsymbol{\theta}} + n\lambda W^{-1}$. The restricted likelihood based on $\boldsymbol{w}_1$ is given in (3.37) with a different M defined in this Section. Following the same arguments as in Section 3.6, we have the GML criterion for λ, $\boldsymbol{\theta}$ and $\boldsymbol{\tau}$ as

$$\mathrm{GML}(\lambda, \boldsymbol{\theta}, \boldsymbol{\tau}) = \frac{\boldsymbol{w}_1^T (Q_2^T M Q_2)^{-1} \boldsymbol{w}_1}{\{\det(Q_2^T M Q_2)^{-1}\}^{\frac{1}{n - q_1 - p}}}, \tag{9.20}$$

where $p = \sum_{k=1}^r p_k$. Similar to (3.41), the REML (GML) estimate of σ^2 is

$$\hat{\sigma}^2 = \frac{n\hat{\lambda} \boldsymbol{w}_1^T (Q_2^T \hat{M} Q_2)^{-1} \boldsymbol{w}_1}{n - q_1 - p}, \tag{9.21}$$

where $\hat{M} = \Sigma_{\hat{\boldsymbol{\theta}}} + n\hat{\lambda} \hat{W}^{-1}$, and $\hat{W}^{-1}$ is the estimate of W^{-1} with $\boldsymbol{\tau}$ being replaced by $\hat{\boldsymbol{\tau}}$.

For clustered data, the leaving-out-one-cluster approach presented in Section 5.2.2 may also be used to estimate the smoothing and variance-covariance parameters.

The SLM model (9.4) reduces to the semiparametric linear regression model (8.4) when the random effects are combined with random errors. Therefore, the methods described in Section 8.2.2 can be used to draw inference about $\boldsymbol{\beta}$ and $\boldsymbol{f}$. Specifically, posterior means and standard deviations can be calculated using formulae (8.17) and (8.18) with $W^{-1} = ZDZ^T + \Lambda$ fixed at its estimate. Bayesian confidence intervals for the overall functions and their components can then be constructed as in Section 8.2.2. Covariances for random effects can be computed using Theorem 1 in Wang (1998a). The bootstrap method may also be used to construct confidence intervals.

9.2.3 The `slm` Function

The connections between SLM models and LME models suggest a relatively simple approach to fitting SLM models using existing software for LME models. Specifically, for $k = 1, \ldots, r$, let $\Sigma_k = Z_{sk} Z_{sk}^T$ be the Cholesky decomposition, where Z_{sk} is a $n \times m_k$ matrix with $m_k = \mathrm{rank}(\Sigma_k)$. Consider the following LME model

$$\boldsymbol{y} = S\boldsymbol{\beta} + T\boldsymbol{d} + \sum_{k=1}^r Z_{sk} \boldsymbol{b}_{sk} + Z\boldsymbol{b} + \boldsymbol{\epsilon}, \tag{9.22}$$

where $\boldsymbol{\beta}$ and $\boldsymbol{d}$ are fixed effects, $\boldsymbol{b}_{sk}$ are random effects with $\boldsymbol{b}_{sk} \sim$ N$(\boldsymbol{0}, \sigma^2\theta_k I_{m_k}/n\lambda)$, $\boldsymbol{b}$ are random effects with $\boldsymbol{b} \sim$ N$(\boldsymbol{0}, \sigma^2 D)$, $\boldsymbol{\epsilon}$ are random errors with $\boldsymbol{\epsilon} \sim$ N$(\boldsymbol{0}, \sigma^2\Lambda)$, and random effects and random errors are mutually independent. Following the same arguments in Section 9.2.2, it can be shown that the PWLS estimates of parameters and nonparametric functions in the SLM model (9.5) correspond to the BLUP estimates in the LME model (9.22). Furthermore, the GML estimates of smoothing and covariance parameters in (9.5) correspond to the REML estimates of covariance parameters in (9.22). Therefore, the PWLS estimates of $\boldsymbol{\beta}$ and $\boldsymbol{f}$, and GML estimates of λ, $\boldsymbol{\theta}$ and $\boldsymbol{\tau}$ in (9.5), can be calculated by fitting the LME model (9.22) with covariance parameters calculated by the REML method. This approach is implemented by the slm function in the assist package where the LME model was fitted using the lme function in the nlme library.

A typical call to the slm function is

```
slm(formula, rk, random)
```

where formula and rk serve the same purposes as those in the ssr function. Combined, they specify the fixed effects. The random argument specifies the random effects the same way as in lme. An object of slm class is returned. The generic function summary can be applied to extract further information. Predictions on different levels of random effects can be computed using the predict function where the nonparametric functions are treated as part of fixed effects. Posterior means and standard deviations for $\boldsymbol{\beta}$ and $\boldsymbol{f}$ can be computed using the intervals function. Examples can be found in Section 9.4.

9.2.4   SS ANOVA Decomposition

We have shown how to build multiple regression models using SS ANOVA decompositions in Chapter 4. The resulting SS ANOVA models have certain modular structures that parallel the classical ANOVA decompositions. In this section we show how to construct similar SS ANOVA decompositions with modular structures that parallel the classical mixed models. We illustrate how to construct SS ANOVA decompositions through two examples. More examples of SS ANOVA decompositions involving random effects can be found in Wang (1998a), Wang and Wahba (1998), and Section 9.4.3. As a general approach, the SS ANOVA decomposition may be employed to build mixed-effects models for other situations.

It is instructive to see how the classical two-way mixed model (9.1) can be derived via an SS ANOVA decomposition. In general, the factor B is considered random since the levels of the factor are chosen at random from a well-defined population of all factor levels. It is of interest to draw an inference about the general population using information from these observed (chosen) levels. Let $\mathcal{X}_1 = \{1, \ldots, a\}$ be the domain of factor A, and $\mathcal{X}_2$ be the population from which the levels of the random factor B are drawn. Assume the following model

$$y_{iwk} = f(i, w) + \epsilon_{iwk}, \quad i \in \mathcal{X}_1; \quad w \in \mathcal{X}_2; \quad k = 1, \ldots, m, \quad (9.23)$$

where $f(i, w)$ is a random variable since w is a random sample from $\mathcal{X}_2$. $f(i, j)$ for $j = 1, \ldots, b$ are realizations of the true mean function defined on $\mathcal{X}_1 \times \mathcal{X}_2$. Let P be the sampling distribution on $\mathcal{X}_2$. Define averaging operators $\mathcal{A}_1^{(1)}$ on $\mathcal{X}_1$ and $\mathcal{A}_1^{(2)}$ on $\mathcal{X}_2$ as

$$\mathcal{A}_1^{(1)} f = \frac{1}{a} \sum_{i=1}^{a} f(i, \cdot),$$

$$\mathcal{A}_1^{(2)} f = \int_{\mathcal{X}_2} f(\cdot, w) dP.$$

$\mathcal{A}_1^{(2)}$ computes population average with respect to the sampling distribution. Let $\mathcal{A}_2^{(1)} = I - \mathcal{A}_1^{(1)}$ and $\mathcal{A}_2^{(2)} = I - \mathcal{A}_1^{(2)}$. An SS ANOVA decomposition can be defined as

$$
\begin{aligned}
f &= \{\mathcal{A}_1^{(1)} + \mathcal{A}_2^{(1)}\}\{\mathcal{A}_1^{(2)} + \mathcal{A}_2^{(2)}\} f \\
&= \mathcal{A}_1^{(1)} \mathcal{A}_1^{(2)} f + \mathcal{A}_2^{(1)} \mathcal{A}_1^{(2)} f + \mathcal{A}_1^{(1)} \mathcal{A}_2^{(2)} f + \mathcal{A}_2^{(1)} \mathcal{A}_2^{(2)} f \\
&\triangleq \mu + \alpha_i + \beta_w + (\alpha\beta)_{iw}.
\end{aligned}
$$

Therefore, the SS ANOVA decomposition leads to the same structure as the classical two-way mixed model (9.1).

Next we discuss how to derive the SS ANOVA decomposition for repeated measures data. As in Section 9.1, suppose we have repeated measurements on a response variable over a period of time or a sequence of doses from multiple individuals. Suppose that individuals are selected at random from a well-defined population $\mathcal{X}_1$. Without loss of generality, denote the time period or dose range as $\mathcal{X}_2 = [0, 1]$. The linear growth curve model (9.2) assumes that responses across time or doses for each individual can be well described by a simple straight line, which may be too restrictive for some applications. Assume the following model

$$y_{wj} = f(w, x_{wj}) + \epsilon_{wj}, \quad w \in \mathcal{X}_1; \quad x_{wj} \in \mathcal{X}_2, \ j = 1, \ldots, n_w, \quad (9.24)$$

where $f(w, x_{wj})$ are random variables since w are random samples from $\mathcal{X}_1$. $f(i, x_{ij})$, $i = 1, \ldots, m$, $j = 1, \ldots, n_i$ are realizations of the true mean functions defined on $\mathcal{X}_1 \times \mathcal{X}_2$. Suppose we want to model the mean function nonparametrically using the cubic spline space $W_2^2[0,1]$ under the construction in Section 2.2. Let P be the sampling distribution on $\mathcal{X}_1$. Define averaging operators

$$\mathcal{A}_1^{(1)}f = \int_{\mathcal{X}_1} f(w,\cdot)dP,$$
$$\mathcal{A}_1^{(2)}f = f(\cdot,0),$$
$$\mathcal{A}_2^{(2)}f = f'(\cdot,0)x.$$

Let $\mathcal{A}_2^{(1)} = I - \mathcal{A}_1^{(1)}$ and $\mathcal{A}_3^{(2)} = I - \mathcal{A}_1^{(2)} - \mathcal{A}_2^{(2)}$. Then

$$\begin{aligned}
f &= \{\mathcal{A}_1^{(1)} + \mathcal{A}_2^{(1)}\}\{\mathcal{A}_1^{(2)} + \mathcal{A}_2^{(2)} + \mathcal{A}_3^{(2)}\}f \\
&= \mathcal{A}_1^{(1)}\mathcal{A}_1^{(2)}f + \mathcal{A}_1^{(1)}\mathcal{A}_2^{(2)}f + \mathcal{A}_1^{(1)}\mathcal{A}_3^{(2)}f + \\
&\quad + \mathcal{A}_2^{(1)}\mathcal{A}_1^{(2)}f + \mathcal{A}_2^{(1)}\mathcal{A}_2^{(2)}f + \mathcal{A}_2^{(1)}\mathcal{A}_3^{(2)}f \\
&= \beta_1 + \beta_2 x + f_2(x) + b_{1w} + b_{2w}x + f_{1,2}(w,x), \qquad (9.25)
\end{aligned}$$

where the first three terms are fixed effects representing the population mean function, and the last three terms are random effects representing the departure of individual w from the population mean function. Since both the first and the last three terms are orthogonal components in $W_2^2[0,1]$, then both the population mean function and the individual departure are modeled by cubic splines.

Based on the SS ANOVA decomposition (9.25), we may consider the following model for observed data

$$y_{ij} = \beta_1 + \beta_2 x_{ij} + f_2(x_{ij}) + b_{1i} + b_{2i}x_{ij} + f_{1,2}(i,x_{ij}) + \epsilon_{ij},$$
$$i = 1,\ldots,m; \quad j = 1,\ldots,n_i.$$

Let $\boldsymbol{b}_i = (b_{1i}, b_{2i})^T$ and assume that $\boldsymbol{b}_i \overset{iid}{\sim} N(0, \sigma^2 D)$. One possible model for the nonparametric random effects $f_{1,2}$ is to assume that $f_{1,2}(i,x)$ is a stochastic process on $[0,1]$ with mean zero and covariance function $\sigma_1^2 R_1(x,y)$, where $R_1(x,y)$ is the RK of $W_2^2[0,1] \ominus \{1,x\}$ defined in (2.4). It is obvious that the linear growth curve model (9.2) is a special case of model (9.25) with $f_2 = 0$ and $\sigma_1^2 = 0$.

9.3 Semiparametric Nonlinear Mixed-Effects Models

9.3.1 The Model

Nonlinear mixed-effects (NLME) models extend LME models by allowing the regression function to depend on fixed and random effects through a nonlinear function. Consider the following NLME model proposed by Lindstrom and Bates (1990) for clustered data:

$$y_{ij} = \psi(\boldsymbol{\phi}_{ij}; \boldsymbol{x}_{ij}) + \epsilon_{ij}, \quad i = 1, \ldots, m; \quad j = 1, \ldots, n_i,$$
$$\boldsymbol{\phi}_{ij} = S_{ij}\boldsymbol{\beta} + Z_{ij}\boldsymbol{b}_i, \quad \boldsymbol{b}_i \stackrel{iid}{\sim} \mathrm{N}(\boldsymbol{0}, \sigma^2\tilde{D}), \tag{9.26}$$

where m is the number of clusters, n_i is the number of observations from the ith cluster, y_{ij} is the jth observation in cluster i, ψ is a known function of a covariate $\boldsymbol{x}$, $\boldsymbol{\phi}_{ij}$ is a q-vector of parameters, ϵ_{ij} are random errors, S_{ij} and Z_{ij} are design matrices for fixed and random effects respectively, $\boldsymbol{\beta}$ is a q_1-vector of population parameters (fixed effects), and $\boldsymbol{b}_i$ is a q_2-vector of random effects for cluster i. Let $\boldsymbol{\epsilon}_i = (\epsilon_{i1}, \ldots, \epsilon_{in_i})^T$. We assume that $\boldsymbol{\epsilon}_i \sim \mathrm{N}(\boldsymbol{0}, \sigma^2\Lambda_i)$, $\boldsymbol{b}_i$ and $\boldsymbol{\epsilon}_i$ are mutually independent, and observations from different clusters are independent.

The first-stage model in (9.26) relates the conditional mean of the response variable to the covariate $\boldsymbol{x}$ and parameters $\boldsymbol{\phi}_{ij}$. The second-stage model relates parameters $\boldsymbol{\phi}_{ij}$ to fixed and random effects. Covariate effects can be incorporated into the second-stage model.

As an extension of the NLME model, Ke and Wang (2001) proposed the following class of *semiparametric nonlinear mixed-effects* (SNM) models:

$$y_{ij} = \mathcal{N}_{ij}(\boldsymbol{\phi}_{ij}, \boldsymbol{f}) + \epsilon_{ij}, \quad i = 1, \ldots, m; \quad j = 1, \ldots, n_i,$$
$$\boldsymbol{\phi}_{ij} = S_{ij}\boldsymbol{\beta} + Z_{ij}\boldsymbol{b}_i, \quad \boldsymbol{b}_i \stackrel{iid}{\sim} \mathrm{N}(\boldsymbol{0}, \sigma^2\tilde{D}), \tag{9.27}$$

where $\boldsymbol{\phi}_{ij}$ is a q-vector of parameters, $\boldsymbol{\beta}$ is a q_1-vector of fixed effects, $\boldsymbol{b}_i$ is a q_2-vector of random effects for cluster i, $\boldsymbol{f} = (f_1, \ldots, f_r)$ are unknown functions, f_k belongs to an RKHS $\mathcal{H}_k$ on an arbitrary domain $\mathcal{X}_k$ for $k = 1, \ldots, r$, and $\mathcal{N}_{ij}$ are known nonlinear functionals on $\mathbb{R}^q \times \mathcal{H}_1 \times \cdots \times \mathcal{H}_r$. As the NLME model, we assume that $\boldsymbol{\epsilon}_i = (\epsilon_{i1}, \ldots, \epsilon_{in_i})^T \sim \mathrm{N}(\boldsymbol{0}, \sigma^2\Lambda_i)$, $\boldsymbol{b}_i$ and $\boldsymbol{\epsilon}_i$ are mutually independent, and observations from different clusters are independent.

It is clear that the SNM model (9.27) is an extension of the SNR model (8.30) with an additional mixed-effects second-stage model. Similar to the SNR model, certain constraints may be required to make an

SNM model identifiable, and often these constraints can be achieved by removing certain components from the model spaces for parameters and $\boldsymbol{f}$. The SLM model (9.26) is a special case when $\mathcal{N}$ is linear in both $\boldsymbol{\phi}$ and $\boldsymbol{f}$.

Let $n = \sum_{i=1}^{m} n_i$, $\boldsymbol{y}_i = (y_{i1}, \ldots, y_{in_i})^T$, $\boldsymbol{y} = (\boldsymbol{y}_1^T, \ldots, \boldsymbol{y}_m^T)^T$, $\boldsymbol{\phi}_i = (\boldsymbol{\phi}_{i1}^T, \ldots, \boldsymbol{\phi}_{in_i}^T)^T$, $\boldsymbol{\phi} = (\boldsymbol{\phi}_1^T, \ldots, \boldsymbol{\phi}_m^T)^T$, $\boldsymbol{\eta}_i(\boldsymbol{\phi}_i, \boldsymbol{f}) = (\mathcal{N}_{i1}(\boldsymbol{\phi}_{i1}, \boldsymbol{f}), \ldots, \mathcal{N}_{in_i}(\boldsymbol{\phi}_{in_i}, \boldsymbol{f}))^T$, $\boldsymbol{\eta}(\boldsymbol{\phi}, \boldsymbol{f}) = (\boldsymbol{\eta}_1^T(\boldsymbol{\phi}_1, \boldsymbol{f}), \ldots, \boldsymbol{\eta}_m^T(\boldsymbol{\phi}_m, \boldsymbol{f}))^T$, $\boldsymbol{\epsilon} = (\boldsymbol{\epsilon}_1^T, \ldots, \boldsymbol{\epsilon}_m^T)^T$, $\boldsymbol{b} = (\boldsymbol{b}_1^T, \ldots, \boldsymbol{b}_m^T)^T$, $\Lambda = \text{diag}(\Lambda_1, \ldots, \Lambda_m)$, $S = (S_{11}^T, \ldots, S_{1n_1}^T, \ldots, S_{m1}^T, \ldots, S_{mn_m}^T)^T$, $Z_i = (Z_{i1}^T, \ldots, Z_{in_i}^T)^T$, $Z = \text{diag}(Z_1, \ldots, Z_m)$, and $D = \text{diag}(\tilde{D}, \ldots, \tilde{D})$. Then model (9.27) can be written in the vector form

$$
\begin{aligned}
\boldsymbol{y} &= \boldsymbol{\eta}(\boldsymbol{\phi}, \boldsymbol{f}) + \boldsymbol{\epsilon}, \quad \boldsymbol{\epsilon} \sim \mathrm{N}(\boldsymbol{0}, \sigma^2 \Lambda), \\
\boldsymbol{\phi} &= S\boldsymbol{\beta} + Z\boldsymbol{b}, \quad \boldsymbol{b} \sim \mathrm{N}(\boldsymbol{0}, \sigma^2 D).
\end{aligned}
\tag{9.28}
$$

Note that model (9.28) is more general than (9.27) in the sense that other SNM models may also be written in this form. We will discuss estimation and inference procedures for the general model (9.28).

9.3.2   Estimation and Inference

Suppose $\mathcal{H}_k = \mathcal{H}_{k0} \oplus \mathcal{H}_{k1}$, where $\mathcal{H}_{k0} = \text{span}\{\phi_{k1}, \ldots, \phi_{kp_k}\}$ and $\mathcal{H}_{k1}$ is an RKHS with RK R_{k1}. Assume that D and Λ depend on an unknown parameter vector $\boldsymbol{\tau}$. We need to estimate $\boldsymbol{\beta}$, $\boldsymbol{f}$, $\boldsymbol{\tau}$, σ^2, and $\boldsymbol{b}$. The marginal likelihood based on model (9.28)

$$
L(\boldsymbol{\beta}, \boldsymbol{f}, \boldsymbol{\tau}, \sigma^2) = (2\pi\sigma^2)^{-\frac{mq_2+n}{2}} |D|^{-\frac{1}{2}} |\Lambda|^{-\frac{1}{2}} \int \exp\left\{-\frac{1}{\sigma^2} g(\boldsymbol{b})\right\} d\boldsymbol{b},
$$

where

$$
g(\boldsymbol{b}) = \frac{1}{2}\left\{(\boldsymbol{y} - \boldsymbol{\eta}(S\boldsymbol{\beta} + Z\boldsymbol{b}, \boldsymbol{f}))^T \Lambda^{-1}(\boldsymbol{y} - \boldsymbol{\eta}(S\boldsymbol{\beta} + Z\boldsymbol{b}, \boldsymbol{f})) + \boldsymbol{b}^T D^{-1} \boldsymbol{b}\right\}.
$$

For fixed $\boldsymbol{\tau}$ and σ^2, we estimate $\boldsymbol{\beta}$ and $\boldsymbol{f}$ as the minimizers of the following penalized likelihood (PL)

$$
l(\boldsymbol{\beta}, \boldsymbol{f}, \boldsymbol{\tau}, \sigma^2) + \frac{n\lambda}{\sigma^2} \sum_{k=1}^{r} \theta_k^{-1} \|P_{k1} f_k\|^2,
\tag{9.29}
$$

where $l(\boldsymbol{\beta}, \boldsymbol{f}, \boldsymbol{\tau}, \sigma^2) = -2 \log L(\boldsymbol{\beta}, \boldsymbol{f}, \boldsymbol{\tau}, \sigma^2)$, P_{k1} is the projection operator onto $\mathcal{H}_{k1}$ in $\mathcal{H}_k$, and $\lambda \theta_k^{-1}$ are smoothing parameters.

The integral in the marginal likelihood is usually intractable because η may depend on $\boldsymbol{b}$ nonlinearly. We now derive an approximation to the log-likelihood using the Laplace method. Let $G = \partial\boldsymbol{\eta}(S\boldsymbol{\beta} + Z\boldsymbol{b}, \boldsymbol{f})/\partial\boldsymbol{b}^T$ and $\tilde{\boldsymbol{b}}$ be the solution to

$$\frac{\partial g(\boldsymbol{b})}{\partial\boldsymbol{b}} = -G^T\Lambda^{-1}(\boldsymbol{y} - \boldsymbol{\eta}(S\boldsymbol{\beta} + Z\boldsymbol{b}, \boldsymbol{f})) + D^{-1}\boldsymbol{b} = 0. \qquad (9.30)$$

Approximating the Hessian by $\partial^2 g(\boldsymbol{b})/\partial\boldsymbol{b}\partial\boldsymbol{b}^T \approx G^T\Lambda^{-1}G + D^{-1}$ and applying the Laplace method for integral approximation, the function l can be approximated by

$$\begin{aligned}
\tilde{l}(\boldsymbol{\beta}, \boldsymbol{f}, \boldsymbol{\tau}, \sigma^2) &= n\log 2\pi\sigma^2 + \log|\Lambda| + \log|I + \tilde{G}^T\Lambda^{-1}\tilde{G}D| \\
&\quad + \frac{1}{\sigma^2}(\boldsymbol{y} - \boldsymbol{\eta}(S\boldsymbol{\beta} + Z\tilde{\boldsymbol{b}}, \boldsymbol{f}))^T\Lambda^{-1}(\boldsymbol{y} - \boldsymbol{\eta}(S\boldsymbol{\beta} + Z\tilde{\boldsymbol{b}}, \boldsymbol{f})) \\
&\quad + \frac{1}{\sigma^2}\tilde{\boldsymbol{b}}^T D^{-1}\tilde{\boldsymbol{b}}, \qquad (9.31)
\end{aligned}$$

where $\tilde{G} = G|_{\boldsymbol{b}=\tilde{\boldsymbol{b}}}$. Replacing l by $\tilde{l}$, ignoring the dependence of $\tilde{G}$ on $\boldsymbol{\beta}$ and $\boldsymbol{f}$, and dropping constant terms, the PL (9.29) reduces to

$$\begin{aligned}
&(\boldsymbol{y} - \boldsymbol{\eta}(S\boldsymbol{\beta} + Z\tilde{\boldsymbol{b}}, \boldsymbol{f}))^T\Lambda^{-1}(\boldsymbol{y} - \boldsymbol{\eta}(S\boldsymbol{\beta} + Z\tilde{\boldsymbol{b}}, \boldsymbol{f})) \\
&+ \tilde{\boldsymbol{b}}^T D^{-1}\tilde{\boldsymbol{b}} + n\lambda\sum_{k=1}^r \theta_k^{-1}\|P_{k1}f_k\|^2. \qquad (9.32)
\end{aligned}$$

It is not difficult to see that estimating $\boldsymbol{b}$ by equation (9.30), and $\boldsymbol{\beta}$ and $\boldsymbol{f}$ by equation (9.32), is equivalent to estimating $\boldsymbol{b}$, $\boldsymbol{\beta}$, and $\boldsymbol{f}$ jointly as minimizers of the following penalized Henderson (hierarchical) likelihood

$$\begin{aligned}
&(\boldsymbol{y} - \boldsymbol{\eta}(S\boldsymbol{\beta} + Z\boldsymbol{b}, \boldsymbol{f}))^T\Lambda^{-1}(\boldsymbol{y} - \boldsymbol{\eta}(S\boldsymbol{\beta} + Z\boldsymbol{b}, \boldsymbol{f})) \\
&+ \boldsymbol{b}^T D^{-1}\boldsymbol{b} + n\lambda\sum_{k=1}^r \theta_k^{-1}\|P_{k1}f_k\|^2. \qquad (9.33)
\end{aligned}$$

Denote the estimates of $\boldsymbol{\beta}$ and $\boldsymbol{f}$ as $\tilde{\boldsymbol{\beta}}$ and $\tilde{\boldsymbol{f}}$. We now discuss the estimation of $\boldsymbol{\tau}$ and σ^2 with $\boldsymbol{\beta}$, $\boldsymbol{f}$, and $\boldsymbol{b}$ fixed at $\tilde{\boldsymbol{\beta}}$, $\tilde{\boldsymbol{f}}$, and $\tilde{\boldsymbol{b}}$. Let $\tilde{W}^{-1} = \Lambda + \tilde{G}D\tilde{G}^T$, $\tilde{\boldsymbol{\eta}} = \boldsymbol{\eta}(S\tilde{\boldsymbol{\beta}} + Z\tilde{\boldsymbol{b}}, \tilde{\boldsymbol{f}})$, and $U = (D^{-1} + \tilde{G}^T\Lambda^{-1}\tilde{G})^{-1}$. Since $\tilde{W} = \Lambda^{-1} - \Lambda^{-1}\tilde{G}U\tilde{G}^T\Lambda^{-1}$ and $\tilde{G}^T\Lambda^{-1}(\boldsymbol{y} - \tilde{\boldsymbol{\eta}}) = D^{-1}\tilde{\boldsymbol{b}}$, then

$$\begin{aligned}
&(\boldsymbol{y} - \tilde{\boldsymbol{\eta}} + \tilde{G}\tilde{\boldsymbol{b}})^T\tilde{W}(\boldsymbol{y} - \tilde{\boldsymbol{\eta}} + \tilde{G}\tilde{\boldsymbol{b}}) \\
&= (\boldsymbol{y} - \tilde{\boldsymbol{\eta}} + \tilde{G}\tilde{\boldsymbol{b}})^T\Lambda^{-1}(\boldsymbol{y} - \tilde{\boldsymbol{\eta}} + \tilde{G}\tilde{\boldsymbol{b}}) \\
&\quad - (\boldsymbol{y} - \tilde{\boldsymbol{\eta}} + \tilde{G}\tilde{\boldsymbol{b}})^T\Lambda^{-1}\tilde{G}U\tilde{G}^T\Lambda^{-1}(\boldsymbol{y} - \tilde{\boldsymbol{\eta}} + \tilde{G}\tilde{\boldsymbol{b}}) \\
&= (\boldsymbol{y} - \tilde{\boldsymbol{\eta}} + \tilde{G}\tilde{\boldsymbol{b}})^T\Lambda^{-1}(\boldsymbol{y} - \tilde{\boldsymbol{\eta}} + \tilde{G}\tilde{\boldsymbol{b}}) \\
&\quad - (\boldsymbol{y} - \tilde{\boldsymbol{\eta}} + \tilde{G}\tilde{\boldsymbol{b}})^T\Lambda^{-1}\tilde{G}U(D^{-1} + \tilde{G}^T\Lambda^{-1}\tilde{G})\tilde{\boldsymbol{b}} \\
&= (\boldsymbol{y} - \tilde{\boldsymbol{\eta}} + \tilde{G}\tilde{\boldsymbol{b}})^T\Lambda^{-1}(\boldsymbol{y} - \tilde{\boldsymbol{\eta}} + \tilde{G}\tilde{\boldsymbol{b}}) - (\boldsymbol{y} - \tilde{\boldsymbol{\eta}} + \tilde{G}\tilde{\boldsymbol{b}})^T\Lambda^{-1}\tilde{G}\tilde{\boldsymbol{b}} \\
&= (\boldsymbol{y} - \tilde{\boldsymbol{\eta}})^T\Lambda^{-1}(\boldsymbol{y} - \tilde{\boldsymbol{\eta}}) + \tilde{\boldsymbol{b}}^T\tilde{G}^T\Lambda^{-1}(\boldsymbol{y} - \tilde{\boldsymbol{\eta}}) \\
&= (\boldsymbol{y} - \tilde{\boldsymbol{\eta}})^T\Lambda^{-1}(\boldsymbol{y} - \tilde{\boldsymbol{\eta}}) + \tilde{\boldsymbol{b}}^T D^{-1}\tilde{\boldsymbol{b}}.
\end{aligned}$$

Note that, based on Theorem 18.1.1 in Harville (1997), $|\tilde{W}^{-1}| = |\Lambda + \tilde{G}D\tilde{G}^T| = |\Lambda||I + \tilde{G}^T\Lambda^{-1}\tilde{G}D|$. Then we can reexpress $\tilde{l}$ in (9.31) as

$$\tilde{l}(\boldsymbol{\beta}, f, \boldsymbol{\tau}, \sigma^2) = n\log 2\pi + \log|\sigma^2\tilde{W}^{-1}| + \frac{1}{\sigma^2}e^T\tilde{W}e, \qquad (9.34)$$

where $e = y - \eta(S\boldsymbol{\beta} + Z\tilde{b}, f) + \tilde{G}\tilde{b}$. Plugging-in estimates of $\boldsymbol{\beta}$ and f in (9.34), profiling with respect to σ^2, and dropping constant terms, we estimate $\boldsymbol{\tau}$ as minimizers of the following approximate negative log-likelihood

$$\log|\tilde{W}^{-1}| + \log(\tilde{e}^T\tilde{W}\tilde{e}), \qquad (9.35)$$

where $\tilde{e} = y - \eta(S\tilde{\boldsymbol{\beta}} + Z\tilde{b}, \tilde{f}) + \tilde{G}\tilde{b}$.

The resulting estimates of $\boldsymbol{\tau}$ are denoted as $\tilde{\boldsymbol{\tau}}$. To account for the loss of degrees of freedom for estimating $\boldsymbol{\beta}$ and f, we estimate σ^2 by

$$\tilde{\sigma}^2 = \frac{\tilde{e}^T\tilde{W}(\tilde{\boldsymbol{\tau}})\tilde{e}}{n - q_1 - df(f)}, \qquad (9.36)$$

where q_1 is the dimension of $\boldsymbol{\beta}$, and $df(f)$ is a properly defined degree of freedom for estimating f. As in Section 8.3.3, we set $df(f) = \text{tr}(\tilde{H}^*)$, where $\tilde{H}^*$ is the hat matrix computed at convergence.

Assume priors (8.12) for f_k, $k = 1, \ldots, r$. Usually the posterior distribution does not have a closed form. Expanding $\boldsymbol{\eta}$ at $\tilde{b}$, we consider the following approximate Bayes model

$$y \approx \eta(S\boldsymbol{\beta} + Z\tilde{b}, f) - \tilde{G}\tilde{b} + \tilde{G}b + \epsilon, \qquad (9.37)$$

where priors for f are given in (8.12), and priors for $\boldsymbol{\beta}$ are $N(\mathbf{0}, \kappa I_{q_1})$. Ignoring the dependence of $\tilde{G}$ on $\boldsymbol{\beta}$ and f and combining random effects with random errors, model (9.37) reduces to an SNR model. Then methods discussed in Section 8.3.3 can be used to draw an inference about $\boldsymbol{\beta}$, $\boldsymbol{\tau}$, and f. In particular, approximate Bayesian confidence intervals can be constructed for f. The bootstrap approach may also be used to draw inference about $\boldsymbol{\beta}$, $\boldsymbol{\tau}$, and f.

9.3.3 Implementation and the snm Function

Since f may interact with $\boldsymbol{\beta}$ and b in a complicated way, it is usually impossible to solve (9.33) and (9.35) directly. The following iterative procedure will be used.

Algorithm for SNM models

1. *Initialize*: Set initial values for $\boldsymbol{\beta}$, f, b, and $\boldsymbol{\tau}$.

2. *Cycle*: Alternate between (a), (b), and (c) until convergence:

   (a) Conditional on current estimates of β, b, and τ, update f by solving (9.33).

   (b) Conditional on current estimates of f and τ, update β and b by solving (9.33).

   (c) Conditional on current estimates of f, β, and b, update τ by solving (9.35).

Note that step (b) corresponds to the *pseudo-data* step, and step (c) corresponds to part of the *LME* step in the Lindstrom–Bates algorithm (Lindstrom and Bates 1990). Consequently, steps (b) and (c) can be implemented by the `nlme` function in the `nlme` library. We now discuss the implementation of step (a). Note that β, b, and τ are fixed as the current estimates, say, β_-, b_- and τ_-. To update f at step (a), we need to fit the first-stage model

$$y = \eta(\phi_-, f) + \epsilon, \quad \epsilon \sim \mathrm{N}(0, \sigma^2 \Lambda_-), \qquad (9.38)$$

where $\phi_- = S\beta_- + Zb_-$, and Λ_- is the covariance matrix with τ fixed at τ_-. First consider the special case when $\mathcal{N}_{ij}$ in (9.27) are linear in f and involve evaluational functionals. Then model (9.27) can be rewritten as

$$y_{ij} = \alpha(\phi_-; x_{ij}) + \sum_{k=1}^{r} \delta_k(\phi_-; x_{ij}) f_k(\gamma_k(\phi_-; x_{ij})) + \epsilon_{ij}, \qquad (9.39)$$

which has the same form as the general SEMOR (self-modeling nonlinear regression) model (8.32). Since ϕ_- and Λ_- are fixed, the procedure described in Section 8.2.2 can be used to update f. In particular, the smoothing parameters may be estimated by the UBR, GCV or GML method at this step. When η is nonlinear in f, the EGN (extended Gauss–Newton) procedure in Section 8.3.3 can be used to update f.

The `snm` function in the `assist` library implements the algorithm when $\mathcal{N}_{ij}$ in (9.27) are linear in f and involve evaluational functionals. A typical call to the `snm` function is

```
snm(formula, func, fixed, random, start)
```

where the arguments `formula` and `func` serve the same purposes as those in the `nnr` function and are specified in the same manner. Following syntax in `nlme`, the `fixed` and `random` arguments specify the fixed, and random effects models in the second-stage model. The option `start` specifies initial values for all parameters in the fixed effects. An object of `snm` class is returned. The generic function `summary` can be applied to extract further information. Predictions at the population level can be computed using the `predict` function. At convergence, approximate Bayesian confidence intervals can be constructed as in Section 8.3.3. Posterior means and standard deviations for f can be computed using the `intervals` function. Examples can be found in Section 9.4.

9.4   Examples

9.4.1   Ozone in Arosa — Revisit

We have fitted SS ANOVA models in Section 4.9.2 and trigonometric spline models with heterogeneous and correlated errors in Section 5.4.2 to the Arosa data. An alternative approach is to consider observations as a long time series. Define a time variable as $t = (\texttt{month} - 0.5)/12 + \texttt{year} - 1$ with domain $[0, b]$, where $b = 45.46$ is the maximum value of t. Denote $\{(y_i, t_i), \ i = 1, \ldots, 518\}$ as observations on the response variable `thick` and time t. Consider the following semiparametric linear regression model

$$y_i = \beta_1 + \beta_2 \sin(2\pi t_i) + \beta_3 \cos(2\pi t_i) + \beta_4 t_i + f(t_i) + \epsilon_i, \qquad (9.40)$$

where $\sin(2\pi t)$ and $\cos(2\pi t)$ model the seasonal trend, $\beta_4 t + f(t)$ models the long-term trend, and ϵ_i are random errors. Note that the space for the parametric component, $\mathcal{H}_0 = \{1, t, \cos 2\pi t, \sin 2\pi t\}$, is the same as the null space of the linear-periodic spline defined in Section 2.11.4 with $\tau = 2\pi$. Therefore, we model the nonparametric function f using the model space $W_2^4[0, b] \ominus \mathcal{H}_0$ (see Section 2.11.4 for more details).

Observations close in time are likely to be correlated. We consider the exponential correlation structure with a nugget effect. Specifically, write $\epsilon_i = \epsilon(t_i)$ and assume that $\epsilon(t)$ is a zero-mean stochastic process with correlation structure

$$\text{Cov}(\epsilon(s), \epsilon(t)) = \begin{cases} \sigma^2(1 - c_0) \exp(-|s - t|/\rho), & s \neq t, \\ \sigma^2, & s = t, \end{cases} \qquad (9.41)$$

where ρ is the range parameter and c_0 is the nugget effect.

In the following we fit model (9.40) with correlation structure (9.41) and compute the overall fit as well as estimates of the seasonal and long-term trends:

```
> Arosa$t <- (Arosa$month-0.5)/12+Arosa$year-1
> arosa.ls.fit3 <- ssr(thick~t+sin(2*pi*t)+cos(2*pi*t),
```

```
  rk=lspline(2*pi*t,type=''linSinCos''), spar=''m'',
  corr=corExp(form=~t,nugget=T), data=Arosa)
> summary(arosa.ls.fit3)
...
GML estimate(s) of smoothing parameter(s) : 19461.83
Equivalent Degrees of Freedom (DF): 4.529238
Estimate of sigma:  17.32503
Correlation structure of class corExp representing
     range     nugget
0.3529361 0.6366844


> tm <- matrix(c(1,1,1,1,1,1,0,1,1,0,0,1,0,0,1), ncol=5,
             byrow=T)
> grid3 <- data.frame(t=seq(0,max(Arosa$t)+0.001,len=500))
> arosa.ls.fit3.p <- predict(arosa.ls.fit3, newdata=grid3,
                    terms=tm)
```

Figure 9.1 shows the overall fit as well as estimates of the seasonal and long-term trends. We can see that the long-term trend is not significantly different from zero.

Sometimes it is desirable to use a stochastic process to model the autocorrelation and regard this process as part of the signal. Then we need to separate this process from other errors and predict it at desired points. Specifically, consider the following SLM model

$$y_i = \beta_1 + \beta_2 \sin(2\pi t_i) + \beta_3 \cos(2\pi t_i) + \beta_4 t_i + f(t_i) + u(t_i) + \epsilon_i, \quad (9.42)$$

where $u(t)$ is a stochastic process independent of ϵ_i with mean zero and $\text{Cov}(u(s), u(t)) = \sigma_1^2 \exp(-|s - t|/\rho)$ with range parameter ρ.

Let $\boldsymbol{t} = (t_1, \ldots, t_{518})^T$ be the vector of design points for the variable t. Let $\boldsymbol{u} = (u(t_1), \ldots, u(t_{518}))^T$ be the vector of the u process evaluated at design points. Then $\boldsymbol{u}$ are random effects, and $\boldsymbol{u} \sim \text{N}(\boldsymbol{0}, \sigma_1^2 D)$, where D is a covariance matrix with the (i, j)th element equals $\exp(-|t_i - t_j|/\rho)$. The SLM model (9.42) cannot be fitted directly using the `slm` function since D depends on the unknown range parameter ρ nonlinearly. We fit model (9.42) in two steps. We first regard $\boldsymbol{u}$ as part of random errors and estimate the range parameter. This is accomplished in `arosa.ls.fit3`. Then we calculate the estimate of D without the nugget effect and regard it as the true covariance matrix. We calculate the Cholesky decomposition of D as $D = ZZ^T$ and transform the random effects $\boldsymbol{u} = Z\boldsymbol{b}$, where $\boldsymbol{b} \sim \text{N}(0, \sigma_1^2 I)$. Then we fit the transformed SLM model and compute the overall fit, estimate of the seasonal trend, and estimate of the long-term trend as follows:

```
> tau <- coef(arosa.ls.fit3$cor.est, F)
```

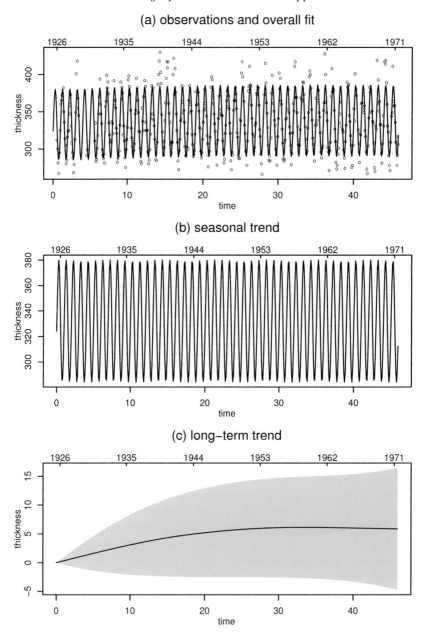

FIGURE 9.1   Arosa data, plots of (a) observations and the overall fit, (b) estimate of the seasonal trend, and (c) estimate of the long-term trend with 95% Bayesian confidence intervals (shaded region). Estimates are based on the fit to model (9.40) with correlation structure (9.41).

```
> D <- corMatrix(Initialize(corExp(tau[1],form=~t),
                 data=Arosa))
> Z <- chol.new(D)
> arosa.ls.fit4 <- slm(thick~t+sin(2*pi*t)+cos(2*pi*t),
    rk=lspline(2*pi*t,type=''linSinCos''),
    random=list(pdIdent(~Z-1)), data=Arosa)
> arosa.ls.fit4.p <- intervals(arosa.ls.fit4,
    newdata=grid3, terms=tm)
```

where `chol.new` is a function in the `assist` library for Cholesky decomposition. Suppose we want to predict u on grid points and denote v as the vector of the u process evaluated at these grid points. Let $R = \text{Cov}(v, u)$. Note that Z is a square invertible matrix since D is invertible. Then $\hat{v} = RD^{-1}\hat{u} = RZ^{-T}\hat{b}$. We compute the prediction as follows:

```
> newdata <- data.frame(t=c(Arosa$t,grid3$t))
> RD <- corMatrix(Initialize(corExp(tau[1],form=~t),
                  data=newdata))
> R <- RD[(length(Arosa$t)+1):length(newdata$t),
          1:length(Arosa$t)]
> b <- as.vector(arosa.ls.fit4$lme.obj$coef$random[[2]])
> u.new <- R%*%t(solve(Z))%*%b
```

Figure 9.2 shows the overall fit, estimate of the seasonal trend, estimate of the long-term trend, and prediction of $u(t)$ (local stochastic trend). The bump during 1940 in Figure 4.10 shows up in the prediction of local stochastic trend.

9.4.2 Lake Acidity — Revisit

We have shown how to investigate geological location effect using an SS ANOVA model in Section 5.4.4. We now describe an alternative approach using random effects. We use the same notations defined in Section 5.4.4. Consider the following SLM model:

$$\text{pH}(x_{i1}, \boldsymbol{x}_{i2}) = f(x_{i1}) + \beta_1 x_{i21} + \beta_2 x_{i22} + u(\boldsymbol{x}_{i2}) + \epsilon(x_{i1}, \boldsymbol{x}_{i2}), (9.43)$$

where $f \in W_2^2(\mathbb{R})$, $u(\boldsymbol{x}_2)$ is a spatial process, and $\epsilon(x_1, \boldsymbol{x}_2)$ are random errors independent of the spatial process. Model (9.43) separates the contribution of the spatial correlation from random errors and regards the spatial process as part of the signal. Assume that $u(\boldsymbol{x}_2)$ is a zero-mean process independent of $\epsilon(x_1, \boldsymbol{x}_2)$ with an exponential correlation structure $\text{Cov}(u(\boldsymbol{x}_2), u(\boldsymbol{z}_2)) = \sigma_1^2 \exp\{-d(\boldsymbol{x}_2, \boldsymbol{z}_2)/\rho\}$ with range parameter ρ, where $d(\boldsymbol{x}_2, \boldsymbol{z}_2)$ represents the Euclidean distance.

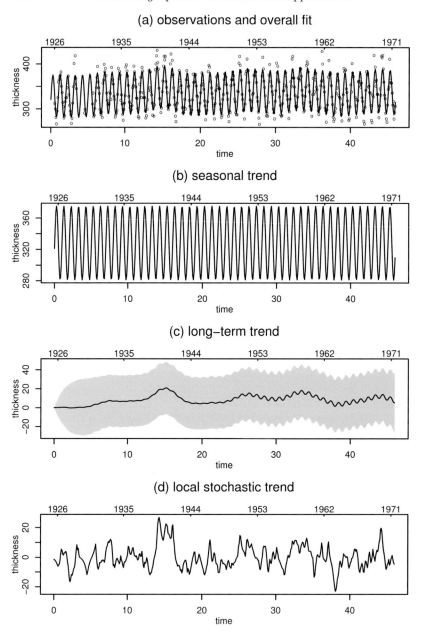

FIGURE 9.2 Arosa data, plots of (a) observations and the overall fit, (b) estimate of the seasonal trend, (c) estimate of the long-term trend with 95% Bayesian confidence intervals (shaded region), and (d) prediction of local stochastic trend. Estimates are based on the fit to model (9.42).

Let $\boldsymbol{u}$ be the vector of the u process evaluated at design points. Then $\boldsymbol{u}$ are random effects and $\boldsymbol{u} \sim N(\boldsymbol{0}, \sigma^2 D)$, where the covariance matrix $D = \{\exp(-d(\boldsymbol{x}_{i2}, \boldsymbol{x}_{j2})/\rho)\}_{i,j=1}^n$ depends on the unknown parameter ρ nonlinearly. Therefore, again, we fit model (9.43) in two steps. We first regard $\boldsymbol{u}$ as part of random errors, estimate the range parameter, and calculate the estimated covariance matrix without the nugget effect:

```
> temp <- ssr(ph~x1+x21+x22, rk=tp(x1), data=acid,
    corr=corExp(form=~x21+x22, nugget=T), spar=''m'')
> tau <- coef(temp$cor.est, F)
> D <- corMatrix(Initialize(corExp(tau[1],form=~x21+x22),
    data=acid))
```

Consider the estimated D as the true covariance matrix. Then we calculate the Cholesky decomposition of D as $D = ZZ^T$ and transform the random effects $\boldsymbol{u} = Z\boldsymbol{b}$, where $\boldsymbol{b} \sim N(\boldsymbol{0}, \sigma_1^2 I)$. Now we are ready to fit the transformed SLM model:

```
> Z <- chol.new(D)
> acid.slm.fit <- slm(ph~x1+x21+x22, rk=tp(x1), data=acid,
                    random=list(pdIdent(~Z-1)))
```

We then calculate the estimated effect of `calcium`:

```
> grid1 <- data.frame(
    x1=seq(min(acid$x1),max(acid$x1),len=100),
    x21=min(acid$x21), x22=min(acid$x22))
> acid.slm.x1.p <- intervals(acid.slm.fit, newdata=grid1,
                    terms=c(0,1,0,0,1))
```

Let $\boldsymbol{v}$ be the vector of the u process evaluated at grid points at which we wish to predict the location effect. Let $R = \mathrm{Cov}(\boldsymbol{v}, \boldsymbol{u})$. Then $\hat{\boldsymbol{v}} = RD^{-1}\hat{\boldsymbol{u}} = RZ^{-T}\hat{\boldsymbol{b}}$.

```
> grid2 <- expand.grid(
    x21=seq(min(acid$x21)-.001,max(acid$x21)+.001,len=20),
    x22=seq(min(acid$x22)-.001,max(acid$x22)+.001,len=20))
> newdata <- data.frame(z1=c(acid$x21,grid2$x21),
                    z2=c(acid$x22,grid2$x22))
> RD <- corMatrix(Initialize(corExp(tau[1], form=~z1+z2),
                    data=newdata))
> R <- RD[(length(acid$x21)+1):length(newdata$z1),
            1:length(acid$x21)]
> b <- as.vector(acid.slm.fit$lme.obj$coef$random[[2]])
> u.new <- R%*%t(solve(Z))%*%b
> acid.slm.x2.p <- u.new+
```

```
acid.slm.fit$lme.obj$coef$fixed[3]*grid2$x21+
acid.slm.fit$lme.obj$coef$fixed[4]*grid2$x22
```

Figure 9.3 plots the estimated calcium (x_1) effect and prediction of the location (x_2) effect.

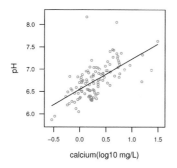 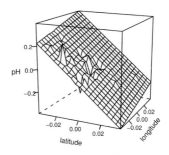

FIGURE 9.3 Lake acidity data, the left panel includes observations and the estimate of the calcium effect f (the constant plus main effect of x_1), and the right panel includes prediction of the location effect $\hat{\beta}_1 x_{21} + \hat{\beta}_{22} x_{22} + \hat{u}(\boldsymbol{x}_2)$.

9.4.3   Coronary Sinus Potassium in Dogs

The dog data set contains measurements of coronary sinus potassium concentration from dogs in four groups: control, extrinsic cardiac denervation three weeks prior to coronary occlusion, extrinsic cardiac denervation immediately prior to coronary occlusion, and bilateral thoracic sympathectomy and stellectomy three weeks prior to coronary occlusion. We are interested in (i) estimating the group (treatment) effects, (ii) estimating the group mean concentration as a function of time, and (iii) predicting response over time for each dog. There are two categorical covariates, group and dog, and a continuous covariate, time. We code the group factor as 1 to 4, and the observed dog factor as 1 to 36. Coronary sinus potassium concentrations for all dogs are shown in Figure 9.4.

Let t be the time variable transformed into $[0, 1]$. We treat group and time as fixed factors. From the design, the dog factor is nested within the group factor. We treat dog as a random factor. For group k, denote

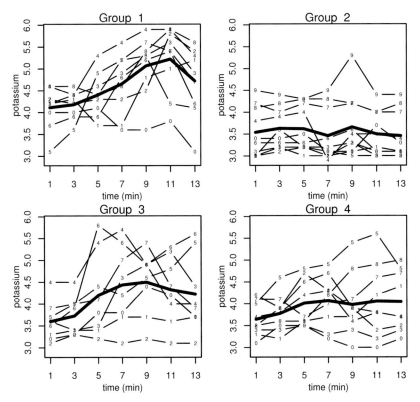

FIGURE 9.4 Dog data, coronary sinus potassium concentrations over time for each dog. Solid thick lines link within group average concentrations at each time point.

$\mathcal{B}_k$ as the population from which the dogs in group k were drawn, and P_k as the sampling distribution. Assume the following model

$$y_{kwj} = f(k, w, t_j) + \epsilon_{kwj}, \quad k = 1, \ldots, 4; \quad w \in \mathcal{B}_k; \quad t_j \in [0, 1], \quad (9.44)$$

where y_{kwj} is the observed potassium concentration at time t_j from dog w in the population $\mathcal{B}_k$, $f(k, w, t_j)$ is the true concentration at time t_j of dog w in the population $\mathcal{B}_k$, and ϵ_{kwj} are random errors. The function $f(k, w, t_j)$ is defined on $\{\{1\} \otimes \mathcal{B}_1, \{2\} \otimes \mathcal{B}_2, \{3\} \otimes \mathcal{B}_3, \{4\} \otimes \mathcal{B}_4\} \otimes [0, 1]$. Note that $f(k, w, j)$ is a random variable since w is a random sample from $\mathcal{B}_k$. What we observe are realizations of this true mean function plus random errors. We use label i to denote dogs we actually observe.

Suppose we want to model the time effect using the cubic spline model space $W_2^2[0, 1]$ under the construction in Section 2.6, and the group effect using the classical one-way ANOVA model space $\mathbb{R}^4$ under the construction in Section 4.3.1. Define the following four averaging operators:

$$\mathcal{A}_2 f = \int_{\mathcal{B}_k} f(k, w, t) dP_k(w),$$

$$\mathcal{A}_1 f = \frac{1}{4} \sum_{k=1}^{4} \mathcal{A}_2 f(k, t),$$

$$\mathcal{A}_3 f = \int_0^1 f(k, w, t) dt,$$

$$\mathcal{A}_4 f = \left(\int_0^1 \frac{\partial f(k, w, t)}{\partial t} dt \right) (t - 0.5).$$

Then we have the following SS ANOVA decomposition:

$$\begin{aligned}
f &= \{\mathcal{A}_1 + (\mathcal{A}_2 - \mathcal{A}_1) + (I - \mathcal{A}_2)\}\{\mathcal{A}_3 + \mathcal{A}_4 + (I - \mathcal{A}_3 - \mathcal{A}_4)\}f \\
&= \mathcal{A}_1 \mathcal{A}_3 f + \mathcal{A}_1 \mathcal{A}_4 f + \mathcal{A}_1 (I - \mathcal{A}_3 - \mathcal{A}_4)f \\
&\quad + (\mathcal{A}_2 - \mathcal{A}_1)\mathcal{A}_3 f + (\mathcal{A}_2 - \mathcal{A}_1)\mathcal{A}_4 f + (\mathcal{A}_2 - \mathcal{A}_1)(I - \mathcal{A}_3 - \mathcal{A}_4)f \\
&\quad + (I - \mathcal{A}_2)\mathcal{A}_3 f + (I - \mathcal{A}_2)\mathcal{A}_4 f + (I - \mathcal{A}_2)(I - \mathcal{A}_3 - \mathcal{A}_4)f \\
&= \mu + \beta(t - 0.5) + s_1(t) + \xi_k + \delta_k(t - 0.5) + s_2(k, t) \\
&\quad + \alpha_{w(k)} + \gamma_{w(k)}(t - 0.5) + s_3(k, w, t), \quad\quad\quad (9.45)
\end{aligned}$$

where μ is a constant, $\beta(t-0.5)$ is the linear main effect of time, $s_1(t)$ is the smooth main effect of time, ξ_k is the main effect of group, $\delta_k(t-0.5)$ is the smooth-linear interaction between time and group, $s_2(k, t)$ is the smooth-smooth interaction between time and group, $\alpha_{w(k)}$ is the main effect of dog, $\gamma_{w(k)}(t-0.5)$ is the smooth-linear interaction between time and dog, and $s_3(k, w, t)$ is the smooth-smooth interaction between time and dog. The overall main effect of time equals $\beta(t - 0.5) + s_1(t)$, the overall interaction between time and group equals $\delta_k(t - 0.5) + s_2(k, t)$, and the overall interaction between time and dog equals $\gamma_{w(k)}(t-0.5) + s_3(k, w, t)$. The first six terms are fixed effects. The last three terms are random effects since they depend on the random variable w. Depending on time only, the first three terms represent the mean curve for all dogs. The middle three terms measure the departure of the mean curve for a particular group from the mean curve for all dogs. The last three terms measure the departure of a particular dog from the mean curve of a population from which the dog was chosen.

Based on the SS ANOVA decomposition (9.45), we will fit the following three models:

- *Model 1* includes the first seven terms in (9.45). It has a different population mean curve for each group plus a random intercept for each dog. We assume that $\alpha_i \overset{iid}{\sim} N(0, \sigma_1^2)$, $\epsilon_{kij} \overset{iid}{\sim} N(0, \sigma^2)$, and the random effects and random errors are mutually independent.

- *Model 2* includes the first eight terms in (9.45). It has a different population mean curve for each group plus a random intercept and a random slope for each dog. We assume that $(\alpha_i, \gamma_i) \overset{iid}{\sim} N(\mathbf{0}, \sigma^2 D_1)$, where D_1 is an unstructured covariance matrix, $\epsilon_{kij} \overset{iid}{\sim} N(0, \sigma^2)$, and the random effects and random errors are mutually independent.

- *Model 3* includes all nine terms in (9.45). It has a different population mean curve for each group plus a random intercept, a random slope, and a smooth random effect for each dog. We assume that $(\alpha_i, \gamma_i) \overset{iid}{\sim} N(\mathbf{0}, \sigma^2 D_1)$, $s_3(k, i, t)$ are stochastic processes that are independent between dogs with mean zero and covariance function $\sigma_2^2 R_1(s, t)$, where R_1 is the cubic spline RK given in Table 2.2, $\epsilon_{kij} \overset{iid}{\sim} N(0, \sigma^2)$, and the random effects and random errors are mutually independent.

Model 1 and Model 2 can be fitted as follows:

```
> data(dog)
> dog.fit1 <- slm(y~time,
    rk=list(cubic(time), shrink1(group),
            rk.prod(kron(time-.5),shrink1(group)),
            rk.prod(cubic(time),shrink1(group))),
    random=list(dog=~1), data=dog)
> dog.fit1
Semi-parametric linear mixed-effects model fit by REML
  Model: y ~ time
  Data: dog
  Log-restricted-likelihood: -180.4784

Fixed: y ~ time
(Intercept)        time
  3.8716210   0.4339335

Random effects:
 Formula: ~1 | dog
        (Intercept)  Residual
StdDev:   0.4980355 0.3924478

GML estimate(s) of smoothing parameter(s) : 0.0002338082
  0.0034079441 0.0038490075 0.0002048518
```

```
Equivalent Degrees of Freedom (DF):   13.00259
Estimate of sigma:   0.3924478
```

```
Number of Observations: 252
```

```
> dog.fit2 <- update(dog.fit1, random=list(dog=~time))
> dog.fit2
Semi-parametric linear mixed-effects model fit by REML
  Model: y ~ time
  Data: dog
  Log-restricted-likelihood: -166.4478
```

```
Fixed: y ~ time
(Intercept)          time
  3.8767107    0.4196788
```

```
Random effects:
 Formula: ~time | dog
 Structure: General positive-definite,
            Log-Cholesky parametrization
            StdDev      Corr
(Intercept) 0.4188186 (Intr)
time        0.5592910 0.025
Residual    0.3403256
```

```
GML estimate(s) of smoothing parameter(s) : 1.674916e-04
  3.286885e-03 5.781563e-03 8.944897e-05
Equivalent Degrees of Freedom (DF):   13.83309
Estimate of sigma:   0.3403256
```

```
Number of Observations: 252
```

To fit Model 3, we need to find a way to specify the smooth (non-parametric) random effect s_3. Let $\boldsymbol{t} = (t_1, \ldots, t_7)^T$ be the time points at which measurements were taken for each dog (note that time points are the same for all dogs), $\boldsymbol{u}_i = (s_3(k, i, t_1), \ldots, s_3(k, i, t_7))^T$, and $\boldsymbol{u} = (\boldsymbol{u}_1^T, \ldots, \boldsymbol{u}_{36}^T)^T$. Then $\boldsymbol{u}_i \overset{iid}{\sim} \mathrm{N}(\boldsymbol{0}, \sigma_2^2 D_2)$, where D_2 is the RK of a cubic spline evaluated at the design points $\boldsymbol{t}$. Let $D_2 = HH^T$ be the Cholesky decomposition of D_2, $D = \mathrm{diag}(D_2, \ldots, D_2)$, and $Z = \mathrm{diag}(H, \ldots, H)$. Then $ZZ^T = D$. We can write $\boldsymbol{u} = Z\boldsymbol{b}_2$, where $\boldsymbol{b}_2 \sim \mathrm{N}(\boldsymbol{0}, \sigma_2^2 I_n)$ and $n = 252$. Then we can specify the random effects $\boldsymbol{u}$ using the matrix Z.

```
> D2 <- cubic(dog$time[1:7])
> H <- chol.new(D2)
```

```
> Z <- kronecker(diag(36), H)
> dog$all <- rep(1,36*7)
> dog.fit3 <- update(dog.fit2,
    random=list(all=pdIdent(~Z-1),dog=~time))
> summary(dog.fit3)
Semi-parametric Linear Mixed Effects Model fit
  Model: y ~ time
  Data: dog

Linear mixed-effects model fit by REML
 Data: dog
       AIC      BIC    logLik
  322.1771 360.9131 -150.0885

Random effects:
 Formula: ~Z - 1 | all
 Structure: Multiple of an Identity
             Z1       Z2       Z3       Z4       Z5       Z6
StdDev: 3.90843 3.90843 3.90843 3.90843 3.90843 3.90843
 ...

 Formula: ~time | dog %in% all
 Structure: General positive-definite, Log-Cholesky
            parametrization
            StdDev     Corr
(Intercept) 0.4671544 (Intr)
time        0.5716972 -0.083
Residual    0.2383448

Fixed effects: y ~ time
               Value  Std.Error  DF  t-value p-value
(Intercept) 3.885270 0.08408051 215 46.20892   0e+00
time        0.404652 0.10952837 215  3.69449   3e-04
 Correlation:
     (Intr)
time -0.224
 ...
Smoothing spline:
 GML estimate(s) of smoothing parameter(s): 8.775858e-05
 1.561206e-03 3.351111e-03 2.870198e-05
 Equivalent Degrees of Freedom (DF):  15.37740
```

The dimension of b_2 associated with the smooth random effects s_3

equals the total number of observations in the above construction. Therefore, computation and/or memory required for larger sample size can be prohibitive. One approach to stabilize and speed up the computation is to use the low rank approximation (Wood 2003). For generality, suppose time points for different dogs may be different. Let t_{ij} for $j = 1, \ldots, n_i$ be the time points for dog i, $\boldsymbol{u}_i = (s_3(k, i, t_{i1}), \ldots, s_3(k, i, t_{in_i}))^T$ and $\boldsymbol{u} = (\boldsymbol{u}_1^T, \ldots, \boldsymbol{u}_m^T)^T$ where m is the number of dogs. Then $\boldsymbol{u}_i \sim \mathrm{N}(\boldsymbol{0}, \sigma_2^2 \Pi_i)$, where Π_i is the RK of a cubic spline evaluated at the design points $\boldsymbol{t}_i = (t_{i1}, \ldots, t_{in_i})$. Let $\Pi_i = U_i \Gamma_i U_i^T$ be the eigendecomposition where U_i is an $n_i \times n_i$ orthogonal matrix, $\Gamma_i = \mathrm{diag}(\gamma_{i1}, \ldots, \gamma_{in_i})$ and $\gamma_{i1} \leq \gamma_{i2} \leq \ldots \leq \gamma_{in_i}$ are eigenvalues. Usually some of the eigenvalues are much smaller than others. Discard $n_i - k_i$ smallest eigenvalues and let $H_{i1} = U_i \Gamma_{i1}$ where Γ_{i1} is an $n_i \times k_i$ matrix with diagonal elements equal $\sqrt{\gamma_{i1}}, \ldots, \sqrt{\gamma_{ik_i}}$ and all other elements equal zero. Then $\Pi_i \approx H_{i1} H_{i1}^T$ and $D = \mathrm{diag}(\Pi_1, \ldots, \Pi_m) \approx Z_1 Z_1^T$ where $Z_1 = \mathrm{diag}(H_{11}, \ldots, H_{m1})$. We can approximate Model 3 using $\boldsymbol{u}_1 = Z_1 \boldsymbol{b}_2$ where $\boldsymbol{b}_2 \sim \mathrm{N}(\boldsymbol{0}, \sigma_2^2 I_K)$ and $K = \sum_{i=1}^m k_i$. The dimension K can be much smaller than n. Low rank approximations for other situations can be constructed similarly. The above approximation procedure is implemented to the dog data as follows:

```
> chol.new1 <- function(Q, cutoff) {
    tmp <- eigen(Q)
    num <- sum(tmp$values<cutoff)
    k <- ncol(Q)-num
    t(t(as.matrix(tmp$vector[,1:k]))*sqrt(tmp$values[1:k]))
    }
> H1 <- chol.new1(D2, 1e-3)
> Z1 <- kronecker(diag(36), H1)
> dog$all1 <- rep(1,nrow(Z1))
> dog.fit3.1 <- update(dog.fit2,
    random=list(all1=pdIdent(~Z1-1), dog=~time))
> summary(dog.fit3.1)
Semi-parametric Linear Mixed Effects Model fit
  Model: y ~ time
  Data: dog

Linear mixed-effects model fit by REML
  Data: dog
        AIC      BIC    logLik
  323.4819 362.2180 -150.7409

Random effects:
```

```
Formula: ~Z1 - 1 | all1
Structure: Multiple of an Identity
            Z11     Z12     Z13     Z14     Z15     Z16
StdDev: 3.65012 3.65012 3.65012 3.65012 3.65012 3.65012
...


Formula: ~time | dog %in% all1
Structure: General positive-definite, Log-Cholesky
  parametrization
            StdDev     Corr
(Intercept) 0.4635580 (Intr)
time        0.5690967 -0.07
Residual    0.2527408

Fixed effects: y ~ time
              Value     Std.Error  DF   t-value  p-value
(Intercept) 3.883828 0.08418382 215 46.13509    0e+00
time        0.407229 0.11063259 215  3.68092    3e-04
 Correlation:
     (Intr)
time -0.228
...
Smoothing spline:
 GML estimate(s) of smoothing parameter(s): 9.964788e-05
   1.764225e-03 3.716363e-03 3.200021e-05
 Equivalent Degrees of Freedom (DF):  15.39633
```

where eigenvalues smaller than 10^{-3} are discarded and $k_i = 2$. The R function `chol.new1` computes the truncated Cholesky decomposition. The fitting criteria and parameter estimates are similar to those from the full model. Another fit based on low rank approximation with the cutoff value 10^{-3} being replaced by 10^{-4} produces almost identical results as those from the full model.

As discussed in Section 9.2.2, Models 1, 2, and 3 are connected with three LME models and these connections are used to fit the SLM models. We can compare these three corresponding LME models as follows:

```
> anova(dog.fit1$lme.obj, dog.fit2$lme.obj,
        dog.fit3$lme.obj)
                 Model df     AIC       BIC      logLik
dog.fit1$lme.obj     1  8 376.9568 405.1285 -180.4784
dog.fit2$lme.obj     2 10 352.8955 388.1101 -166.4478
dog.fit3$lme.obj     3 11 322.1771 360.9131 -150.0885
```

```
                           Test   L.Ratio  p-value
dog.fit1$lme.obj
dog.fit2$lme.obj              1 vs 2 28.06128 <.0001
dog.fit3$lme.obj             2 vs 3 32.71845 <.0001
```

Even though they do not compare three SLM models directly, these comparison results seem to indicate that Model 3 is more favorable. More research on model selection and inference for SLM and SNM models is necessary.

We can calculate estimates of the population mean curves for four groups based on Model 3 as follows:

```
> dog.grid <- data.frame(time=rep(seq(0,1,len=50),4),
                group=as.factor(rep(1:4,rep(50,4))))
> e.dog.fit3 <- intervals(dog.fit3, newdata=dog.grid,
                terms=rep(1,6))
```

Figure 9.5 shows the estimated mean curves and 95% Bayesian confidence intervals based on the fit `dog.fit3`.

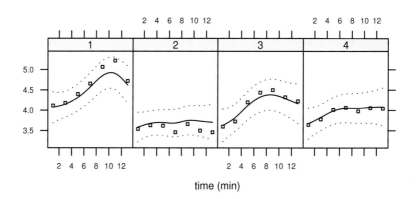

FIGURE 9.5 Dog data, estimates of the group mean response curves (solid lines) with 95% Bayesian confidence intervals (dotted lines) based on the fit `dog.fit3`. Squares are within group average concentrations.

We have shrunk the group mean curves toward the overall mean curve. That is, we have penalized the **group** main effect ξ_k and the smooth-linear **group-time** interaction $\delta_k(t - 0.5)$ in the SS ANOVA decomposition (9.45). From Figure 9.5 we can see that the estimated population mean curve for group 2 is biased upward, while the estimated population mean curves for group 1 is biased downward. This is because responses from group 2 are smaller, while responses from group 1 are larger than those from groups 3 and 4. Thus, their estimates are pulled toward the overall mean. Shrinkage estimates in this case may not be advantageous since `group` only has four levels. One may want to leave ξ_k and $\delta_k(t-0.5)$ terms unpenalized to reduce biases. We can rewrite the fixed effects in (9.45) as

$$
\begin{aligned}
f_k(t) &\triangleq \mu + \beta(t - 0.5) + s_1(t) + \xi_k + \delta_k(t - 0.5) + s_2(k, t) \\
&= \{\mu + \xi_k\} + \{\beta(t - 0.5) + \delta_k(t - 0.5)\} + \{s_1(t) + s_2(k, t)\} \\
&= \tilde{\xi}_k + \tilde{\delta}_k(t - 0.5) + \tilde{s}_2(k, t), \qquad\qquad (9.46)
\end{aligned}
$$

where $f_k(t)$ is the mean curve for group k. Assume that $f_k \in W_2^2[0, 1]$. Define penalty as $\int_0^1 (f_k''(t))^2 dt = ||\tilde{s}_2(k, t)||^2$. Then the constant term $\tilde{\xi}_k$ and the linear term $\tilde{\delta}_k(t - 0.5)$ are not penalized. We can refit *Model 1*, *Model 2*, and *Model 3* under this new form of penalty as follows:

```
> dog.fit4 <- slm(y~group*time,
    rk=list(rk.prod(cubic(time),kron(group==1)),
            rk.prod(cubic(time),kron(group==2)),
            rk.prod(cubic(time),kron(group==3)),
            rk.prod(cubic(time),kron(group==4))),
    random=list(dog=~1), data=dog)
> dog.fit5 <- update(dog.fit4, random=list(dog=~time))
> dog.fit6 <- update(dog.fit5,
    random=list(all=pdIdent(~Z-1),dog=~time))
> e.dog.fit6 <- intervals(dog.fit6, newdata=dog.grid,
                          terms=rep(1,12))
```

Figure 9.6 shows the estimated mean curves and 95% Bayesian confidence intervals based on the fit `dog.fit6`. The estimated mean functions are less biased.

We now show how to calculate predictions for all dogs based on the fit `dog.fit6`. Predictions based on other models may be derived similarly. For a particular dog i in group k, its prediction at time z can be computed as $\tilde{\xi}_k + \tilde{\delta}_k(z - 0.5) + \tilde{s}_2(k, z) + \hat{\alpha}_i + \hat{\gamma}_i(z - 0.5) + \hat{s}_3(k, i, z)$. Prediction of the fixed effects can be computed using the `prediction` function. Predictions of random effects $\hat{\alpha}_i$ and $\hat{\gamma}_i$ can be extracted from the fit. Therefore, we only need to compute $\hat{s}_3(k, i, z)$. Suppose we want to predict s_3 for dog i in group k on a vector of points $z_i = (z_{i1}, \ldots, z_{ig_i})^T$. Let $v_i = (s_3(k, i, z_{i1}), \ldots, s_3(k, i, z_{ig_i}))^T$ and $C_i = \text{Cov}(v_i, u_i) = \{R_1(z_{ik}, t_j)\}_{k=1}^{g_i}{}_{j=1}^{7}$ for $i = 1, \ldots, 36$, where $R_1(z, t)$ is the cubic spline RK given in Table 2.2. Let $v = (v_1^T, \ldots, v_{36}^T)^T$, $R =$

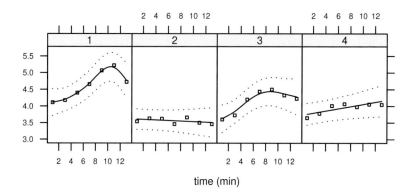

time (min)

FIGURE 9.6   Dog data, estimates of the group mean response curves (solid lines) with 95% Bayesian confidence intervals (dotted lines) based on the fit `dog.fit6`. Squares are within group average responses.

$\mathrm{diag}(C_1, \ldots, C_{36})$, and $\hat{\boldsymbol{u}}$ be the prediction of $\boldsymbol{u}$. We then can compute the prediction for all dogs as $\hat{\boldsymbol{v}} = RD^{-1}\hat{\boldsymbol{u}}$. The smallest eigenvalue of D is close to zero, and thus D^{-1} cannot be calculated precisely. We will use an alternative approach that does not require inverting D. Note that $\boldsymbol{u} = Z\boldsymbol{b}_2$ and denote the estimate of $\boldsymbol{b}_2$ as $\hat{\boldsymbol{b}}_2$. If we can find a vector $\boldsymbol{r}$ (need not to be unique) such that

$$Z^T \boldsymbol{r} = \hat{\boldsymbol{b}}_2, \tag{9.47}$$

then

$$\hat{\boldsymbol{v}} = RD^{-1}\hat{\boldsymbol{u}} = RD^{-1}Z\hat{\boldsymbol{b}}_2 = RD^{-1}ZZ^T\boldsymbol{r} = R\boldsymbol{r}.$$

So the task now is to solve (9.47). Let

$$Z = (Q_1 \ Q_2) \begin{pmatrix} V \\ \boldsymbol{0} \end{pmatrix}$$

be the QR decomposition of Z. We consider $\boldsymbol{r}$ in the space spanned by Q_1: $\boldsymbol{r} = Q_1\boldsymbol{\alpha}$. Then, from (9.47), $\boldsymbol{\alpha} = V^{-T}\hat{\boldsymbol{b}}_2$. Thus, $\boldsymbol{r} = Q_1 V^{-T}\hat{\boldsymbol{b}}_2$ is a solution to (9.47). This approach also applies to the situation when D is singular. In the following we calculate predictions for all 36 dogs on a set of grid points. Note that groups 1, 2, 3, and 4 have 9, 10, 8, and 9 dogs, respectively.

```
> dog.grid2 <- data.frame(time=rep(seq(0,1,len=50),36),
                          dog=rep(1:36,rep(50,36)))
```

```
> R <- kronecker(diag(36),
                 cubic(dog.grid2$time[1:50],dog$time[1:7]))
> b1 <- dog.fit6$lme.obj$coef$random$dog
> b2 <- as.vector(dog.fit6$lme.obj$coef$random$all)
> Z.qr <- qr(Z)
> r <- qr.Q(Z.qr)%*%solve(t(qr.R(Z.qr)))%*%b2
> tmp1 <- c(rep(e.dog.fit6$fit[dog.grid$group==1],9),
            rep(e.dog.fit6$fit[dog.grid$group==2],10),
            rep(e.dog.fit6$fit[dog.grid$group==3],8),
            rep(e.dog.fit6$fit[dog.grid$group==4],9))
> tmp2 <- as.vector(rep(b1[,1],rep(50,36)))
> tmp3 <- as.vector(kronecker(b1[,2],dog.grid2$time[1:50]))
> u.new <- as.vector(R%*%r)
> p.dog.fit6 <- tmp1+tmp2+tmp3+u.new
```

Predictions for dogs 1, 2, 26, and 27 are shown in Figure 9.7.

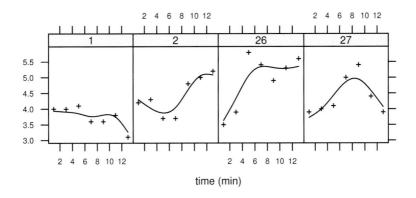

FIGURE 9.7 Dog data, predictions for dogs 1, 2, 26, and 27. Pluses are observations and solid lines are predictions.

9.4.4 Carbon Dioxide Uptake

The carbon dioxide data set contains five variables: `plant` identifies each plant; `Type` has two levels, Quebec and Mississippi, indicating the origin of the plant; `Treatment` indicates two treatments, nonchilling and chilling; `conc` gives ambient carbon dioxide concentrations (mL/L); and `uptake` gives carbon dioxide uptake rates (umol/m^2 sec). Figure 9.8

shows the CO_2 uptake rates for all plants.

The objective of the experiment was to evaluate the effect of plant type and chilling treatment on CO_2 uptake. Pinheiro and Bates (2000) gave detailed analyses of this data set based on NLME models. They reached the following NLME model:

$$\texttt{uptake}_{ij} = e^{\phi_{1i}}\{1 - e^{-e^{\phi_{2i}}(\texttt{conc}_j - \phi_{3i})}\} + \epsilon_{ij},$$
$$\phi_{1i} = \beta_{11} + \beta_{12}\texttt{Type} + \beta_{13}\texttt{Treatment} + \beta_{14}\texttt{Treatment:Type} + b_i,$$
$$\phi_{2i} = \beta_{21},$$
$$\phi_{3i} = \beta_{31} + \beta_{32}\texttt{Type} + \beta_{33}\texttt{Treatment} + \beta_{34}\texttt{Treatment:Type},$$
$$i = 1, \ldots, 12; \; j = 1, \ldots, 7, \qquad\qquad (9.48)$$

where $\texttt{uptake}_{ij}$ denotes the CO_2 uptake rate of $\texttt{plant}$ i at CO_2 ambient concentration $\texttt{conc}_j$; $\texttt{Type}$ equals 0 for plants from Quebec and 1 for plants from Mississippi, $\texttt{Treatment}$ equals 0 for chilled plants and 1 for control plants; $e^{\phi_{1i}}$, $e^{\phi_{2i}}$, and ϕ_{3i} denote, respectively, the asymptotic uptake rate, the uptake growth rate, and the maximum ambient CO_2 concentration at which no uptake is verified for plant i; random effects $b_i \overset{iid}{\sim} N(0, \sigma_b^2)$; and random errors $\epsilon_{ij} \overset{iid}{\sim} N(0, \sigma^2)$. Random effects and random errors are mutually independent. Note that we used exponential transformations to enforce the positivity constraints.

```
> data(CO2)
> co2.nlme <-
    nlme(uptake~exp(a1)*(1-exp(-exp(a2)*(conc-a3))),
         fixed=list(a1+a2~Type*Treatment,a3~1),
         random=a1~1, groups=~Plant, data=CO2,
         start=c(log(30),0,0,0,log(0.01),0,0,0,50))
> summary(co2.nlme)
Nonlinear mixed-effects model fit by maximum likelihood
  Model: uptake ~ exp(a1)*(1-exp(-exp(a2)*(conc - a3)))
 Data: CO2
       AIC      BIC    logLik
  393.2869 420.0259 -185.6434
...
```

Fits of model (9.48) are shown in Figure 9.8 as dotted lines. Based on model (9.48), one may conclude that the CO_2 uptake is higher for plants from Quebec, and that chilling, in general, results in lower uptake, and its effect on Mississippi plants is much larger than on Quebec plants.

We use this data set to demonstrate how to fit an SNM model and how to check if an NLME model is appropriate. As an extension of (9.48),

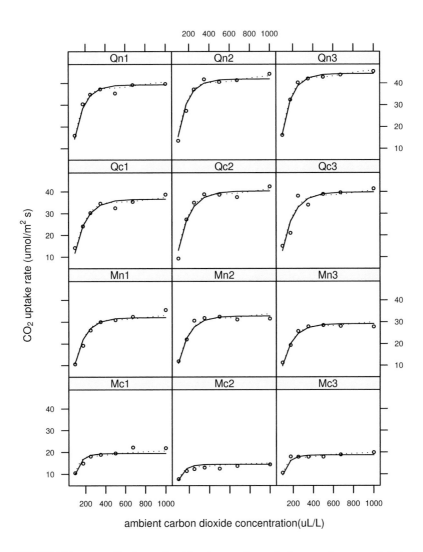

FIGURE 9.8 Carbon dioxide data, plot of observations, and fitted curve for each plant. Circles are CO_2 uptake rates. Solid lines represent SNM model fits from `co2.smm`. Dotted lines represent NLME model fits from `co2.nlme`. Strip names represent IDs of plants with "Q" indicating Quebec, "M" indicating Mississippi, "c" indicating chilled, "n" indicating nonchilled, and "1", "2", "3" indicating the replicate numbers.

we consider the following SNM model

$$\text{uptake}_{ij} = e^{\phi_{1i}} f(e^{\phi_{2i}}(\text{conc}_j - \phi_{3i})) + \epsilon_{ij},$$
$$\phi_{1i} = \beta_{12}\text{Type} + \beta_{13}\text{Treatment} + \beta_{14}\text{Treatment:Type} + b_i,$$
$$\phi_{2i} = \beta_{21},$$
$$\phi_{3i} = \beta_{31} + \beta_{32}\text{Type} + \beta_{33}\text{Treatment} + \beta_{34}\text{Treatment:Type},$$
$$i = 1, \ldots, 12; \; j = 1, \ldots, 7, \tag{9.49}$$

where $f \in W_2^2[0, b]$ for some fixed $b > 0$, and the second-stage model is similar to that in (9.48). In order to test if the parametric model (9.48) is appropriate, we use the exponential spline introduced in Section 2.11.2 with $\gamma = 1$. Then $\mathcal{H}_0 = \text{span}\{1, \exp(-x)\}$, and $\mathcal{H}_1 = W_2^2[0, b] \ominus \mathcal{H}_0$ with RK given in (2.58). Note that β_{11} in (9.48) is excluded from (9.49) to make f free of constraint on the vertical scale. We need the side conditions that $f(0) = 0$ and $f(x) \neq 0$ for $x \neq 0$ to make β_{31} identifiable with f. The first condition reduces $\mathcal{H}_0$ to $\tilde{\mathcal{H}}_0 = \text{span}\{1 - \exp(-x)\}$ and is satisfied by all functions in $\mathcal{H}_2$. We do not enforce the second condition because it is satisfied by all reasonable estimates. Thus the model space for f is $\tilde{\mathcal{H}}_0 \oplus \mathcal{H}_1$. It is clear that the NLME model is a special case of the SNM model with $f \in \tilde{\mathcal{H}}_0$. In the following we fit the SNM model (9.49) with initial values chosen from the NLME fit. The procedure converged after five iterations.

```
> M <- model.matrix(~Type*Treatment, data=CO2)[,-1]
> co2.snm <- snm(uptake~exp(a1)*f(exp(a2)*(conc-a3)),
    func=f(u)~list(~I(1-exp(-u))-1,lspline(u,type=''exp'')),
    fixed=list(a1~M-1,a3~1,a2~Type*Treatment),
    random=list(a1~1), group=~Plant, verbose=T,
    start=co2.nlme$coe$fixed[c(2:4,9,5:8)], data=CO2)
> summary(co2.snm)
Semi-parametric Nonlinear Mixed Effects Model fit
  Model: uptake ~ exp(a1) * f(exp(a2) * (conc - a3))
  Data: CO2

       AIC       BIC      logLik
  406.4865  441.625  -188.3760
  ...

GCV estimate(s) of smoothing parameter(s): 1.864814
Equivalent Degrees of Freedom (DF):   4.867183


Converged after 5 iterations
```

Fits of model (9.49) are shown in Figure 9.8 as solid lines. Since the data set is small, different initial values may lead to different estimates.

However, the overall fits are similar. We also fitted models with AR(1) within-subject correlations and covariate effects on ϕ_3. None of these models improve fits significantly. The estimates are comparable to the NLME fit, and the conclusions remain the same as those based on (9.48).

To check if the parametric NLME model (9.48) is appropriate, we calculate the posterior means and standard deviations using the function `intervals`. Note that the `intervals` function returns an object of class called "bCI" to which the generic function `plot` can be applied directly.

```
> co2.grid2 <- data.frame(u=seq(0.3, 11, len=50))
> co2.ci <- intervals(co2.snm, newdata=co2.grid2,
    terms=matrix(c(1,1,1,0,0,1), ncol=2, byrow=T))
> plot(co2.ci,
    type.name=c(''overall'',''parametric'',''smooth''))
```

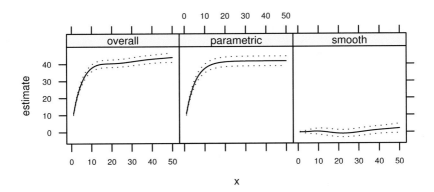

FIGURE 9.9 Carbon dioxide data, estimate of the overall function f (left), its projection onto $\tilde{\mathcal{H}}_0$ (center), and $\mathcal{H}_1$ (right). Solid lines are fitted values. Dash lines are approximate 95% Bayesian confidence intervals.

Figure 9.9 shows the estimate of f and its projection onto $\tilde{\mathcal{H}}_0$ and $\mathcal{H}_1$. The zero line is inside the Bayesian confidence intervals for the projection onto $\mathcal{H}_1$ (smooth component), which suggests that the parametric NLME model (9.48) is adequate.

9.4.5 Circadian Rhythm — Revisit

We have fitted an SIM (8.59) for normal subjects in Section 8.4.6 where parameters β_{i1}, β_{i2}, and β_{i3} are deterministic. The fixed effects SIM (8.59) has several drawbacks: (1) potential correlations among observations from the same subject are ignored; (2) the number of deterministic parameters is large; and (3) it is difficult to investigate covariate effects on parameters and/or the common shape function. In this section we show how to fit the hormone data using SNM models. More details can be found in Wang, Ke and Brown (2003).

We first consider the following mixed-effects SIM for a single group

$$\texttt{conc}_{ij} = \mu + b_{1i} + \exp(b_{2i}) f(\texttt{time}_{ij} - \text{alogit}(b_{3i})) + \epsilon_{ij},$$
$$i = 1, \ldots, m, \quad j = 1, \ldots, n_i, \tag{9.50}$$

where the fixed effect μ represents 24-hour mean of the population, the random effects b_{1i}, b_{2i}, and b_{3i} represent deviations in 24-hour mean, amplitude, and phase of subject i. We assume that $f \in W_2^2(per) \ominus \{1\}$ and $\boldsymbol{b}_i = (b_{1i}, b_{2i}, b_{3i})^T \overset{iid}{\sim} \text{N}(\boldsymbol{0}, \sigma^2 D)$, where D is an unstructured positive-definite matrix. The assumption of zero population mean for amplitude and phase, and the removal of constant functions from the periodic spline space, take care of potential confounding between amplitude, phase, and the nonparametric common shape function f. We fit model (9.50) to cortisol measurements from normal subjects as follows:

```
> nor <- horm.cort[horm.cort$type==''normal'',]
> nor.snm.fit <- snm(conc~b1+exp(b2)*f(time-alogit(b3)),
    func=f(u)~list(periodic(u)),
    data=nor, fixed=list(b1~1), random=list(b1+b2+b3~1),
    start=c(mean(nor$conc)), groups=~ID, spar=''m'')
> summary(nor.snm.fit)
Semi-parametric Nonlinear Mixed Effects Model fit
  Model: conc ~ b1 + exp(b2) * f(time - alogit(b3))
  Data: nor

       AIC      BIC    logLik
  176.1212 224.1264 -70.07205

Random effects:
  Formula: list(b1 ~ 1, b2 ~ 1, b3 ~ 1)
  Level: ID
  Structure: General positive-definite,
  Log-Cholesky parametrization
          StdDev    Corr
```

```
b1        0.2462385 b1       b2
b2        0.1803665 -0.628
b3        0.2486114  0.049 -0.521
Residual 0.3952836
```

```
Fixed effects: list(b1 ~ 1)
        Value  Std.Error DF  t-value p-value
b1 1.661412 0.07692439 98 21.59799       0
```

```
GML estimate(s) of smoothing parameter(s): 0.0001200191
Equivalent Degrees of Freedom (DF):  9.988554
```

```
Converged after 10 iterations
```

We compute predictions for all subjects evaluated at grid points:

```
> nor.grid <- data.frame(ID=rep(unique(nor$ID),rep(50,9)),
                         time=rep(seq(0,1,len=50),9))
> nor.snm.p <- predict(nor.fit, newdata=nor.grid)
```

The predictions are shown in Figure 9.10. In the following we also fit model (9.50) to depression and Cushing groups. Observations and predictions based on model (9.50) for these two groups are shown in Figures 9.11 and 9.12, respectively.

```
> dep <- horm.cort[horm.cort$type==''depression'',]
> dep.snm.fit <- snm(conc~b1+exp(b2)*f(time-alogit(b3)),
    func=f(u)~list(periodic(u)),
    data=dep, fixed=list(b1~1), random=list(b1+b2+b3~1),
    start=c(mean(dep$conc)), groups=~ID, spar=''m'')
> cush <- horm.cort[horm.cort$type==''cushing'',]
> cush.snm.fit <- snm(conc~b1+exp(b2)*f(time-alogit(b3)),
    func=f(u)~list(periodic(u)),
    data=cush, fixed=list(b1~1), random=list(b1+b2+b3~1),
    start=c(mean(cush$conc)), groups=~ID, spar=''m'')
```

We calculate the posterior means and standard deviations of the common shape functions for all three groups:

```
> ci.grid <- data.frame(time=seq(0,1,len=50))
> nor.ci <- intervals(nor.snm.fit, newdata=ci.grid)
> dep.ci <- intervals(dep.snm.fit, newdata=ci.grid)
> cush.ci <- intervals(cush.snm.fit, newdata=ci.grid)
```

The estimated common shape functions and 95% Bayesian confidence intervals for three groups are shown in Figure 9.13.

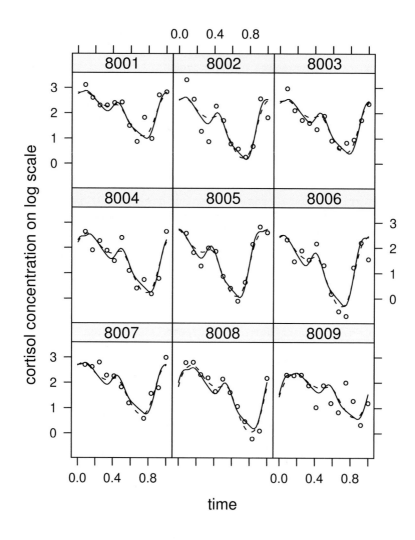

FIGURE 9.10 Hormone data, normal subjects, plots of cortisol concentrations (circles), and fitted curves based on model (9.50) (solid lines) and model (9.53) (dashed lines). Subjects' ID are shown in the strip.

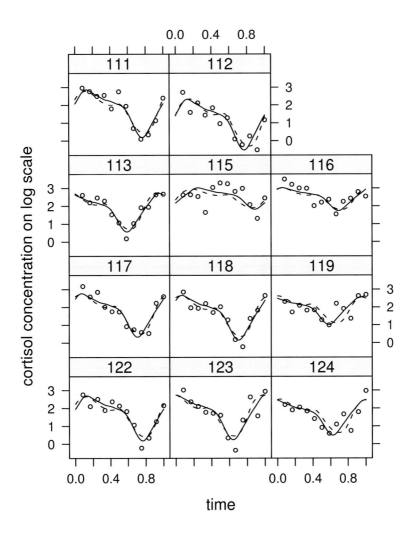

FIGURE 9.11 Hormone data, depressed subjects, plots of cortisol concentrations (circles), and fitted curves based on model (9.50) (solid lines) and model (9.53) (dashed lines). Subjects' ID are shown in the strip.

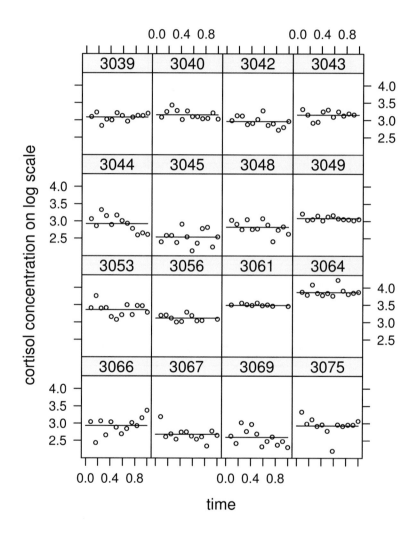

FIGURE 9.12 Hormone data, subjects with Cushing's disease, plots of cortisol concentrations (circles), and fitted curves based on model (9.50) (solid lines). Subjects' ID are shown in the strip.

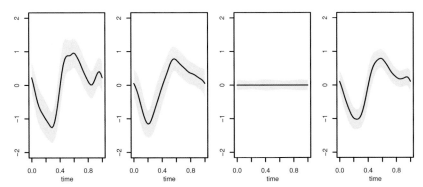

FIGURE 9.13 Hormone data, estimates of the common shape function f (lines), and 95% Bayesian confidence intervals (shaded regions). The lefts three panels are estimates based on model (9.50) for normal, depression, and Cushing groups, respectively. The right panel is the estimate based on model (9.53) for combined data from normal and depression groups.

It is obvious that the common function for the Cushing group is almost zero, which suggests that, in general, circadian rhythms are lost for Cushing patients. It seems that the shape functions for normal and depression groups are similar. We now test the hypothesis that the shape functions for normal and depression groups are the same by fitting data from these two groups jointly. Consider the following model

$$\text{conc}_{ijk} = \mu_k + b_{1ik} + \exp(b_{2ik})f(k, \text{time}_{ijk} - \text{alogit}(b_{3ik})) + \epsilon_{ijk},$$
$$i = 1, \ldots, m, \quad j = 1, \ldots, n_i, \quad k = 1, 2, \tag{9.51}$$

where k represents group factor with $k = 1$ and $k = 2$ corresponding to the depression and normal groups, respectively; fixed effect μ_k is the population 24-hour mean of group k; random effects b_{1ik}, b_{2ik}, and b_{3ik} represent the ith subject's deviation of 24-hour mean, amplitude, and phase. Note that subjects are nested within groups. We allow different variances for the random effects in each group. That is, we assume that $\boldsymbol{b}_{ik} = (b_{1ik}, b_{2ik}, b_{3ik})^T \overset{iid}{\sim} \text{N}(\boldsymbol{0}, \sigma_k^2 D)$, where D is an unstructured positive-definite matrix. We assume different common shape functions for each group. Thus f is a function of both group (denoted as k) and time. Since f is periodic in time, we model f using the tensor product space $\mathbb{R}^2 \otimes W_2^2(per)$. Specifically, consider the SS ANOVA decomposition (4.22). The constant and main effect of group are removed for identifiability with μ_k. Therefore, we assume the following model for f:

$$f(k, \text{time}) = f_1(\text{time}) + f_{12}(k, \text{time}), \tag{9.52}$$

where $f_1(\text{time})$ is the main effect of time, and $f_{12}(k, \text{time})$ is the interaction between group and time. The hypothesis $H_0 : f(1, \text{time}) = f(2, \text{time})$ is equivalent to $H_0 : f_{12}(k, \text{time}) = 0$ for all values of the time variable. Model (9.51) is fitted as follows:

```
> nordep <- horm.cort[horm.cort$type!=''cushing'',]
> nordep$type <- as.factor(as.vector(nordep$type))
> nordep.fit1 <- snm(conc~b1+exp(b2)*f(type,time-alogit(b3)),
    func=f(g,u)~list(list(periodic(u),
                     rk.prod(shrink1(g),periodic(u)))),
    data=nordep, fixed=list(b1~type), random=list(b1+b2+b3~1),
    groups=~ID, weights=varIdent(form=~1|type),
    spar=''m'', start=c(1.8,-.2))
> summary(nordep.fit1)
Semi-parametric Nonlinear Mixed Effects Model fit
 Model: conc ~ b1 + exp(b2) * f(type, time - alogit(b3))
 Data: nordep

       AIC      BIC    logLik
  441.4287 542.7464 -191.5463

Random effects:
 Formula: list(b1 ~ 1, b2 ~ 1, b3 ~ 1)
 Level: ID
 Structure: General positive-definite,
 Log-Cholesky parametrization
               StdDev    Corr
b1.(Intercept) 0.3403483 b1.(I) b2
b2             0.2936284 -0.781
b3             0.2962941  0.016 -0.159
Residual       0.4741110

Variance function:
 Structure: Different standard deviations per stratum
 Formula: ~1 | type
 Parameter estimates:
depression    normal
 1.000000   0.891695
Fixed effects: list(b1 ~ type)
                   Value  Std.Error  DF  t-value  p-value
b1.(Intercept)  1.8389689 0.08554649 218 21.496719  0.0000
b1.typenormal  -0.1179035 0.12360018 218 -0.953911  0.3412
 Correlation:
```

```
              b1.(I)
b1.typenormal -0.692

GML estimate(s) of smoothing parameter(s): 4.022423e-04
                                           2.256957e+02
Equivalent Degrees of Freedom (DF):  19.16807

Converged after 15 iterations
```

The smoothing parameter for the interaction term $f_{12}(k, \texttt{time})$ is large, indicating that the interaction is negligible. We compute posterior mean and standard deviation of the interaction term:

```
> u <- seq(0,1,len=50)
> nordep.inter <- intervals(nordep.fit1, terms=c(0,1),
    newdata=data.frame(g=rep(c(''normal'',''depression''),
    c(50,50)),u=rep(u,2)))
> range(nordep.inter$fit)
 -9.847084e-06  1.349702e-05
> range(nordep.inter$pstd)
  0.001422883 0.001423004
```

The posterior means are on the magnitude of 10^{-5}, while the posterior standard deviations are on the magnitude of 10^{-3}. The estimate of f_{12} is essentially zero. Therefore, it is appropriate to assume the same shape function for normal and depression groups.

Under the assumption of one shape function for both normal and depression groups, we now can investigate differences of 24-hour mean, amplitude, and phase between two groups. For this purpose, consider the following model

$$\begin{aligned}
\texttt{conc}_{ijk} = \mu_k + b_{1ik} + \exp(b_{2ik} + \delta_{k,2}d_1) \times \\
f(\texttt{time}_{ijk} - \text{alogit}(b_{3ik} + \delta_{k,2}d_2)) + \epsilon_{ijk}, \\
i = 1, \dots, m, \quad j = 1, \dots, n_i, \quad k = 1, 2,
\end{aligned} \qquad (9.53)$$

where $\delta_{k,2}$ is the Kronecker delta, and parameters d_1 and d_2 account for the differences in amplitude and phase, respectively, between normal and depression groups.

```
> nordep.fit2 <- snm(conc~b1+exp(b2+d1*I(type==''normal''))
    *f(time-alogit(b3+d2*I(type==''normal'')))),
    func=f(u)~list(periodic(u)), data=nordep,
    fixed=list(b1~type,d1+d2~1), random=list(b1+b2+b3~1),
    groups=~ID, weights=varIdent(form=~1|type),
```

```
   spar=''m'', start=c(1.9,-0.3,0,0))
> summary(nordep.fit2)
Semi-parametric Nonlinear Mixed Effects Model fit
 Model: conc ~ b1 + exp(b2 + d1 * I(type == ''normal'')) *
 f(time - alogit(b3 +      d2 * I(type == ''normal'')))
 Data: nordep

     AIC      BIC    logLik
 429.9391 503.4998 -193.7516


Random effects:
 Formula: list(b1 ~ 1, b2 ~ 1, b3 ~ 1)
 Level: ID
 Structure: General positive-definite,
           Log-Cholesky parametrization
                StdDev    Corr
b1.(Intercept) 0.3309993 b1.(I) b2
b2             0.2841053 -0.781
b3             0.2901979  0.030 -0.189
Residual       0.4655115


Variance function:
 Structure: Different standard deviations per stratum
 Formula: ~1 | type
 Parameter estimates:
depression      normal
 1.0000000   0.8908655
Fixed effects: list(b1 ~ type, d1 + d2 ~ 1)
                 Value   Std.Error  DF   t-value   p-value
b1.(Intercept)  1.8919482 0.08590594 216 22.023485  0.0000
b1.typenormal  -0.2558220 0.14361590 216 -1.781293  0.0763
d1              0.2102017 0.10783207 216  1.949343  0.0525
d2              0.0281460 0.09878700 216  0.284916  0.7760
 Correlation:
              b1.(I) b1.typ d1
b1.typenormal -0.598
d1             0.000 -0.509
d2             0.000  0.023 -0.159

GML estimate(s) of smoothing parameter(s): 0.0004723142
Equivalent Degrees of Freedom (DF): 9.217902


Converged after 9 iterations
```

The differences of 24-hour mean and amplitude are borderline significant, while the difference of phase is not. We refit without the d_2 term:

```
> nordep.fit3 <- snm(conc~b1+
    exp(b2+d1*I(type==``normal''))*f(time-alogit(b3)),
    func=f(u)~list(periodic(u)), data=nordep,
    fixed=list(b1~type,d1~1), random=list(b1+b2+b3~1),
    groups=~ID, weights=varIdent(form=~1|type),
    spar=``m'', start=c(1.9,-0.3,0))
> summary(nordep.fit3)
Semi-parametric Nonlinear Mixed Effects Model fit
  Model: conc ~ b1 + exp(b2 + d1 * I(type == ``normal'')) *
  f(time - alogit(b3))
  Data: nordep

       AIC      BIC    logLik
  425.2350 495.3548 -192.4077


Random effects:
  Formula: list(b1 ~ 1, b2 ~ 1, b3 ~ 1)
  Level: ID
  Structure: General positive-definite,
  Log-Cholesky parametrization
                  StdDev    Corr
b1.(Intercept) 0.3302233 b1.(I) b2
b2             0.2835421 -0.780
b3             0.2898165  0.033 -0.192
Residual       0.4647148


Variance function:
  Structure: Different standard deviations per stratum
  Formula: ~1 | type
  Parameter estimates:
depression     normal
 1.0000000  0.8902948
Fixed effects: list(b1 ~ type, d1 ~ 1)
                   Value  Std.Error  DF  t-value  p-value
b1.(Intercept)  1.8919931 0.08574693 217 22.064849  0.0000
b1.typenormal  -0.2567236 0.14327117 217 -1.791872  0.0745
d1              0.2148962 0.10620988 217  2.023316  0.0443
 Correlation:
              b1.(I) b1.typ
b1.typenormal -0.598
```

```
d1              0.000 -0.512
```

```
GML estimate(s) of smoothing parameter(s): 0.0004742399
Equivalent Degrees of Freedom (DF):  9.20981
```

```
Converged after 8 iterations
```

The predictions based on the final fit are shown in Figures 9.10 and 9.11. The estimate of the common shape function f is shown in the right panel of Figure 9.13. Data from two groups are pooled to estimate the common shape function, which leads to narrower confidence intervals. The final model suggests that the depressed subjects have their mean cortisol level elevated and have less profound circadian rhythm than normal subjects.

To take a closer look, we extract estimates of 24-hour mean levels and amplitudes for all subjects in normal and depression groups and perform binary recursive partitioning using these two variables:

```
> nor.mean <- nordep.fit3$coef$fixed[1]+
              nordep.fit3$coef$fixed[2]+
              nordep.fit3$coef$random$ID[12:20,1]
> dep.mean <- nordep.fit3$coef$fixed[1]+
              nordep.fit3$coef$random$ID[1:11,1]
> nor.amp <- exp(nordep.fit3$coef$fixed[3]+
                 nordep.fit3$coef$random$ID[12:20,2])
> dep.amp <- exp(nordep.fit3$coef$random$ID[1:11,2])
> u <- c(nor.mean, dep.mean)
> v <- c(nor.amp, dep.amp)
> s <- c(rep(''n'',9),rep(''d'',11))
> library(rpart)
> prune(rpart(s~u+v), cp=.1)
n= 20

node), split, n, loss, yval, (yprob)
      * denotes terminal node

1) root 20 9 d (0.5500000 0.4500000)
  2) u>=1.639263 12 2 d (0.8333333 0.1666667) *
  3) u< 1.639263 8 1 n (0.1250000 0.8750000) *
```

Figure 9.14 shows the estimated 24-hour mean levels plotted against the estimated amplitudes. There is a negative relationship between the 24-hour mean and amplitude. The estimate of correlation between b_{1ik} and b_{2ik} equals -0.781. The normal subjects and depressed patients can be well separated by the 24-hour mean level.

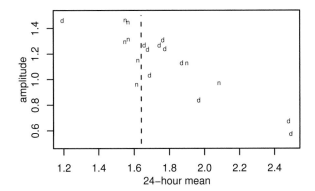

FIGURE 9.14 Hormone data, plot of the estimated 24-hour mean levels against amplitudes. Normal subjects and depressed patients are marked as "n" and "d", respectively. The dotted line represents partition based on the tree method.

Appendix A

Data Sets

In this appendix we describe data sets used for illustrations in this book. Table A.1 lists all data sets.

TABLE A.1   List of all data sets.

Air quality	New York air quality measurements
Arosa	Monthly ozone measurements from Arosa
Beveridge	Beveridge wheat price index
Bond	Treasury and GE bonds
Canadian weather	Monthly temperature and precipitation from 35 Canadian stations
Carbon dioxide	Carbon dioxide uptake in grass plants
Chickenpox	Monthly chickenpox cases in New York City
Child growth	Height of a child over one school year
Dog	Coronary sinus potassium in dogs
Geyser	Old Faithful geyser data
Hormone	Cortisol concentrations
Lake acidity	Acidity measurements from lakes
Melanoma	Melanoma incidence rates in Connecticut
Motorcycle	Simulated motorcycle accident data
Paramecium caudatum	Growth of paramecium caudatum population
Rock	Measurements on petroleum rock samples
Seizure	IEEG segments from a seizure patient
Star	Magnitude of the Mira variable R Hydrae
Stratford weather	Daily maximum temperatures in Stratford
Superconductivity	Superconductivity magnetization
Texas weather	Texas historical climate data
Ultrasound	Ultrasound imaging of the tongue shape
USA climate	Average winter temperatures in USA
Weight loss	Weight loss of an obese patient
WESDR	Wisconsin Epidemiological Study of Diabetic Retinopathy
World climate	Global average winter temperature

A.1   Air Quality Data

This data set contains daily air quality measurements in New York City, May to September of 1973. Four variables were measured: mean ozone in parts per billion from 1300 to 1500 hours at Roosevelt Island (denoted as `Ozone`), average wind speed in miles per hour at 0700 and 1000 hours at LaGuardia Airport (denoted as `Wind`), maximum daily temperature in degrees Fahrenheit at LaGuardia Airport (denoted as `Temp`), and solar radiation in Langleys in the frequency band 4000-7700 Angstroms from 0800 to 1200 hours at Central Park (denoted as `Solar.R`). The data set is available in R with the name `airquality`.

A.2   Arosa Ozone Data

This is a data set in Andrews and Herzberg (1985) contains monthly mean ozone thickness (Dobson units) in Arosa, Switzerland, from 1926 to 1971. It consists of 518 observations on three variables: `thick` for ozone thickness, `month`, and `year`. The data set is available in library `assist` with the name `Arosa`.

A.3   Beveridge Wheat Price Index Data

The data set contains Beveridge wheat price index averaged over many locations in western and central Europe from 1500 to 1869. The data set is available in library `tseries` with the name `bev`.

A.4   Bond Data

144 GE (General Electric Company) bonds and 78 Treasury bonds were collected from Bloomberg. The data set contains four variables: `name` of a bond, current `price`, `payment` at future `times`, and `type` of the bond. The number of payments till due ranges from 1 to 27 with a median 3

for the GE bonds, and from 1 to 58 with a median 2.5 for Treasury. The maximum time to maturity is 14.8 years for GE bonds and 28.77 years for Treasury bonds. The data set is available in library `assist` with the name `bond`.

A.5 Canadian Weather Data

The data set contains mean monthly temperature and precipitation at 35 Canadian weather stations (Ramsay and Silverman 2005). It consists of 420 observations on five variables: `temp` for temperature in Celsius, `prec` for precipitation in millimeters, `station` code number, geological `zone`, and `month`. The data set is available in library `fda` with the name `CanadianWeather`.

A.6 Carbon Dioxide Data

This data set comes from a study of cold tolerance of a C_4 grass species, *Echinochloa crus-galli*. A total of 12 four-week-old `plants` were used in the study. There were two `types` of plants: six from Quebec and six from Mississippi. Two `treatments`, nonchilling and chilling, were assigned to three plants of each type. Nonchilled plants were kept at 26^oC, and chilled plants were subject to 14 hours of chilling at 7^oC. After 10 hours of recovery at 20^oC, CO_2 `uptake` rates (in umol/m^{2}s) were measured for each plant at seven `concentrations` of ambient CO_2 in increasing, consecutive order. More details can be found in Pinheiro and Bates (2000). The data set is available in library `nlme` with the name `CO2`.

A.7 Chickenpox Data

The data set, downloaded at `http://robjhyndman.com/tsdldata/` `epi/chicknyc.dat`, contains monthly number of reported cases of chickenpox in New York City from 1931 to the first 6 months of 1972. It consists of 498 observations on three variables: count, month, and year. The data set is available in library assist with the name chickenpox.

A.8 Child Growth Data

Height measurements of a child were recorded over one school year (Ramsay 1998). The data set contains 83 observations on two variables: height in centimeters and day in a year. The data set is available in library fda with the name onechild.

A.9 Dog Data

A total of 36 dogs were assigned to four groups: control, extrinsic cardiac denervation 3 weeks prior to coronary occlusion, extrinsic cardiac denervation immediately prior to coronary occlusion, and bilateral thoracic sympathectomy and stellectomy 3 weeks prior to coronary occlusion. Coronary sinus potassium concentrations (milliliter equivalents per liter) were measured on each dog every 2 minutes from 1 to 13 minutes after occlusion. This data was originally presented by Grizzle and Allen (1969). It is available in library assist with the name dog.

A.10 Geyser Data

This data set contains 272 measurements from the Old Faithful geyser in Yellowstone National Park. Two variables were recorded: duration as the eruption time in minutes, and waiting as the waiting time in minutes to the next eruption. The data set is available in R with the name faithful.

A.11 Hormone Data

In an experiment to study immunological responses in humans, blood samples were collected every two hours for 24 hours from 9 healthy normal volunteers, 11 patients with major depression and 16 patients with Cushing's syndrome. These blood samples were analyzed for parameters that measure immune functions and hormones of the Hypothalamic-Pituitary-Adrenal axis (Kronfol, Nair, Zhang, Hill and Brown 1997). We will concentrate on hormone cortisol. The data set contains four variables: ID for subject index, time for time points when blood samples were taken, type as a group indicator for subjects and conc for cortisol concentration on $\log_{10}$ scale. The variable time is scaled into the interval $[0, 1]$. The data set is available in library assist with the name horm.cort.

A.12 Lake Acidity Data

This data set was derived by Douglas and Delampady (1990) from the Eastern Lakes Survey of 1984. The study involved measurements of 1789 lakes in three Eastern US regions: Northeast, Upper Midwest, and Southeast. We use a subset of 112 lakes in the southern Blue Ridge mountains area. The data set contains 112 observations on four variables: ph for water pH level, t1 for calcium concentration in $\log_{10}$ milligrams per liter, x1 for latitude, and x2 for longitude. The data set is available in library assist with the name acid.

A.13 Melanoma Data

This is a data set in Andrews and Herzberg (1985) that contains numbers of melanoma cases per 100,000 in the state of Connecticut during 1936–1972. It consists of 37 observations on two variables: cases for numbers of melanoma cases per 100,000, and year. The data set is available in library fda with the name melanoma.

A.14   Motorcycle Data

These data come from a simulated motorcycle crash experiment on the efficacy of crash helmets. The data set contains 133 measurements on two variables: `accel` as head acceleration in g of a subject and `time` as time after impact in milliseconds. The data set is available in library `MASS` with the name `mcycle`.

A.15   *Paramecium caudatum* Data

This is a data set in Gause (1934) that contains growth of paramecium caudatum population in the medium of Osterhout. It consists of 25 observations on two variables: `days` since the start of the experiment, and `density` representing the mean number of individuals in 0.5 milliliter of medium of four different cultures started simultaneously. The data set is available in library `assist` with the name `paramecium`.

A.16   Rock Data

This data set contains measurements on 48 rock samples collected from a petroleum reservoir. Four variables were measured: `area` of pores space in pixels out of 256 by 256, perimeter in pixels (denoted as `peri`), shape in perimeter/area$^{1/2}$, and permeability in milli-Darcies (denoted as `perm`). The data set is available in R with the name `rock`.

A.17   Seizure Data

This data set, provided by Li Qin, contains two 5-minute intracranial electroencephalograms (IEEG) segments from a seizure patient: `base` includes the baseline segment extracted at least 4 hours before the seizure's onset, and `preseizure` includes the segment right before a seizure's clinical onset. The sampling rate is 200 Hertz. Therefore, there are 60,000 time points in each segment. The data set is available in library `assist` with the name `seizure`.

A.18   Star Data

This data set, provided by Marc G. Genton, contains magnitude (brightness) of the Mira variable R Hydrae during 1900–1950. It consists of two variables: `magnitude` and `time` in days. The data set is available in library `assist` with the name `star`.

A.19   Stratford Weather Data

This is part of a climate data set downloaded from the Carbon Dioxide Information Analysis Center at `http://cdiac.ornl.gov/ftp/ndp070`. Daily maximum temperatures from the station in Stratford, Texas, in the year 1990 were extracted. The year was divided into 73 five-day periods, and measurements on the third day in each period were selected as observations. Therefore, the data set consists of 73 observations on two variables: y as the observed maximum temperature in Fahrenheit, and x as the time scaled into $[0, 1]$. The data set is available in library `assist` with the name `Stratford`.

A.20   Superconductivity Data

The data come from a study involving superconductivity magnetization modeling conducted by the National Institute of Standard and Technology. The data set contains 154 observations on two variables: `magnetization` in ampere×meter^2/kilogram, and `log time` in minutes. Temperature was fixed at 10 degrees Kelvin. The data set is available in library `NISTnls` with the name `Bennett5`.

A.21   Texas Weather Data

The data set contains average monthly `temperatures` during 1961–1990 from 48 weather `stations` in Texas. It also contains geological locations of these stations in terms of longitude (`long`) and latitude (`lat`). The data set is available in library `assist` with the name `TXtemp`.

A.22   Ultrasound Data

Ultrasound imaging of the tongue provides real-time information about the shape of the tongue body at different stages in articulation. This data set comes from an experiment conducted in the Phonetics and Experimental Phonology Lab of New York University led by Professor Lisa Davidson. Three Russian speakers produced the consonant sequence, /gd/, in three different linguistic environments:

2words: the g was at the end of one word followed by d at the beginning of the next word. For example, the Russian phrase *pab^{j}eg* d*amoj*;

cluster: the g and d were both at the beginning of the same word. For example, the phrase *xot* gd*amam*;

schwa: the g and d were at the beginning of the same word but are separated by the short vowel schwa (indicated by [∂]). For example, the phrase *pr∂tajitat^{j}g∂da'voj*".

Details about the ultrasound experiment can be found in Davidson (2006). We use a subset from a single subject, with three replications for each environment, 15 points recorded from each of 9 slices of tongue curves separated by 30 ms (milliseconds). The data set contains four variables: `height` as tongue height in mm (millimeters), `length` as tongue length in mm, `time` as the time in ms and `env` as the environment with three levels: 2words, cluster, and schwa. The data set is available in library `assist` with the name `ultrasound`.

A.23   USA Climate Data

The data set contains average winter (December, January, and February) temperatures (`temp`) in 1981 from 1214 stations in the United States. It also contains geological locations of these stations in terms of longitude (`long`) and latitude (`lat`). The data set is available in library `assist` with the name `USAtemp`.

A.24   Weight Loss Data

The data set contains 52 observations on two variables, `Weight` in kilograms of a male obese patient, and `Days` since the start of a weight rehabilitation program. The data set is available in library `MASS` with the name `wtloss`.

A.25   WESDR Data

Wisconsin Epidemiological Study of Diabetic Retinopathy (WESDR) is an epidemiological study of a cohort of diabetic patients receiving their medical care in an 11-county area in Southern Wisconsin. A number of medical, demographic, ocular, and other covariates were recorded at the baseline and later examinations along with a retinopathy score for each eye. Detailed descriptions of the study were given in Klein, Klein, Moss, Davis and DeMets (1988) and Klein, Klein, Moss, Davis and DeMets (1989). This subset contains 669 observations on five variables: `num` for subject ID, `dur` for duration of diabetes at baseline, `gly` for glycosylated hemoglobin, `bmi` for body mass index (weight in kilograms/(height in meters)^2), and `prg` for progression status of diabetic retinopathy at the first follow-up (1 for progression and 0 for nonprogression). The data set is available in library `assist` with the name `wesdr`.

A.26   World Climate Data

The data were obtained from the Carbon Dioxide Information and Analysis Center at Oak Ridge National Laboratory. The data set contains average winter (December, January, and February) temperatures (`temp`) in 1981 from 725 stations around the globe, and geological locations of these stations in terms of longitude (`long`) and latitude (`lat`). The data set is available in library `assist` with the name `climate`.

Appendix B

Codes for Fitting Strictly Increasing Functions

B.1   C and R Codes for Computing Integrals

The following functions k2 and k4 compute the scaled Bernoulli polynomials $k_2(x)$ and $k_4(x)$ in (2.27), and the function rc computes the RK of the cubic spline in Table 2.2:

```
static double
k2(double x) {
  double value;
  x = fabs(x);
  value = x - 0.5;
  value *= value;
  value = (value-1./12.)/2.;
  return(value);
}

static double
k4(double x) {
  double val;
  x = fabs(x);
  val = x - 0.5;
  val *= val;
  val = (val * val - val/2. + 7./240.)/24.;
  return(val);
}

static double
rc(double x, double y) {
  double value;
  value = k2(x) * k2(y)- k4 (x - y);
  return(value);
}
```

The following functions `integral_s`, `integral_f`, and `integral_1` compute three-point Gaussian quadrature approximations to integrals $\int_0^x f(s)ds$, $\int_0^x f(s)R_1(s,y)ds$, and $\int_0^x \int_0^y f(s)f(t)R_1(s,t)dsdt$, respectively, where R_1 is the RK of the cubic spline in Table 2.2:

```c
void integral_s(double *f, double *x, long *n, double *res)
{
  long i;
  double sum=0.0;

  for(i=0; i< *n; i++){
    sum += (x[i+1]-x[i])*(0.2777778*(f[3*i]+f[3*i+2])+
    0.4444444*f[3*i+1]);
    res[i] = sum;
  }
}

void integral_f(double *x, double *y, double *f,
                long *nx, long *ny, double *res)
{
  long i, j;
  double x1, y1, sum=0.0;

  for(i=0;i< *ny; i++){
    sum = 0.0;
    for(j=0; j< *nx; j++){
      x1 = x[j+1]-x[j];
      sum += x1*(0.2777778*(f[3*j]*
      rc(x[j]+x1*0.1127017, y[i])+
      f[3*j+2]*rc(x[j]+x1*0.8872983, y[i]))+
      0.4444444*f[3*j+1]*rc(x[j]+x1*0.5,y[i]));
      res[i*(*nx)+j] = sum;
    }
  }
}

void integral_1(double *x, double *y, double *f,
                long *n1, long *n2, double *res)
{
  long i, j, t, s;
  double x1, y1, sum=0.0, sum_tmp;

  for(i=0; i< *n1; i++){
```

```
  x1 = x[i+1]-x[i];
  sum = 0.0;
  for(j=0; j< *n2; j++){
   y1 = y[j+1]-y[j];
   sum_tmp = 0.2777778*0.2777778*(f[3*i]*f[3*j])*
    rc(x[i]+x1*0.1127017, y[j]+y1*0.1127017)+
    0.2777778*0.4444444*((f[3*i]*f[3*j+1])*
    rc(x[i]+x1*0.1127017, y[j]+y1*0.5)+
    (f[3*i+1]*f[3*j])*rc(x[i]+x1*0.5,y[j]+y1*0.1127017));
   sum_tmp += 0.4444444*0.4444444*((f[3*i+1]*f[3*j+1])*
    rc(x[i]+x1*0.5,y[j]+y1*0.5))+
    0.2777778*0.2777778*((f[3*i+2]*f[3*j+2])*
    rc(x[i]+x1*0.8872983, y[j]+ y1*0.8872983));
   sum_tmp += 0.2777778*0.2777778*((f[3*i]*f[3*j+2])*
    rc(x[i]+x1*0.1127017, y[j]+y1*0.8872983)+
    (f[3*i+2]*f[3*j])*
    rc(x[i]+x1*0.8872983,y[j]+y1*0.1127017))+
    0.4444444*0.2777778*((f[3*i+1]*f[3*j+2])*
    rc(x[i]+x1*0.5, y[j]+y1*0.8872983)+
    (f[3*i+2]*f[3*j+1])*rc(x[i]+x1*0.8872983, y[j]+y1*0.5));
   sum += sum_tmp*x1*y1;
   res[i*(*n2)+j] = sum;
  }
 }
}
```

The following R functions provide interface with the C functions
integral_s, integral_f, and integral_1:

```
int.s <- function(f, x, low=0) {
 n <- length(x)
 x <- c(low, x)
 .C(''integral_s'', as.double(f), as.double(x),
  as.integer(n), val = double(n))$val
}

int.f <- function(x, y, f, low=0) {
 nx <- length(x)
 ny <- length(y)
 x <- c(low, x)
 res <- .C(''integral_f'', as.double(x),
  as.double(y), as.double(f), as.integer(nx),
  as.integer(ny), val=double(nx*ny))$val
 matrix(res, ncol=ny, byrow=F)
```

```
}

int1 <- function(x, f.val, low=0) {
 n <- length(x)
 x <- c(low, x)
 if(length(f.val) != 3 * n) stop(''input not match'')
 res <- matrix(.C(''integral_1'', as.double(x),
  as.double(x), as.double(f.val), as.integer(n),
  as.integer(n), val = double(n * n))$val, ncol = n)
 apply(res, 1, cumsum)
}
```

B.2   R Function inc

The following R function implements the EGN procedure for model (7.3).

```
inc <- function(y, x, spar=''v'', grid=x, limnla=c(-6,0),
                prec=1.e-6, maxit=50, verbose=F)
{
 n <- length(x)
 org.ord <- match(1:n, (1:n)[order(x)])
 s.x <- sort(x)
 s.y <- y[order(x)]
 x1 <- c(0, s.x[-n])
 x2 <- s.x-x1
 q.x <- as.vector(rep(1,3)%o%x1+
  c(0.1127017,0.5,0.8872983)%o%x2)

# function for computing derivatives
 k1 <- function(x) x-.5
 k2 <- function(x) ((x-.5)^2-1/12)/2
 dk2 <- function(x) x-.5
 dk4 <- function(x)
  sign(x)*((abs(x)-.5)^3/6-(abs(x)-.5)/24)
 drkcub <- function(x,z) dk2(x)%o%k2(z)-
  dk4(x%o%rep(1,length(z))-rep(1,length(x))%o%z)

# compute starting value
 ini.fit <- ssr(s.y~I(s.x-.5), cubic(s.x))
 g.der <- ini.fit$coef$d[2]+drkcub(q.x,x)%*%ini.fit$coef$c
```

```
h.new <- abs(g.der)+0.005

# begin iteration
iter <- cover <- 1
h.old <- h.new
repeat {
 if(verbose) cat(''\n Iteration: '', iter)
 yhat <- s.y-int.s(h.new*(1-log(h.new)),s.x)
 smat <- cbind(int.s(h.new, s.x), int.s(h.new*q.x,s.x))
 qmat <- int1(s.x,h.new)
 fit <- ssr(yhat~smat, qmat, spar=spar, limnla=limnla)
 if(verbose)
  cat(''\nSmoothing parameter: '', fit$rkpk$nlaht)
 dd <- fit$coef$d
 cc <- as.vector(fit$coef$c)
 h.new <- as.vector(exp(cc%*%int.f(s.x,q.x,h.new)+
  dd[2]+dd[3]*q.x))
 cover <- mean((h.new-h.old)^2)
 h.old <- h.new
 if(verbose)
  cat(''\nConvergent Criterion: '', cover, ''\n'')
 if(cover<prec || iter>(maxit-1)) break
 iter <- iter + 1
}
if(iter>=maxit) print(''convergence not achieved!'')
y.fit <- (smat[,1]+fit$rkpk$d[1])[org.ord]
f.fit <- as.vector(cc%*%int.f(s.x,grid,h.new)+
 dd[2]+dd[3]*grid)
x1 <- c(0, grid[-length(grid)])
x2 <- grid-x1
q.x <- as.vector(rep(1,3)%o%x1+
 c(0.1127017,0.5,0.8872983)%o%x2)
h.new <- as.vector(exp(cc%*%int.f(s.x,q.x,h.new)+
 dd[2]+dd[3]*q.x))
y.pre <- int.s(h.new,grid)+fit$rkpk$d[1]
sigma <- sqrt(sum((y-y.fit)^2)/(length(y)-fit$df))
list(fit=fit, iter=c(iter, cover),
      pred=list(x=grid,y=y.pre,f=f.fit),
      y.fit=y.fit, sigma=sigma)
}
```

where x and y are vectors of the independent and dependent variables; grid is a vector of grid points of the x variable used for assessing convergence and prediction; and options `spar` and `limnla` are similar to those in the `ssr` function. Let $h(x) = g'(x) = \exp\{f(x)\}$. To get the initial value for the function f, we first fit a cubic spline to model (7.1). Denote the fitted function as g_0. We then use $\log(|g_0'(x)| + \delta)$ as the initial value for f, where $\delta = 0.005$ is a small positive number for numerical stability. Since $g_0(x) = d_1\phi_1(x) + d_2\phi_2(x) + \sum_{i=1}^{n} c_i R_1(x_i, x)$, we have $g_0'(x) = d_2 + \sum_{i=1}^{n} c_i \partial R_1(x_i, x)/\partial x$, where $\partial R_1(x_i, x)/\partial x$ is computed by the function `drkcub`. Functions `k1` and `k2` compute scaled Bernoulli polynomials defined in (2.27), and functions `dk2` and `dk4` compute $k_2'(x)$ and $k_4'(x)$, respectively. As a by-product, the line starting with `g.der` shows how to compute the first derivative for a cubic spline fit.

Appendix C

Codes for Term Structure of Interest Rates

C.1 C and R Codes for Computing Integrals

The following rc function computes the RK of the cubic spline in Table 2.1:

```
static double
rc(double x, double y) {
  double val, tmp;
  tmp = (x+y-fabs(x-y))/2.0;
  val = (tmp)*(tmp)*(3.0*(x+y-tmp)-tmp)/6.0;
  return(val);
}
```

In addition to the functions integral_s, integral_f, and integral_1 presented in Appendix B, we need the following function integral_2 for computing three-point Gaussian quadrature approximations to the integral $\int_0^x \int_0^y f_1(s)f_2(t)R_1(s,t)dsdt$, where R_1 is the RK of the cubic spline in Table 2.1:

```
void integral_2(double *x, double *y, double *fx,
      double *fy, long *n1, long *n2, double *res)
{
  long i,j, t, s;
  double x1, y1, sum=0.0, sum_tmp;

  for(i=0; i< *n1; i++){
   x1 = x[i+1]-x[i];
   sum = 0.0;
   for(j=0; j< *n2; j++){
    y1 = y[j+1]-y[j];
    sum_tmp = 0.2777778*0.2777778*(fx[3*i]*fy[3*j])*
     rc(x[i]+x1*0.1127017, y[j]+y1*0.1127017)+
```

```
      0.2777778*0.4444444*((fx[3*i]*fy[3*j+1])*
      rc(x[i]+x1*0.1127017, y[j]+y1*0.5)+
      (fx[3*i+1]*fy[3*j])*rc(x[i]+
      x1*0.5, y[j]+y1*0.1127017));
    sum_tmp += 0.4444444*0.4444444*((fx[3*i+1]*fy[3*j+1])*
      rc(x[i]+x1*0.5, y[j]+y1*0.5)) + 0.2777778*0.2777778*
      ((fx[3*i+2]*fy[3*j+2])*
      rc(x[i]+x1*0.8872983, y[j]+y1*0.8872983));
    sum_tmp += 0.2777778*0.2777778*((fx[3*i]*fy[3*j+2])*
      rc(x[i]+x1*0.1127017, y[j]+y1*0.8872983)+
      (fx[3*i+2]*fy[3*j])*
      rc(x[i]+x1*0.8872983, y[j]+y1*0.1127017))+
      0.4444444*0.2777778*((fx[3*i+1]*fy[3*j+2])*
      rc(x[i]+x1*0.5, y[j]+y1*0.8872983)+
      (fx[3*i+2]*fy[3*j+1])*
      rc(x[i]+x1*0.8872983, y[j]+y1*0.5));
    sum += sum_tmp*x1*y1;
    res[i*(*n2)+j] = sum;
    }
  }
}
```

Note that the rc function called inside integral_f, integral_1, and integral_2 in this Appendix computes the RK R_1 of the cubic spline in Table 2.1.

The following R functions provide interface with the C function integral_2:

```
int2 <- function(x, y, fx, fy, low.x=0, low.y=0) {
nx <- length(x)
ny <- length(y)
if((length(fx) != 3 * nx) || (length(fy) != 3 * ny))
  stop(''input not match'')
x <- c(low.x, x)
y <- c(low.y, y)
res <- matrix(.C(''integral_2'', as.double(x),
  as.double(y), as.double(fx), as.double(fy),
  as.integer(nx), as.integer(ny), val=double(nx*ny))$val,
  ncol=ny, byrow=T)
apply(res, 2, cumsum)
}
```

C.2   R Function for One Bond

The following one.bond function implements the EGN algorithm to fit model (7.35):

```
one.bond <- function(price, payment, time, name,
spar=''m'', limnla=c(-3,6)) {
# pre-processing the data
# the data has to be sorted by the name
group <- as.vector(table(name))
y <- price[cumsum(group)]
n.time <- length(time)

# create variables for 3-point Gaussian quadrature
s.time <- sort(time)
x1.y <- c(0, s.time[-n.time])
x2.y <- s.time-x1.y
org.ord <- match(1:n.time, (1:n.time)[order(time)])
q.time <- as.vector(rep(1,3)%o%x1.y+
 c(0.1127017,0.5,0.8872983)%o%x2.y)

# initial values for f
f0 <- function(x) rep(0.04, length(x))
f.old <- f.val <- f0(q.time)

# create s and q matrices
S <- cbind(time,time*time/2.0)
Lambda <- int1(s.time,
 rep(1,3*length(s.time)))[org.ord,org.ord]
Lint <- int.f(s.time,q.time,
 rep(1,3*length(s.time)))[org.ord,]

# begin iteration
iter <- cover <- 1
repeat {
 fint <- int.s(f.val,s.time)[org.ord]
 X <- assist:::diagComp(matrix(payment*exp(-fint),nrow=1),
                     group)
 ytilde <- X%*%(1+fint)-y
 T <- X%*%S; Q <- X%*%Lambda%*%t(X)
 fit <- ssr(ytilde~T-1, Q, spar=spar, limnla=limnla)
 dd <- fit$coef$d; cc <- fit$coef$c
```

```
    f.val <- as.vector((cc%*%X)%*%Lint+dd[1]+dd[2]*q.time)
    cover <- mean((f.val-f.old)^2)
    if(cover<1.e-6 || iter>20) break
    iter<- iter+1; f.old <- f.val
  }
  tmp <- -int.s(f.val,s.time)[org.ord]
  yhat <- apply(assist:::diagComp(matrix(payment*exp(tmp),
    nrow=1),group),1,sum)
  sigma <- sqrt(sum((y-yhat)^2)/(length(y)-fit$df))
  list(fit=fit, iter=c(iter, cover), call=match.call(),
       f.val=f.val, q.time=q.time, dc=exp(tmp),
       y=list(y=y,yhat=yhat), sigma=sigma)
}
```

where variable names are self-explanatory.

C.3   R Function for Two Bonds

The following `two.bond` function implements the nonlinear Gauss–Seidel algorithm to fit model (7.36):

```
two.bond <- function(price, payment, time, name, type,
  spar=''m'', limnla=c(-3,6), prec=1.e-6, maxit=20) {
  # pre-processing the data
  # the data in each group has to be sorted by the name
  group1 <- as.vector(table(name[type==''govt'']))
  y1 <- price[type==''govt''][cumsum(group1)]
  time1 <- time[type==''govt'']
  n1.time <- length(time1)
  payment1 <- payment[type==''govt'']
  group2 <- as.vector(table(name[type==''ge'']))
  y2 <- price[type==''ge''][cumsum(group2)]
  time2 <- time[type==''ge'']
  n2.time <- length(time2)
  payment2 <- payment[type==''ge'']
  y <- c(y1, y2)
  group <- c(group1, group2)
  payment <- c(payment1, payment2)
  error <- 0

  # create variables for 3-point Gaussian quadrature
```

```
s.time1 <- sort(time1)
x1.y1 <- c(0, s.time1[-n1.time])
x2.y1 <- s.time1-x1.y1
org.ord1 <- match(1:n1.time, (1:n1.time)[order(time1)])
q.time1 <- as.vector(rep(1,3)%o%x1.y1+
 c(0.1127017,0.5,0.8872983)%o%x2.y1)
s.time2 <- sort(time2)
x1.y2 <- c(0, s.time2[-n2.time])
x2.y2 <- s.time2-x1.y2
org.ord2 <- match(1:n2.time, (1:n2.time)[order(time2)])
q.time2 <- as.vector(rep(1,3)%o%x1.y2+
 c(0.1127017,0.5,0.8872983)%o%x2.y2)

# initial values for f
f10 <- function(x) rep(0.04, length(x))
f20 <- function(x) rep(0.01, length(x))
f1.val1 <- f10(q.time1)
f1.val2 <- f10(q.time2)
f1.old <- c(f1.val1,f1.val2)
f2.val2 <- f20(q.time2)
f2.old <- f2.val2

# create s and q matrices
S1 <- cbind(time1,time1*time1/2.0)
S2 <- cbind(time2,time2*time2/2.0)
L1 <- int1(s.time1,
 rep(1,3*length(s.time1)))[org.ord1,org.ord1]
L2 <- int1(s.time2,
 rep(1,3*length(s.time2)))[org.ord2,org.ord2]
L12 <- int2(s.time1,s.time2,rep(1,3*length(s.time1)),
 rep(1,3*length(s.time2)))
L12 <- L12[org.ord1,org.ord2]
Lambda <- rbind(cbind(L1,L12),cbind(t(L12),L2))
L1int <- int.f(s.time1,c(q.time1,q.time2),
 rep(1,3*length(s.time1)))[org.ord1,]
L2int <- int.f(s.time2,c(q.time1,q.time2),
 rep(1,3*length(s.time2)))[org.ord2,]
Lint <- rbind(L1int,L2int)
L2int2 <- int.f(s.time2,q.time2,
 rep(1,3*length(s.time2)))[org.ord2,]

# begin iteration
iter <- cover <- 1
```

```
repeat {
 # update f1
 f1int1 <- int.s(f1.val1,s.time1)[org.ord1]
 f1int2 <- int.s(f1.val2,s.time2)[org.ord2]
 f2int2 <- int.s(f2.val2,s.time2)[org.ord2]
 X <- assist:::diagComp(matrix(payment*
  exp(-c(f1int1,f1int2+f2int2)),nrow=1),group)
 ytilde1 <- X%*%(1+c(f1int1,f1int2))-y
 T <- X%*%rbind(S1,S2)
 Q <- X%*%Lambda%*%t(X)
 fit1 <- try(ssr(ytilde1~T-1,Q,spar=spar,limnla=limnla))
 if (class(fit1)==``try-error'') {error=1; break}
 if (class(fit1)!=``try-error'') {
  dd <- fit1$coef$d; cc <- fit1$coef$c
  f1.val <- as.vector((cc%*%X)%*%Lint+dd[1]+
   dd[2]*c(q.time1,q.time2))
  f1.val1 <- f1.val[1:(3*n1.time)]
  f1.val2 <- f1.val[-(1:(3*n1.time))]
 }

 # update f2
 f1int2 <- int.s(f1.val2,s.time2)[org.ord2]
 X2 <- assist:::diagComp(matrix(payment2*
  exp(-f1int2-f2int2),nrow=1),group2)
 ytilde2 <- X2%*%(1+f2int2)-y2
 T2 <- X2%*%S2
 Q22 <- X2%*%L2%*%t(X2)
 fit2 <- try(ssr(ytilde2~T2-1,Q22,spar=spar,
  limnla=limnla))
 if (class(fit2)==``try-error'') {error=1; break}
 if (class(fit2)!=``try-error'') {
  dd <- fit2$coef$d; cc <- fit2$coef$c
  f2.val2 <- as.vector((cc%*%X2)%*%L2int2+
   dd[1]+dd[2]*q.time2)
 }
 cover <- mean(((c(f1.val1,f1.val2,f2.val2)-
  c(f1.old, f2.old))^2)
 if(cover<prec || iter>maxit) break
 iter<- iter + 1
 f1.old <- c(f1.val1,f1.val2)
 f2.old <- f2.val2
}
tmp1 <- -int.s(f1.val1,s.time1)[org.ord1]
```

```
tmp2 <- -int.s(f1.val2+f2.val2,s.time2)[org.ord2]
yhat <- apply(assist:::diagComp(matrix(payment*
 exp(c(tmp1,tmp2)),nrow=1),group),1,sum)
sigma <- NA
if (error==0) sigma <- sqrt(sum((y-yhat)^2)/
 (length(y)-fit1$df-fit2$df))
list(fit1=fit1, fit2=fit2, iter=c(iter, cover, error),
 call=match.call(),
 f.val=list(f1=f1.val1,f2=f1.val2+f2.val2),
 f2.val=f2.val2,
 q.time=list(q.time1=q.time1,q.time2=q.time2),
 dc=list(dc1=exp(tmp1),dc2=exp(tmp2)),
 y=list(y=y,yhat=yhat), sigma=sigma)
}
```

The matrices S1, S2, T, T2, L1, L2, L12, and Q represent S_1, S_2, T, T_2, Λ_1, Λ_2, Λ_{12}, and Σ respectively, in the description about the Gauss–Seidel algorithm for model (7.36) in Section 7.6.3. In the output, f1 and f2 in the list f.val contain estimated forward rates for two bonds evaluated at time points q.time1 and q.time2 in the list q.time; f2.val contains estimated credit spread evaluated at time points q.time2; and dc1 and dc2 in the list dc contain estimated discount rates for two bonds evaluated at the observed time points.

References

Abramowitz, M. and Stegun, I. A. (1964). *Handbook of Mathematical Functions with Formulas, Graphs, and Mathematical Tables*, Washington, DC: National Bureau of Standards.

Andrews, D. F. and Herzberg, A. M. (1985). *Data: A Collection of Problems From Many Fields for the Student and Research Worker*, Springer, Berlin.

Aronszajn, N. (1950). Theory of reproducing kernels, *Transactions of the American Mathematics Society* **68**: 337–404.

Bennett, L. H., Swartzendruber, L. J. Turchinskaya, M. J., Blendell, J. E., Habib, J. M. and Seyoum, H. M. (1994). Long-time magnetic relaxation measurements on a quench melt growth YBCO superconductor, *Journal of Applied Physics* **76**: 6950–6952.

Berlinet, A. and Thomas-Agnan, C. (2004). *Reproducing Kernel Hilbert Spaces in Probability and Statistics*, Kluwer Academic, Norwell, MA.

Breslow, N. E. and Clayton, D. G. (1993). Approximate inference in generalized linear mixed models, *Journal of the American Statistical Association* **88**: 9–25.

Carroll, R. J., Fan, J., Gijbels, I. and Wand, M. P. (1997). Generalized partial linear single-index models, *Journal of the American Statistical Association* **92**: 477–489.

Coddington, E. A. (1961). *An Introduction to Ordinary Differential Equations*, Prentice-Hall, NJ.

Cox, D. D., Koh, E., Wahba, G. and Yandell, B. (1988). Testing the (parametric) null model hypothesis in (semiparametric) partial and generalized spline model, *Annals of Statistics* **16**: 113–119.

Cox, D. R. and Hinkley, D. V. (1974). *Theoretical Statistics*, Chapman and Hall, London.

Craven, P. and Wahba, G. (1979). Smoothing noisy data with spline functions, *Numerische Mathematik* **31**: 377–403.

Dalzell, C. J. and Ramsay, J. O. (1993). Computing reproducing kernels with arbitrary boundary constraints, *SIAM Journal on Scientific Computing* **14**: 511–518.

Davidson, L. (2006). Comparing tongue shapes from ultrasound imaging using smoothing spline analysis of variance., *Journal of the Acoustical Society of America* **120**: 407–415.

Davies, R. B. (1980). The distribution of a linear combination of χ^2 random variables, *Applied Statistics* **29**: 323–333.

Debnath, L. and Mikusiński, P. (1999). *Introduction to Hilbert Spaces with Applications*, Academic Press, London.

Douglas, A. and Delampady, M. (1990). *Eastern Lake Survey — Phase I: documentation for the data base and the derived data sets*, SIMS Technical Report 160. Department of Statistics, University of British Columbia, Vancouver.

Duchon, J. (1977). Spline minimizing rotation-invariant semi-norms in Sobolev spaces, pp. 85–100. In *Constructive Theory of Functions of Several Variables*, W. Schemp and K. Zeller eds., Springer, Berlin.

Earn, D. J. D., Rohani, P., Bolker, B. M. and Gernfell, B. T. (2000). A simple model for complex dynamical transitions in epidemics, *Science* **287**: 667–670.

Efron, B. (2001). Selection criteria for scatterplot smoothers, *Annals of Statistics* **29**: 470–504.

Efron, B. (2004). The estimation of prediction error: covariance penalties and cross-validation (with discussion), *Journal of the American Statistical Association* **99**: 619–632.

Eubank, R. (1988). *Spline Smoothing and Nonparametric Regression*, Dekker, New York.

Eubank, R. (1999). *Nonparametric Regression and Spline Smoothing, 2nd ed.*, Dekker, New York.

Evans, M. and Swartz, T. (2000). *Approximating Integrals via Monte Carlo and Deterministic Methods*, Oxford University Press, Oxford, UK.

Fisher, M. D., Nychka, D. and Zervos, D. (1995). *Fitting the term structure of interest rates with smoothing spline*, Working Paper 95-1, Finance and Eonomics Discussion Series, Federal Reserve Board.

Flett, T. M. (1980). *Differential Analysis*, Cambridge University Press, London.

Friedman, J. H. and Stuetzle, W. (1981). Projection pursuit regression, *Journal of the American Statistical Association* **76**: 817–823.

Gause, G. F. (1934). *The Struggle for Existence*, Williams & Wilkins, Baltimore, MD.

Genton, M. G. and Hall, P. (2007). Statistical inference for evolving periodic functions, *Journal of the Royal Statistical Society B* **69**: 643–657.

Green, P. J. and Silverman, B. W. (1994). *Nonparametric Regression and Generalized Linear Models: A Roughness Penalty Approach*, Chapman and Hall, London.

Grizzle, J. E. and Allen, D. M. (1969). Analysis of growth and dose response curves, *Biometrics* **25**: 357–381.

Gu, C. (1992). Penalized likelihood regression: A Bayesian analysis, *Statistica Sinica* **2**: 255–264.

Gu, C. (2002). *Smoothing Spline ANOVA Models*, Springer, New York.

Guo, W., Dai, M., Ombao, H. C. and von Sachs, R. (2003). Smoothing spline ANOVA for time-dependent spectral analysis, *Journal of the American Statistical Association* **98**: 643–652.

Hall, P., Kay, J. W. and Titterington, D. M. (1990). Asymptotically optimal difference-based estimation of variance in nonparametric regression, *Biometrika* **77**: 521–528.

Harville, D. (1976). Extension of the Gauss-Markov theorem to include the estimation of random effects, *Annals of Statistics* **4**: 384–395.

Harville, D. A. (1997). *Matrix Algebra From A Statistician's Perspective*, Springer, New York.

Hastie, T. and Tibshirani, R. (1990). *Generalized Additive Models*, Chapman and Hall, London.

Hastie, T. and Tibshirani, R. (1993). Varying coefficient model, *Journal of the Royal Statistical Society B* **55**: 757–796.

Heckman, N. (1997). *The theory and application of penalized least squares methods or reproducing kernel Hilbert spaces made easy*, University of British Columbia Statistics Department Technical Report number 216.

Heckman, N. and Ramsay, J. O. (2000). Penalized regression with model-based penalties, *Canadian Journal of Statistics* **28**: 241–258.

Jarrow, R., Ruppert, D. and Yu, Y. (2004). Estimating the term structure of corporate debt with a semiparametric penalized spline model, *Journal of the American Statistical Association* **99**: 57–66.

Ke, C. and Wang, Y. (2001). Semi-parametric nonlinear mixed-effects models and their applications (with discussion), *Journal of the American Statistical Association* **96**: 1272–1298.

Ke, C. and Wang, Y. (2004). Nonparametric nonlinear regression models, *Journal of the American Statistical Association* **99**: 1166–1175.

Kimeldorf, G. S. and Wahba, G. (1971). Some results on Tchebycheffian spline functions, *Journal of Mathematical Analysis and Applications* **33**: 82–94.

Klein, R., Klein, B. E. K., Moss, S. E., Davis, M. D. and DeMets, D. L. (1988). Glycosylated hemoglobin predicts the incidence and progression of diabetic retinopathy, *Journal of the American Medical Association* **260**: 2864–2871.

Klein, R., Klein, B. E. K., Moss, S. E., Davis, M. D. and DeMets, D. L. (1989). Is blood pressure a predictor of the incidence or progression of diabetic retinopathy, *Archives of Internal Medicine* **149**: 2427–2432.

Kronfol, Z., Nair, M., Zhang, Q., Hill, E. and Brown, M. (1997). Circadian immune measures in healthy volunteers: Relationship to hypothalamic-pituitary-adrenal axis hormones and sympathetic neurotransmitters, *Psychosomatic Medicine* **59**: 42–50.

Lawton, W. H., Sylvestre, E. A. and Maggio, M. S. (1972). Self-modeling nonlinear regression, *Technometrics* **13**: 513–532.

Lee, Y., Nelder, J. A. and Pawitan, Y. (2006). *Generalized Linear Models with Random Effects: Unified Analysis via H-likelihood*, Chapman and Hall, London.

Li, K. C. (1986). Asymptotic optimality of C_L and generalized cross-validation in ridge regression with application to spline smoothing, *Annals of Statistics* **14**: 1101–1112.

Lindstrom, M. J. and Bates, D. M. (1990). Nonlinear mixed effects models for repeated measures data, *Biometrics* **46**: 673–687.

Liu, A. and Wang, Y. (2004). Hypothesis testing in smoothing spline models, *Journal of Statistical Computation and Simulation* **74**: 581–597.

Liu, A., Meiring, W. and Wang, Y. (2005). Testing generalized linear models using smoothing spline methods, *Statistica Sinica* **15**: 235–256.

Liu, A., Tong, T. and Wang, Y. (2007). Smoothing spline estimation of variance functions, *Journal of Computational and Graphical Statistics* **16**: 312–329.

Ma, X., Dai, B., Klein, R., Klein, B. E. K., Lee, K. and Wahba, G. (2010). *Penalized likelihood regression in reproducing kernel Hilbert spaces with randomized covariate data*, University of Wisconsin Statistics Department Technical Report number 1158.

McCullagh, P. and Nelder, J. (1989). *Generalized Linear Models*, Chapman and Hall, London.

Meinguet, J. (1979). Multivariate interpolation at arbitrary points made simple, *Journal of Applied Mathematics and Physics (ZAMP)* **30**: 292–304.

Neal, D. (2004). *Introduction to Population Biology*, Cambridge University Press, Cambridge, UK.

Nychka, D. (1988). Bayesian confidence intervals for smoothing splines, *Journal of the American Statistical Association* **83**: 1134–1143.

Opsomer, J. D., Wang, Y. and Yang, Y. (2001). Nonparametric regression with correlated errors, *Statistical Science* **16**: 134–153.

O'Sullivan, F. (1986). A statistical perspective on ill-posed inverse problems (with discussion), *Statistical Science* **4**: 502–527.

Parzen, E. (1961). An approach to time series analysis, *Annals of Mathematical Statistics* **32**: 951–989.

Pinheiro, J. and Bates, D. M. (2000). *Mixed-effects Models in S and S-plus*, Springer, New York.

Qin, L. and Wang, Y. (2008). Nonparametric spectral analysis with applications to seizure characterization using EEG time series, *Annals of Applied Statistics* **2**: 1432–1451.

Ramsay, J. O. (1998). Estimating smooth monotone functions, *Journal of the Royal Statistical Society B* **60**: 365–375.

Ramsay, J. O. and Silverman, B. W. (2005). *Functional Data Analysis*, 2nd ed., Springer, New York.

Rice, J. A. (1984). Bandwidth choice for nonparametric regression, *Annals of Statistics* **12**: 1215–1230.

Robinson, G. K. (1991). That BLUP is a good thing: The estimation of random effects (with discussion), *Statistical Science* **6**: 15–51.

Ruppert, D., Wand, M. P. and Carroll, R. J. (2003). *Semiparametric Regression*, Cambridge, New York.

Schumaker, L. L. (2007). *Spline Functions: Basic Theory, 3rd ed.*, Cambridge University Press, Cambridge, UK.

Smith, M. and Kohn, R. (2000). Nonparametric seemingly unrelated regression, *Journal of Econometrics* **98**: 257–281.

Speckman, P. (1995). Fitting curves with features: semiparametric change-point methods, *Computing Science and Statistics* **26**: 257–264.

Stein, M. (1990). A comparison of generalized cross-validation and modified maximum likelihood for estimating the parameters of a stochastic process, *Annals of Statistics* **18**: 1139–1157.

Tibshirani, R. and Knight, K. (1999). The covariance inflation criterion for adaptive model selection, *Journal of the Royal Statistical Society B* **61**: 529–546.

Tong, T. and Wang, Y. (2005). Estimating residual variance in nonparametric regression using least squares, *Biometrika* **92**: 821–830.

Venables, W. N. and Ripley, B. D. (2002). *Modern Applied Statistics with S, 4th ed.*, Springer, New York.

Wahba, G. (1980). Automatic smoothing of the log periodogram, *Journal of the American Statistical Association* **75**: 122–132.

Wahba, G. (1981). Spline interpolation and smoothing on the sphere, *SIAM Journal on Scientific Computing* **2**: 5–16.

Wahba, G. (1983). Bayesian confidence intervals for the cross-validated smoothing spline, *Journal of the Royal Statistical Society B* **45**: 133–150.

Wahba, G. (1985). A comparison of GCV and GML for choosing the smoothing parameters in the generalized spline smoothing problem, *Annals of Statistics* **4**: 1378–1402.

Wahba, G. (1987). Three topics in ill posed inverse problems, pp. 37–51. In *Inverse and Ill-Posed Problems*, M. Engl and G. Groetsch, eds. Academic Press, New York.

Wahba, G. (1990). *Spline Models for Observational Data*, SIAM, Philadelphia, PA. CBMS-NSF Regional Conference Series in Applied Mathematics, Vol. 59.

Wahba, G. and Wang, Y. (1995). Behavior near zero of the distribution of GCV smoothing parameter estimates for splines, *Statistics and Probability Letters* **25**: 105–111.

Wahba, G., Wang, Y., Gu, C., Klein, R. and Klein, B. E. K. (1995). Smoothing spline ANOVA for exponential families, with application to the Wisconsin Epidemiological Study of Diabetic Retinopathy, *Annals of Statistics* **23**: 1865–1895.

Wang, Y. (1994). *Smoothing Spline Analysis of Variance of Data From Exponential Families*, Ph.D. Thesis, University of Wisconsin-Madison, Department of Statistics.

Wang, Y. (1997). GRKPACK: fitting smoothing spline analysis of variance models to data from exponential families, *Communications in Statistics: Simulation and Computation* **26**: 765–782.

Wang, Y. (1998a). Mixed-effects smoothing spline ANOVA, *Journal of the Royal Statistical Society B* **60**: 159–174.

Wang, Y. (1998b). Smoothing spline models with correlated random errors, *Journal of the American Statistical Association* **93**: 341–348.

Wang, Y. and Brown, M. B. (1996). A flexible model for human circadian rhythms, *Biometrics* **52**: 588–596.

Wang, Y. and Ke, C. (2009). Smoothing spline semi-parametric nonlinear regression models, *Journal of Computational and Graphical Statistics* **18**: 165–183.

Wang, Y. and Wahba, G. (1995). Bootstrap confidence intervals for smoothing splines and their comparison to Bayesian confidence intervals, *Journal of Statistical Computation and Simulation* **51**: 263–279.

Wang, Y. and Wahba, G. (1998). Discussion of "Smoothing Spline Models for the Analysis of Nested and Crossed Samples of Curves" by Brumback and Rice, *Journal of the American Statistical Association* **93**: 976–980.

Wang, Y., Guo, W. and Brown, M. B. (2000). Spline smoothing for bivariate data with applications to association between hormones, *Statistica Sinica* **10**: 377–397.

Wang, Y., Ke, C. and Brown, M. B. (2003). Shape invariant modelling of circadian rhythms with random effects and smoothing spline ANOVA decomposition, *Biometrics* **59**: 804–812.

Wang, Y., Wahba, G., Chappell, R. and Gu, C. (1995). Simulation studies of smoothing parameter estimates and Bayesian confidence intervals in Bernoulli SS ANOVA models, *Communications in Statistics: Simulation and Computation* **24**: 1037–1059.

Wong, W. (2006). Estimation of the loss of an estimate. In *Frontiers in Statistics*, J. Fan and H. L. Koul eds. Imperial College Press, London.

Wood, S. N. (2003). Thin plate regression splines, *Journal of the Royal Statistical Society B* **65**: 95–114.

Xiang, D. and Wahba, G. (1996). A generalized approximate cross validation for smoothing splines with non-Gaussian data, *Statistica Sinica* **6**: 675–692.

Yang, Y., Liu, A. and Wang, Y. (2005). Detecting pulsatile hormone secretions using nonlinear mixed effects partial spline models, *Biometrics* pp. 230–238.

Ye, J. M. (1998). On measuring and correcting the effects of data mining and model selection, *Journal of the American Statistical Association* **93**: 120–131.

Yeshurun, Y., Malozemoff, A. P. and Shaulov, A. (1996). Magnetic relaxation in high-temperature superconductors, *Reviews of Modern Physics* **68**: 911–949.

Yorke, J. A. and London, W. P. (1973). Recurrent outbreaks of measles, chickenpox and mumps, *American Journal of Epidemiology* **98**: 453–482.

Yu, Y. and Ruppert, D. (2002). Penalized spline estimation for partially linear single index models, *Journal of the American Statistical Association* **97**: 1042–1054.

Yuan, M. and Wahba, G. (2004). Doubly penalized likelihood estimator in heteroscedastic regression, *Statistics and Probability Letters* **69**: 11–20.

Author Index

Subject Index